PLANKTON

A Guide to Their Ecology and Monitoring for Water Quality

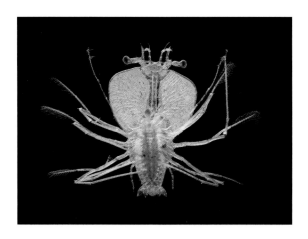

EDITORS: IAIN M. SUTHERS, DAVID RISSIK
AND ANTHONY J. RICHARDSON

CSIRO

PUBLISHING

CRC Press
Taylor & Francis Group
Boca Raton London New York

CRC Press is an imprint of the
Taylor & Francis Group, an **informa** business

A catalogue record for this book is available from the National Library of Australia.

ISBN: 9781486308798 (hbk.)
ISBN: 9781486308804 (epdf)
ISBN: 9781486308811 (epub)

Published exclusively in Australia and New Zealand by:
CSIRO Publishing
Locked Bag 10
Clayton South VIC 3169
Australia

Telephone: +61 3 9545 8400
Email: publishing.sales@csiro.au
Website: www.publish.csiro.au

Published exclusively throughout the world (excluding Australia and New Zealand)
by CRC Press, with ISBN 978-0-367-03016-2

CRC Press
6000 Broken Sound Parkway NW, Suite 300,
Boca Raton, Florida 33487
USA
Tel: 800-272-7737
Website: www.crcpress.com

Front cover: Various plankton species. Photos: Julian Uribe Palomino.
All illustrations are by the authors unless otherwise specified.

Set in 10.5/14 Palatino and Optima
Edited by Peter Storer
Cover design by Andrew Weatherill
Typeset by Thomson Digital
Printed in China by Asia Pacific Offset

Dec18_01

Contents

Preface

Many local councils and estuary managers collect phytoplankton and zooplankton in response to the increasing incidence of algal (phytoplankton) blooms in estuaries and coastal waters. Recent studies have shown that the biomass of algae is a better indicator of nutrient stress in waterways than nutrient concentrations. Unfortunately, there has been a lack of consistency and scientific rigour in the methodologies used for sampling, which has often resulted in unresolved outcomes. Monitoring studies are often poorly designed and are *ad hoc* – making it difficult to identify an appropriate management response. We wish to provide a guide for those preparing or maintaining a water-quality program, as well as to educate people about plankton. By increasing the general awareness about the inhabitants of our water, we can tackle many water quality issues.

The objectives of this guide are to introduce plankton as indices of water quality to managers and students. We hope to enable people to design, implement and conduct a meaningful plankton sampling program, which may accommodate future changes in technology and respond to new concepts, needs and ideas.

This guide is intended for those concerned with water quality and resource management in state government, local governments (council engineers, town planners and landscape architects), community groups (Landcare, Rivercare), environmental consultants and teachers. The information provided is also of value to those working with or studying phytoplankton and zooplankton in freshwater and coastal marine waters, including those in the aquaculture industry. Management concerns and case studies are key features of this guide to demonstrate the utility of plankton studies for water quality management and other applications including aquaculture, fisheries and ecosystem assessments. Our target readership includes those without large budgets, operating from small boats, and who may have limited experience with marine or freshwater sampling programs.

In this second edition we welcome Anthony Richardson who heavily revised Chapter 8 on marine zooplankton. Anthony also brought his experience with indicator species and climate change (Chapter 3 Use of plankton for management) and his enthusiasm for plankton ('Fun facts on plankton' in Chapter 1).

The other major change was to replace the models in management component (Chapter 9) with plankton and teaching. For this, we welcome Tim Roe, a teacher with the Moreton Bay Environmental Education Centre, who provides some useful plankton exercises for people of all ages.

List of contributors

Penelope Ajani
Climate Change Cluster, University of Technology Sydney, Sydney, Australia

Ian A.E. Bayly
Flinders Island, Tasmania, Australia

Lee Bowling
Honorary Research Fellow, Centre for Ecosystem Science, University of New South Wales, Sydney, Australia

Frank Coman
Commonwealth Scientific and Industrial Research Organisation (CSIRO) Oceans and Atmosphere, Brisbane, Australia

Ruth Eriksen
Commonwealth Scientific and Industrial Research Organisation (CSIRO) Oceans and Atmosphere, Hobart, Australia

Mark Gibbs
Institute of Sustainable Futures, Queensland University of Technology, Brisbane, Australia

Tsuyoshi Kobayashi
Science Division, Office of Environment and Heritage NSW, Sydney, Australia

Anthony G. Miskiewicz
Adjunct, School of Biological, Earth & Environmental Sciences, University of New South Wales, Sydney, Australia

Kylie Pitt
Griffith School of Environment and Science, Griffith University, Brisbane, Australia

Anna M Redden
Acadia Centre for Estuarine Research, Acadia University, Nova Scotia, Canada

Anthony J. Richardson
Commonwealth Scientific and Industrial Research Organisation (CSIRO) Oceans and Atmosphere and School of Mathematics, University of Queensland, Brisbane, Australia

David Rissik
BMT and National Climate Change Adaptation Research Facility, Griffith University, Brisbane, Australia

Timothy Roe
Moreton Bay Environmental Education Centre, Department of Education, Queensland, Brisbane, Australia

Peter Rothlisberg
Commonwealth Scientific and Industrial Research Organisation (CSIRO) Oceans and Atmosphere, Brisbane, Australia

Russell J. Shiel
School of Biological Sciences, University of Adelaide, Adelaide, Australia

Anita Slotwinski
Commonwealth Scientific and Industrial Research Organisation (CSIRO) Oceans and Atmosphere, Brisbane, Australia

Iain M. Suthers
Sydney Institute of Marine Science, and School of Biological, Earth & Environmental Sciences, University of New South Wales, Sydney, Australia

Julian Uribe-Palomino
Commonwealth Scientific and Industrial Research Organisation (CSIRO) Oceans and Atmosphere, Brisbane, Australia

Jock Young
Commonwealth Scientific and Industrial Research Organisation (CSIRO) Oceans and Atmosphere, Hobart, Australia

Acknowledgements

We acknowledge the pioneering work in marine plankton identification in Australia by William Dakin, Alan Colefax and Isobel Bennett during the 1940s, and the others that followed.

This work would not have been possible without the support of our home institutions, noted under the list of contributors. We acknowledge the many wonderful people of the Integrated Marine Observing System (IMOS) – Australia's national ocean observing system; and the Marine National Facility; and the financial support of the Australian Research Council.

We also acknowledge the different organisations who have recognised the role of plankton in management and have used plankton to underpin management decisions. They have helped enthuse us to promote plankton as an essential component of managing aquatic systems.

We thank our past co-authors of the first edition, including Mark Baird, Michael N Dawson, William Froneman, Anthony Jakeman, Alison J. King, Daniel Large, Rebecca Letcher, Lachlan T.H. Newham, Gina Newton, Murray Root, Brian Sanderson, and Stephanie Wallace. Over the years, many others have given us advice including Gustaaf Hallegraeff and David McKinnon; many students and colleagues assisted with figures including Chapter 2 (Adrien Greene and Sylvia Dove) and Fig. 6.5 (Rick van den Enden, AAD). For Chapter 7 we thank Professors W. Foissner, J.J. Gilbert, D. J. Patterson and B.V. Timms for comments on the early manuscript. We thank Emeritus Professor Patrick De Deckker for comments on ostracods and providing a scanning electron photomicrograph of *Newhamia fenestrata*. Similarly we acknowledge the comments by Michelle Burford and Chantelle Rissik for their reviews of Chapters 8 and 9.

And by no means least, we thank our partners – Karen Whitehead, Chantelle Rissik and Glynnis Richardson – for accommodating our many hours working with plankton and writing this book.

1

The importance of plankton

Iain M. Suthers, Anthony J. Richardson and David Rissik

Phytoplankton and zooplankton – tiny drifting plants and animals – are vital components of aquatic systems. It is sobering to think that all of the large and charismatic animals in aquatic systems that we are familiar with – the fish, seabirds and mammals – are minor components of the food web when the biomass of different groups is considered (Fig. 1.1). Aquatic systems are dominated by very small organisms that we rarely see: bacteria, phytoplankton and zooplankton. This huge biomass of very small organisms means that most ecosystem services provided by aquatic systems are provided by plankton. Plankton not only provide the food for higher trophic levels such as fish, seabirds, penguins, seals and sharks, but produce oxygen, cycle nutrients, process many of the pollutants that humans dispose of through our waterways, and help to remove carbon dioxide from our atmosphere. Without the diverse roles of plankton, our waterways and oceans would be virtually devoid of life and our planet would be very different.

Being at the base of the food web and in such huge numbers, plankton are strongly influenced by water quality because they cannot isolate themselves as oysters do by closing their shells in adverse conditions. Plankton are effectively our aquatic 'canaries-in-a-coal mine', providing an indication of the effects of hourly changes in water quality integrated over days and weeks.

Management of water quality can be supported by having a broad understanding of plankton and their interaction with the environment. Phytoplankton respond rapidly within days to changes in light, nutrients, pollution or sediment load, changes in water flow or estuarine flushing, and in response to grazing by larger zooplankton. Therefore, from a manager's perspective, the response time of plankton is comparable to changes in water quality, which contrasts with changes in the benthic community or in fish that respond over broader scales of months or many kilometres (Fig. 1.2).

The amount and type of phytoplankton present in the water can inform managers about the health of the waterways and where management actions may be required. High biomass of phytoplankton often reflects excessive nutrient inputs (eutrophication) and this can cause problems when the phytoplankton blooms die and decay, depleting oxygen levels in the water. The types of plankton present in the water are also important, because several phytoplankton species are toxic and can be harmful to humans, but not necessarily to the vectors of the toxin, such as oysters or fish. It is important to know about harmful phytoplankton species to manage causes of blooms.

1.1 What are plankton?

Plankton may be defined as any organisms that cannot swim against a current. Most plankton can swim or adjust their position by changing buoyancy but they lack the power to swim against a persistent current. Most plankton are microscopic,

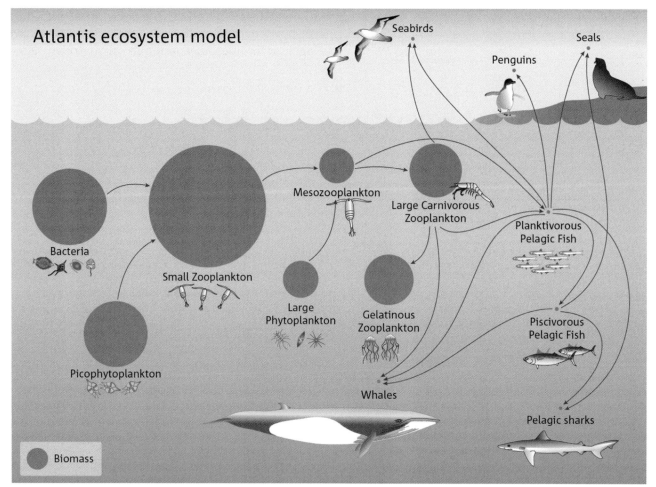

Fig. 1.1. Diagram of a marine food web off south-east Australia, with the size of spheres proportional to the biomass of each group. Plankton dominate the biomass; the large species that we know well such as fish, whales, seabirds and penguins have a minuscule biomass in comparison. The biomass is estimated from a balanced ecosystem model based on available biomass data for each marine group ('Atlantis', Fulton *et al.* 2004). This diagram shows only biomass, and if presented as production of biomass over a year, would show plankton and bacteria to massively dominate the ecosystem even more than is shown by biomass alone.

which affects their swimming ability, but some zooplankton such as jellyfish can be huge: up to 2 m in diameter (Chapter 8).

Phytoplankton, such as diatoms and dinoflagellates, grow in the presence of sunlight and nutrients such as nitrogen and phosphorus. These single-celled organisms are the 'grasses of the sea' and are the basis of ocean productivity. Many of these 'plants' – but not all – are in turn grazed by zooplankton, which is dominated by small crustaceans such as copepods, shrimps and their larvae, and by smaller single-celled microzooplankton. The amount of phytoplankton in the water column

reflects the influence of several environmental factors and processes. These competing processes may be summed up as 'bottom-up', such as those concerning nutrients and light, which drive primary production, or 'top-down', such as predation by copepods or other grazers.

Phytoplankton contain photosynthetically active pigments such as chlorophyll, which enable them to use energy from sunlight to convert carbon dioxide into complex organic molecules, such as sugar or protein (i.e. they are autotrophs). Chlorophyll is used as an estimate of phytoplankton biomass. The majority of chlorophyll in tropical and subtropical

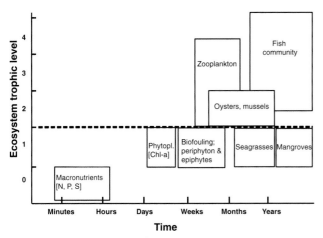

Fig. 1.2. Range of possible estuarine health and water quality indicators available, illustrating the higher trophic level and intermediate integrating period of zooplankton. Phytoplankton (Phytopl.) is often quantified as the concentration of chlorophyll-*a*, the primary photosynthetic pigment (Chl-a).

coastal waters is found in the very smallest of cells – the size of bacteria, (~0.001 mm or 1 μm). These very small cells have high surface-area-to-volume ratio, allowing them to out-compete larger phytoplankton cells in the race for nutrients.

Exceptions abound where some of these single-celled 'plants' do not fix their own carbon, but engulf and consume other plant cells (i.e. they are heterotrophic like an animal and have no photosynthetic pigments). Other single-celled organisms both photosynthesise (like a plant) and they eat other organisms (like an animal). If you work on phytoplankton, the distinction between plants and animals becomes blurred. Other phytoplankton have the potential to form harmful algal blooms (HABs) – producing red tides or toxic algae – but there are only a few species responsible (just a fraction of a per cent of all phytoplankton species may be harmful; see Chapter 5 and 6). Most phytoplankton are enormously beneficial, such as those used in the aquaculture industry as food for young fish and shellfish. There are distinct forms and different sizes of the major phytoplankton groups and this book will guide you through their identification.

Zooplankton refers to the small multicellular animal life, dominated by crustaceans and some types of gelatinous animals. Zooplankton includes

representatives of nearly all of the 34 major groups or phyla (a phylum is a discrete evolutionary lineage) of multicellular animals alive today. Zooplankton includes the larvae of many familiar animals that spend only a portion of their life as plankton – fish, crabs, lobsters, oysters, mussels, jellyfish and starfish – and are known as meroplankton (Chapter 2). Holoplankton spend their entire life in the plankton and include copepods, ctenophores, arrow worms and salps. Some typical benthic animals, such as snails, marine worms and even tiny fish, have some holoplanktonic species with fascinating specialised body forms.

The most abundant animals on the planet are copepods (Sections 1.2.9, 8.3.1), and may comprise over 95% of zooplankton abundance and biomass (Fig. 1.1). Only occasionally will jellyfish, ctenophores or salps predominate. There are over 12 000 species of copepods (yet only 78 species of krill!), and each species of copepod develops by moulting through six larval (naupliar) stages and five juvenile (copepodite) forms until they reach the final and sixth adult stage (Fig. 2.5). Many zooplankton species have young stages that look very different to their adults, making the study of zooplankton interesting, complex and challenging. This book will guide you through this complexity and discuss the traditional and a few modern ways one can study zooplankton.

1.2 Fun facts about plankton

With most plankton being microscopic, the amazing roles of plankton are largely hidden from us. Here we present some fun facts about plankton to highlight their many diverse, yet critical roles, which go totally unnoticed.

1.2.1 Did you know that our society is based on plankton?

You might be surprised to learn that our cars run on plankton, and many parts of cars are even built from plankton! Plankton also make most products we use in everyday life, from plumbing materials to our clothes. Here's why. Our society is based on petroleum products – our cars, planes and trains that keep us moving, the roads we drive on, and the plastics that we use to make our everyday products. Petroleum is formed by dead zooplankton and phytoplankton sinking to an ancient seafloor. Under low oxygen conditions, plankton is not broken down by bacteria, but is buried by sediment. Over time, pressure and heat convert the plankton and sediment into sedimentary rock. If there is sufficient organic content (i.e. plankton) and the right temperature (90–160°C), then oil and natural gas can form. After the petroleum is extracted, it is then distilled into fractions to separate it into its constituents (liquefied petroleum gas, gasoline, jet fuel, kerosene and diesel). These petroleum products are then used to produce many products including asphalt, nitrogen fertilisers and plastics. We use plastics in everything from our cars (up to ~20%), synthetics such as nylon, acrylic, polyester. So, our petroleum-based society literally runs on plankton!

1.2.2 Plankton shaped early human society

It might seem far-fetched, but plankton also shaped early human society. It was the naturalist Charles Darwin who said that fire was one of the most significant achievements of humanity. To make fire, early humans used flint: a type of rock commonly formed from silica-rich plankton such as diatoms and radiolarians. Flint was used for many stone age tools, including weapons, but it was the ability of a flint edge to produce sparks when struck against the rock pyrite that was revolutionary. The ability to produce fire provided early hominids protection from predators, a method for hunting, the ability to cook food, and the capacity to expand activities into the darker and colder hours of night. These cultural innovations changed our diet and behaviour, allowing humans to disperse across the world. Without the ability to use flint formed from plankton to produce fire, it is difficult to see that human society would be where it is today.

1.2.3 Plankton in the movies

Plankton and Karen are well known characters to those who enjoyed the animated TV series

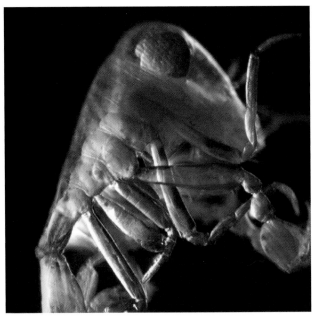

Fig. 1.3.   The hyperiid amphipod *Phronima*: the likely inspiration for the alien in the movie of the same name (photo: Anita Slotwinski).

SpongeBob SquarePants. Plankton is a copepod, while Karen is his supercomputer. But when sorting through your plankton samples and seeing what different plankton species look like close up, you'll realise that plankton has probably inspired many characters in the movie industry.

One of the most iconic movie monsters in film history – the antagonist in the film 'Alien' – is thought to be inspired by plankton. *Phronima* is a large planktonic amphipod, up to ~40 mm in size with a head like a praying mantis insect. It has huge eyes and large predatory arms (Fig. 1.3). However, the most remarkable aspect is that some species parasitise salps (Box 1.1): translucent barrel-shaped plankton. *Phronima* uses the salp for protection and as a flotation device, swimming it through the water, feeding on its host and other plankton as it goes, and rearing its young inside. Finally, the young eventually emerge from the salp, like the Alien erupted from a human host in the movie!

1.2.4 Amazing single-celled plankton inspire architects and engineers!

The structure of many plankton groups such as diatoms, coccolithophores, silicoflagellates, tintinnids,

Box 1.1 Salps, larvaceans and climate change

Salps and the appendicularians have been described as the fastest growing metazoans (multi-cellular animals) on the planet (Hopcroft and Roff 1995). They consume tiny phytoplankton and bacteria that are up to eight orders of magnitude smaller than themselves (a much greater size difference than the copepod diet), and produce dense faecal pellets that rapidly sink. Therefore salps have the potential to alter regional food webs and even global fluxes of carbon via their faecal pellets (Henschke *et al.* 2016). The most common salp off south-east Australia, *Thalia democratica,* can reproduce both sexually and asexually (Henschke *et al.* 2013). An individual or solitary may produce a chain of individual clones, resulting in the population doubling or more per day (Heron 1972). Salps compete with other zooplankton such as copepods and krill. In the Southern Ocean, for example, a decrease in krill populations over the last 50 years has been accompanied by an increase in salp populations (Atkinson *et al.* 2004). In subtropical waters, the relative abundance of salps in the zooplankton community could alter the balance between those predator species that avoid salps and those fish for whom salps are an important component of their diet.

radiolarians, foraminiferans and acantharians are wildly elegant and diverse, being spinose, ribbed, geometric, geodesic, perforated, fluted, ornamented and stellate. This variety of forms was made famous by early ecologists and palaeontologists such as Earnst Haeckl in 1904 in his *Kunstformen der Nature ('Art Forms of Nature')*. These stunning shapes inspired Art Nouveau architecture and design, including René Binet's design for the Printemps department store. Such diverse planktonic forms are now inspiring architects and engineers through biomimetics (Pohl and Nachtigall 2015). Biomimetics is the field of imitation of natural systems and elements to solve human problems. The similarity between how nature solves problems and how architects do can be illustrated by the central dome of Galleria Vittorio Emmanuele in Milan, Italy. This glass dome is supported by a structure of radial and concentric ribs structurally and functionally reminiscent of the valve of the radial centric diatom *Arachnoidiscus* (Fig. 1.4).

Fig. 1.4.   The valve of the radial centric diatom *Arachnoidiscus* (left) and the central dome of Galleria Vittorio Emmanuele in Milan, Italy (right).

Over 100 000 different plankton species have evolved, with different geometries forming a suite of templates for structures in different industries. Most remarkably, plankton have evolved different composite materials to make their structure strong and lightweight, but from very few based materials such as silica and calcium carbonate and mixing these with organic compounds. Composite materials are increasingly used in the building industry because they are strong and lightweight. Scientists are now investigating how plankton lay down composite materials in different orientations to resist stresses in particular directions.

A recent example of biomimetics was the use of radiolarians in the design of support towers for wind turbines (Pohl and Nachtigall 2015). The design criterion was that the support tower had to have three legs, with these joining to a central tube, and had to be 25% lighter and stronger than existing structures. Diatoms were considered, but their structure has too many small pores to optimise light and nutrient transfer in photosynthesis and they are basically enclosed boxes, whereas radiolarians feed on small organisms and therefore have open skeletal frameworks within a tetrahedral design space that make them ideal as a template for support structures. Several radiolarian genera were considered, analysed using computer aided design for how well they cope with stresses and their ease of fabrication, and ultimately the genus *Chlathrocorys* was chosen. This was because of its homogeneous stress distribution, simple design in spite of complex geometry, its ability to prevent buckling by tensile structure, and its relative ease to be built in steel.

1.2.5 Bluebottles and the plankton collective

The Portuguese man o'war (*Physalia*) or bluebottles are the scourge of summer surf, delivering a painful sting to the unwary swimmer (see Box 9.1). Their typical habitat is the offshore ocean surface where they capture zooplankton on their dangling tentacles, which may extend up to 5 m long. Small fish may school around the tentacles and it is possible that over days of association the fish slime gains certain compounds, so that *Physalia* is tricked

Fig. 1.5.   An impressively large late stage phyllosoma of a southern (or eastern) rock lobster.

into not stinging them. *Physalia* is often driven onshore by the afternoon sea breeze in summer, along with the predatory sea slug *Glaucus*. Amazingly, *Glaucus* not only eat *Physalia*, but they can transfer the stinging cells from *Physalia* to their own tissue and store them for their own defence! *Physalia* is a type of jellyfish known as a siphonophore, which is composed of a colony of genetically identical individual clones similar to those which make up coral. A siphonophore is truly a collective, as its hundreds of individuals each perform specific functions such as stinging, digestion, fishing, reproduction or flotation. For example, the characteristic bluebottle float is actually composed of a single individual. Siphonophore colonies can be astonishingly long: the siphonophore *Praya dubia* can grow to 50 m long in the benign deep ocean waters, making it one of the longest animals on Earth.

1.2.6 The bizarre life of lobsters in the plankton

Like most crustaceans, female lobster release tiny larvae (nauplii), which go through many stages by moulting their exoskeleton. But, unlike most crustaceans that have a larval duration of several weeks, the larval duration of lobsters can be several years! Lobster larvae are commonly called phyllosoma and are perfectly adapted to floating, as they have a thin, flat, transparent body, with long legs (Fig. 1.5; see also Section 8.3.2.1, Fig. 8.12M). A secret of their survival is to avoid predators by being transparent and living in offshore eddies where there are relatively few predators. The diet of phyllosoma was unknown until recently when genetic analysis of their stomach contents revealed that they commonly eat jellyfish, salps and chaetognaths (arrow worms). Some species of phyllosoma eat jellyfish while they hitchhike a ride on them, using them both as food and as a flotation device. These species have modified appendages for grooming, because the mucus produced by jellyfish promotes bacterial growth (Kamio *et al.* 2015). Scientists at sea have also observed phyllosoma capturing arrow worms (chaetognaths) with their 10 chelate limbs, and eating them like a carrot. Phyllosoma feed and build up a reserve of fat (lipid) that sustains them for their final moult into a miniature lobster (puerulus) and the long swim back to the coast.

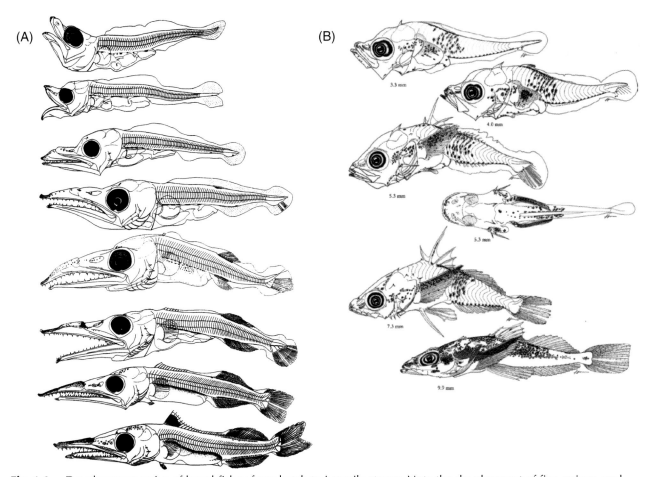

Fig. 1.6. Development series of larval fishes from hatch to juvenile stages. Note the development of fins, spines, and dorsal flexion of the notochord to form the tail. **(A)** Wahoo *Acanthocybium solandri*, from after hatching (2.8 mm) to 13.2 mm standard length (SL); family Scombridae, order Perciformes. Wahoo is a globally distributed fish and important in recreational fishing. Reproduced from Richards 1989. **(B)** Dusky flathead (*Platycephalus fuscus*, family Platycephalidae, order Scorpaeniformes) from 3.3 to 9.9 mm. Dusky flathead are a popular coastal species of eastern Australia (reproduced with permission from Neira *et al.* 1998).

1.2.7 Remarkable larval fish – transformers and killers

Like adults, larval fish have a very diverse morphology including body shape, size and patterns of head and fin spination and pigmentation. Larvae of live-bearers such as seahorses can hatch looking like miniature adults, while larvae hatching from benthic eggs (i.e. attached to rocks, like a Nemo fish) usually have well-developed eyes, mouth and fins. Most larval fish hatch from pelagic eggs and barely have any fins, unpigmented eyes and a little yolk or lipid to sustain them until they begin feeding (see Chapters 7 and 8). As they grow, the morphology changes as the fins develop and the larvae transforms into a juvenile fish (Fig. 1.6A,B). Some species can have a dramatic transformation/metamorphosis from larval to juvenile stages with development and loss of elongate head and fin spines and eye position and shape. Larval flounder hatch with symmetrical heads and eyes like any other larval fish, but approaching metamorphosis one eye migrates around the skull to the topside as they prepare for adult life and rest one side on the substrate.

Life is difficult as a larval fish, and the natural mortality rate is large (as much as 20% per day) due to starvation or predation by copepods, krill, jellyfish and arrow worms. It is a major goal of fisheries science to understand the dynamics of this larval mortality so future fish numbers can be forecast for fisheries management. Most larval fish hatch from eggs without much to live on. They barely have any fins and only unpigmented eyes and a little yolk or lipid to sustain them until their jaws develop and they can begin to feed on larval copepods (nauplii). As larvae grow and their mouth gape increases, they can ingest a greater variety and size of prey. Some species of tunas and mackerel become voracious predators of other larval fishes, including their siblings, and seem to be all jaws and no tail (Fig. 1.6A).

1.2.8 Phytoplankton supply every second breath you take

Land plants and phytoplankton use the green pigment chlorophyll to harness sunlight to produce organic compounds from carbon dioxide, and release oxygen: a process known as photosynthesis. On land, most photosynthesis is performed by huge land plants, but in the ocean it is the microscopic phytoplankton that are the key players. The main groups of phytoplankton are the diatoms, dinoflagellates and cyanobacteria (bacteria that photosynthesise). While far too small to be easily observed (less than 1 μm in size), the biomass and significance of the cyanobacteria is remarkable. During the 1980s, several groups of marine cyanobacteria were discovered, including *Prochlorococcus*. *Prochlorococcus* is tiny (0.6 μm) and dominates the huge subtropical and tropical oceans. Only one species has been described, *Prochlorococcus marinus*, so it is likely to be the most abundant species on the plant, typically with over 100 million cells per litre, equating to 3 octillion (10^{27}) individuals: more than the number of sand grains on Earth (Flombaum *et al.* 2013). There are different strains of this species at different depths and light regimes in the ocean. This single species produces about one-fifth of all the oxygen on Earth. Together with the diatoms, dinoflagellates and other phytoplankton, they produce nearly 50% of all oxygen generated by primary producers on Earth, so nearly every second breath we take! We call this one of the many 'ecosystem services' provided by plankton. We get the oxygen produced by phytoplankton for free and we take it for granted, but how much would you pay for a lungful of air?

1.2.9 Red tides and Noctiluca

A red tide is a colloquial name for a bloom of phytoplankton that turns the water reddish brown. A red tide may be a HAB but not always. A common red-tide-forming organism is the dinoflagellate *Noctiluca scintillans*, sometimes known as sea-sparkle (Fig. 3.2). *Noctiluca* is also bioluminescent (Section 1.2.12), producing a greenish glow during the night by activating some commensal bioluminescent bacteria. *Noctiluca* has no photosynthetic pigments and feeds at night on other phytoplankton, small zooplankton and their eggs. It contains no toxins, other than a dilute solution of ammonium chloride, which, in large quantities, can irritate the skin and cause localised fish kills. During the final senescent stages of its life, the cell swells up to a comparatively large size of 2 mm diameter and becomes buoyant, thus concentrating at the surface as a reddish, or even bright pink, stain. Estuaries and coastlines around the world with high nutrient concentration frequently have *Noctiluca* blooms.

Off the west coast of South Africa, upwelling brings nutrients to the surface waters, and this can lead to huge phytoplankton blooms, including *Noctiluca* in the middle of summer. If upwelling is followed by a long period of calm, then *Noctiluca* blooms can be concentrated inshore. When the blooms dies, it sinks to the bottom where it is decomposed by bacteria, which can strip the water of oxygen. This low oxygen water can stretch over tens of kilometres, threatening marine life. During summer, the oxygen can be so low in the water that rock lobster, which live on the seafloor, move into the surf zone to obtain more oxygen. When the tide goes out, they can be left stranded. The largest event was in South Africa in 1997, when 2000 tonnes of rock lobster walked out of the water and died on the beach.

The sequence of events that lead to problematic *Noctiluca* blooms illustrate the complexity of plankton ecology. Problems associated with *Noctiluca* could increase in the future: the warming of coastal waters, especially during El Nino years off Australia; an increasingly more environmentally aware public; and the suspicion that *Noctiluca* may have been transported around the world including to Australia by ballast water (McLeod *et al.* 2012). With warming and the strengthening of the warm, poleward flowing East Australian Current, *Noctiluca* was seen for the first time in the Southern Ocean in 2010.

1.2.10 How does zooplankton survive when water dries up in the desert?

Many freshwater zooplankton have developed the capacity to withstand periods of adverse environmental conditions such as droughts and hot temperatures common in deserts. Probably the best known example is the brine shrimp (*Artemia* or 'sea monkeys', Fig. 1.7). They achieve this through a process known as diapause. Diapause of eggs occurs when, following fertilisation, eggs remain viable for long periods, and do not hatch until environmental conditions are appropriate. Diapause can also occur at different life history stages of some copepods, when, in response to environmental signals, they either 'hibernate' for a period or create a cyst that sinks to the bottom of the water column and stays viable in unfavourable conditions, 'hatching' once positive environmental conditions return. This enables diversity to return to desert wetlands and rivers after long periods of drought. Diapause of eggs has been very important for aquaculture in that eggs in diapause are collected and added to aquaculture tanks where they hatch and are used as a food source.

1.2.11 Are copepods the world's most abundant animal?

Sir Alister Hardy, founder of the Continuous Plankton Recorder survey in the North Atlantic, stated that copepods were the most abundant multi-celled animal in the world. Is he right? Well, insects are also strong contenders. Insects far outnumber other terrestrial fauna, with estimates of a trillion (10^{18}) insects on Earth (Schminke 2007). A simple calculation follows for total copepod abundance.

Fig. 1.7. *Artemia salina* (photo: Hans Hillewaert/ Wikimedia Commons, CC BY-SA 4.0).

Assuming that there is one copepod in every litre of sea water, and there is 1347 million km^3 of ocean, then there would 1.35×10^{21} individual copepods in the water column. If benthic and parasitic copepods were included, this figure could be tripled to 4×10^{21} individuals (Schminke 2007), suggesting there are 1000 times more copepods than insects. Finally, if the live weight of a single copepod is 0.036 mg, the biomass of these copepods would be 1.5×10^{10} tonnes, or 500 times the biomass of the whole human population on Earth (Schminke 2007). So yes, copepods could be the most abundant multi-celled animals on the planet!

1.2.12 Bioluminescence in the ocean

Have you ever been at the beach at night, and noticed that when the water is disturbed, perhaps by a breaking wave, that the water lights up? This is probably caused by bacteria or dinoflagellate phytoplankton emitting light as a result of a chemical reactions in their body. Most of the colours are blue-green, but almost the entire visible spectrum can be emitted. The ability to emit light has evolved many separate times in different marine groups. Over 700 marine genera are able to emit light, including many planktonic species within the dinoflagellates, copepods, krill, jellyfish, amphipods and arrow worms (Widder 2010). There are many possible reasons for bioluminescence. For example, the burglar alarm hypothesis is thought to explain bioluminescence in the carnivorous dinoflagellate *Noctiluca* (Section 1.2.9). When water is agitated, such as when a copepod is nearby and is trying to eat *Noctiluca*, it emits light. The light is thought to attract fish, which could then stop the copepod from feeding because it is now concerned with being eaten itself! Thus, *Noctiluca* emits light (the burglar alarm) that attracts the police (fish) and deters burglars (copepods), protecting the innocent victims (*Noctiluca*). Other species that exhibit bioluminescence include krill, which have light organs known as photophores. When krill swarms, their light shows might serve to attract mates or deter predators, or both. Many species of copepods are also bioluminescent. The marine copepod *Pleuromamma* spends a lot of time in the deep, dark ocean and discharges bioluminescent material when threatened. Because predators at that depth have developed a strong sensitivity to light, they are temporarily blinded by the discharged bioluminescent material and the copepod is able to escape. This would be like in the movies when someone is wearing night vision goggles and the light is turned on!

1.2.13 How can the biggest animals that have ever lived eat microscopic zooplankton?

Have you ever wondered how the largest animals that have ever lived – the whales – survive on some of the smallest animals, the microscopic krill and other zooplankton that float in the water? This contrasts starkly with the largest land animals that have ever lived – the elephants and dinosaurs – that eat huge immobile trees. To resolve this seeming paradox of the smallest organisms in the ocean supporting the largest ones, we need to know some biological oceanography. On land and in the ocean, photosynthetic organisms are at the base of the food web, supporting higher trophic levels. In the ocean, these photosynthetic organisms are seagrasses, algae such as kelp, and phytoplankton. As light in the ocean is only available in the top 100 m or less, seagrass and kelp that are anchored to the seafloor must be in water less that this depth (and usually a lot shallower). However, phytoplankton are single-celled and neutrally buoyant, so they easily float in the sun-lit upper layer, not only in coastal waters but throughout the vast expanse of the ocean. Thus, phytoplankton do not need the energetically expensive structures for support such as the stems and trunks of land-dwelling plants. Being single-celled, phytoplankton also have the advantage that light readily penetrates their cell, and nutrients dissolved in the water are directly taken up through its membrane. Thus, phytoplankton do not need leaves or roots. Therefore, their simple single-celled structure and floating lifestyle enable phytoplankton to not only colonise the entire surface layer of the ocean but to grow

Fig. 1.8. A manta ray feeding on zooplankton (photo: Asia Armstrong).

incredibly quickly compared with land plants – phytoplankton have lifespans of days to a week, whereas land plants have lifespans of months to centuries. So, phytoplankton growth rates are more than an order of magnitude faster than land plants. This means they can support large populations of higher trophic levels that eat them because they are replenished rapidly.

Interestingly, though, large marine animals do not eat phytoplankton directly because most are less than 0.05 mm in size and it is very energy intensive to move a very fine sieve through water to capture them. So, the largest marine animals such as blue whales (up to 30 m long), whale sharks (up to 20 m long), and manta rays (up to 7 m wide, Fig. 1.8) have solved this problem by developing a filter-feeding system with a coarser sieve to remove zooplankton (bigger than 1 mm), and they leave the challenge of capturing phytoplankton to the zooplankton.

1.2.14 Phytoplankton as biofuels and food

During photosynthesis, phytoplankton fix carbon dioxide dissolved in water and convert it into carbon-rich organic compounds. In fact, some phytoplankton species produce carbon-rich oils, which makes them ideal candidates for the production of biofuels. They can be converted into a range of different fuels including biodiesel and bioethanol using thermochemical and biochemical methods.

The classic example of this in fresh waters is *Botryococcus*, which is a large phytoplankton consisting of colonies or compound colonies. Hydrocarbons account for up to 40% of the dry weight of *B. braunii* (Wake and Hillen 1981). Additionally, because phytoplankton grow so quickly and take up carbon from the atmosphere, they offer a multi-pronged approach to tackling climate change. The colonial cyanobacterium 'Spirulina' (*Arthrospira platensis*) often forms enormous blooms in alkaline saline inland waters in Australia, Africa and South America. These blooms are highly nutritious source of protein, and not only sustain large populations of flamingos (in Africa), but are directly harvested for human consumption (*Spirulina* tablets and powder are available in health shops). Several initiatives are underway around the globe growing freshwater and marine phytoplankton to provide biofuels. There are still challenges that need to be overcome. These include sustainable water supplies, treatment of wastewater and availability of nutrients.

1.2.15 Plankton and the Gaia hypothesis

Plankton are a key part of taking up CO_2, fixing it via photosynthesis and ultimately having it removed from the surface waters and to the abyssal deep by the sinking of their remains and faecal pellets. This is not quite as simple as it sounds, but pelagic tunicates such as salps, larvaceans and pyrosomes do have potential to harvest small phytoplankton and transfer it to the sea floor (Henschke *et al.* 2013, 2016; Box 1.1).

Plankton has a key role in the various Gaia hypotheses or philosophy, proposed by Dr James Lovelock and others decades ago, which recognised the interrelationships of the oceans, plankton and the atmosphere and treats the Earth as a living organism. Plankton play a key role in the planet's ability to regulate its climate just as an organism regulates its body temperature (homeostasis). For example, when the planet goes through a warm and drying phase, the wind-driven dust deposits iron into the ocean, which stimulates the production of phytoplankton. Some abundant phytoplankton groups such as diatoms and coccolithophores produce a gas dimethylsulphio-propionate (DMSP). This is broken down to dimethylsulphide (DMS) in the ocean and this outgasses to the atmosphere. DMS is oxidised in the atmosphere to sulphur dioxide, and this forms cloud nucleation centres. These cloud nucleation centres generate clouds, which in turn cause rain but they also reflect more sunlight back into space, having a cooling effect on the climate. While there are many critics of the hypothesis, it is based on the sound science of biogeochemistry, the carbon cycle, and the biological pump, and plankton are the main character.

1.2.16 Plankton could amplify climate change: the anti-Gaia hypothesis

The role of plankton in the global carbon cycle in terms of feedback to the atmosphere is not included in any models used in our current global assessments. This is of concern because the magnitude of projected future climate change could be greater than we currently think because plankton are likely to amplify climate change. Here's how.

Phytoplankton abundance is regulated by nutrients in the top sunlit layer of the ocean. This top 100 m of water is warmed by the sun, and remains near the surface because it is less dense than colder waters below. Phytoplankton rapidly strip nutrients in this top water layer. Thus, phytoplankton increase in abundance when nutrients from cold, deeper water upwell into the surface layer. As the Earth warms with climate change, the surface ocean layer warms more than deeper waters, resulting in stronger stratification. This means that less nutrients will be injected into surface layers, reducing phytoplankton abundance, and thus reducing the food available for fish, birds and dolphins. Current models predict 10–20% fewer fish in our oceans in the future.

Of potentially greater concern is that the impact of climate change on phytoplankton will weaken the biological pump. The biological pump is where atmospheric carbon dioxide dissolved in the ocean is taken up by phytoplankton, which then dies and sinks, removing the fixed carbon from surface layers and locking it up in the deep ocean for centuries to millions of years. Therefore, with a lower abundance of phytoplankton because of increased ocean stratification, the biological pump weakens and removes less carbon dioxide from the atmosphere, meaning that more of the carbon dioxide humans release remains there. This is a positive feedback, as more carbon dioxide remaining in the atmosphere leads to more warming and a weaker biological pump, and again more carbon dioxide remaining in the atmosphere and more warming. This positive feedback is enhanced by the increased stratification leading to smaller phytoplankton cells. Smaller cells do not sink as rapidly as larger heavier cells, so less carbon dioxide is removed to the deep ocean. With warmer upper ocean temperatures, these twin outcomes of reduced phytoplankton abundance and smaller cell size reduce the amount of carbon dioxide removed to the deep ocean, and thus accelerates climate change. It is troubling that no global climate models at present include this positive feedback.

A controversial solution by commercial organisations is to increase the ocean's uptake of CO_2 by fertilising the ocean with dissolved iron, which in some oceans is in limited supply (see Section 2.2). But who is willing to pay? And could we add too much iron and induce another ice-age? It is sobering to think that plankton radically changed the planet's atmosphere 2 billion years ago, and could be harnessed to do so again.

1.3 Plankton, water quality and natural resource management

In this book, we have a focus on water quality management because plankton are sensitive indicators of the state of our waterways and the coastal ocean. The major limiting nutrients for phytoplankton are nitrogen, in the form of ammonia (or in its ionic form NH_4^+), nitrate (NO_3^-), nitrite (NO_2^-) and phosphate (PO_4^{3-}). Nitrogen tends to be the limiting nutrient in marine systems, while phosphate is usually the limiting nutrient in freshwater systems. Nitrogen and phosphorus are needed for cell membranes and for proteins such as enzymes. These two nutrients are therefore of prime importance in water quality, and also because human activities usually enhance their abundance via sewage leaks, land clearing, and agricultural use of fertilisers (Box 1.2). Ammonia in particular is indicative of nutrients from human or animal sewage, while nitrate is indicative of nutrients from oceanographic upwelling. In high concentrations, ammonium (NH_4^+) is very toxic to plankton and fish, but in low concentrations, is more easily assimilated by phytoplankton than nitrate. Two other nutrients – silica (Si) and iron (Fe) – are also limiting nutrients for some phytoplankton and are usually derived from the natural weathering of rocks. Therefore, a useful benchmark is the ratio of N: Si or P: Si in coastal regions, which is used as a measure of human: natural nutrient sources (see Box 1.2).

Water quality and the extent of eutrophication have been assessed for decades by many management authorities from the analysis of water samples for nutrients and chlorophyll content. Such analyses are expensive, and quality control of the chemical analysis and the sampling design has often been inadequate. Compared with oceanographic sampling, nutrients in enclosed waters tend to be higher and vary rapidly over time, requiring collections particularly around rainfall events and with adequate replication (Fig. 1.2). Nutrients may behave in chemically and biologically complex ways – for example, phytoplankton may take up nutrients within hours and simply sequester them, waiting for warmer temperatures. Nutrient samples require stringent conditions for collection (such as wearing rubber gloves and controls to allow for the effect of boat engine exhaust) and laboratory analyses require particular attention to quality control. Many managers and scientists argue that nutrient analyses provide a low value for the environmental dollar, and do not achieve the managers' aims. The frequency and spatial replication that water samples should be collected usually exceeds existing budgets.

Many water quality agencies are now in a position to assess their historical data, and find that it is not adequate to determine if water quality has declined in recent years. There is now the added effect of climate change on urbanisation of waterways and on water quality. Some studies have failed not through lack of funds, but by a sampling design

Box 1.2 Eutrophication and the effects of excess nitrogen

Nutrient ratios provide fundamental insights into the nutrient status in a system and have a long history in aquatic systems. The Redfield ratio provides an indication of the typical nutrient concentrations in marine life, which reflects the elemental ratios in coastal and oceanic waters. It was originally proposed by Alfred Redfield and is the ratio of carbon:nitrogen:phosphorus (C:N:P = 117:14:1). The Redfield ratio has also been extended to include silicate, an important element for some phytoplankton (e.g. diatoms, C:Si:N:P = 106:15:16:1). Sewage and excessive use of fertilisers significantly alter this ratio, which alters the natural species composition of phytoplankton. Therefore, not only the concentration of nutrients but also any changes to the ratio of nutrients can increase the predominance of a single group or species – and some phytoplankton may begin to produce toxins under altered nutrient ratios.

Compared with phytoplankton, seagrass growth needs less nitrogen relative to carbon to manufacture cellulose for structural support. Phytoplankton – and the algae that grows on seagrass – requires proportionally more nitrogen because their cells have little structural support. Consequently, seagrasses thrive in clear, low-nutrient waters and can out-compete algae, taking up the sparse nutrients. When humans release nutrients into waterways, phytoplankton are no longer constrained and begin to shade the seagrass, and algae begin to grow on the seagrass blades. This results in a downward spiral – a positive feedback – because, with slower growth, the seagrass blades become further covered in algae, which further retards their growth and encourages dieback, exposing the sediments and releasing more nutrients. Seagrasses are a useful indicator species of water quality, but they take time to warn of an impending crisis (Fig. 1.2).

that did not have adequate controls or replication. Investment in unreplicated estuarine samples at regular monthly intervals would be better served by concentrating the same sampling effort at replicated sites during the summer and around rainfall events (see Chapter 4). In the presence of surplus nutrients, zooplankton grazers may be overwhelmed by rapid exponential growth of some phytoplankton ('bloom') over and above what the ecosystem can assimilate, becoming eutrophic (Box 1.2).

It is important to remember that many phytoplankton blooms occur naturally; they may be stimulated during the spring, or by natural events such as rainfall or upwelling. Often in subtropical and temperate regions, phytoplankton and zooplankton bloom during the spring to late summer period, prompting public concern. And yet springtime blooms of the blue-green phytoplankton and gelatinous salps are usually examples of natural events.

Nutrient assimilation by plankton, and nutrient 'accountability' to determine if the event is natural or induced by humans highlights the need for using plankton in a study of water quality. Phytoplankton may assimilate surplus nutrients, which may be grazed by zooplankton and productively pass them up the food chain to fish. Some species of phytoplankton or zooplankton can be indicator species of environmental health by integrating the conditions of the past few days or weeks (Fig. 1.2). Furthermore, a few phytoplankton – mainly cyanobacteria and dinoflagellates – produce toxins that become concentrated in filter-feeding animals such as oysters, mussels and even fish.

Through coastal urbanisation and population growth, estuaries and coastal waters have unsavoury swimming conditions, poor fishing and bad press, which translate into reduced spending by tourists and reduced community pride. Natural resource management is a rapidly expanding field, which is increasingly underpinned by rigorous science. In addition to understanding more about the systems we manage, it is also important to measure the performance or outcomes of management decisions and practices. What is the environmental

dollar value for an artificial wetland versus more river bank fencing? This can be achieved by undertaking well-designed, hypothesis-based, monitoring programs.

1.4 Plankton in management

Although the focus of this book is on water quality, plankton are used in management in many other ways and have many other interactions with humans (Table 1.1). We use and rely upon plankton much more than most people realise. It is worth reading through Table 1.1 because it summarises much of the information in this book and it also highlights the critical roles that plankton play in our society. We have also provided links to the various parts of the book that describe the management issues and human interactions in more detail.

1.5 Outline of this book

This book draws together disparate literature and views to convey a modern, pragmatic approach to the study of plankton and water quality. We are writing for non-specialists, particularly those concerned with the quality of waterways and our coastal ocean. The study of plankton is not only a curiosity or a class exercise, but a practical, integrated measure of water quality (Fig. 1.2). We use management issues with examples and discuss logistics to maximise the utility of this guide.

Plankton size is a persistent theme throughout this book. It is the first feature that a beginner can use and it is a pervasive feature in many plankton models of nutrient uptake, growth, longevity and grazing. No plankton keys are provided – instead we provide images and sketches as a guide to the common types and, where applicable, provide a reference to detailed guides.

Chapter 2 provides an overview of plankton ecology and planktonic habitats for the non-specialist. It describes the life cycles of plankton, plankton behaviour including vertical migration, and the effect of tides on plankton and on sampling strategies.

Table 1.1.   Summary of the management issues and human interactions associated with plankton

Issues-interactions	Section of book
Water quality monitoring	Through coastal urbanisation and population growth, the water quality of estuaries and coastal waters can diminish, affecting tourism and the community. Water quality monitoring by natural resource managers includes key plankton indicators such as chlorophyll-*a*, a measure of phytoplankton biomass, a sensitive index of eutrophication (Box 1.2, Box 3.3). Ratios of abundance of key phytoplankton groups such as diatoms and dinoflagellates can provide insight into the health of waterways (Section 3.5). Phosphorus (Box 3.4) and nitrogen (Box 3.5) are usually the nutrients responsible. Monitoring can also help to elucidate when natural processes have caused blooms and associated community angst (e.g. Box 6.7).
Human health and harmful algal blooms	Phytoplankton are also monitored because some species are toxic. Freshwater cyanobacterial blooms can cause skin and eye irritations and digestive upsets in bathers (Sections 3.6, 3.8). Cyanobacteria produce toxins that cause the breakdown of liver cells and attack the nervous system (Box 3.6). Cyanobacterial blooms can also cause death of people or livestock by drinking contaminated water (Section 5.2). Another major issue is that shellfish filter toxic phytoplankton and retain the toxin (Section 3.7), which can result in several syndromes. Zooplankton have also been used widely as indicators to monitor pollution including acidification, eutrophication, pesticide pollution, algal toxins and climate warming (Section 3.8).
Human health and jellyfish	Jellyfish blooms cause major socio-economic impacts on coastal infrastructure and commercial fishing, aquaculture and tourism industries. Jellyfish may clog the cooling water intakes of coastal power plants or ships and the seawater intakes of desalination plants, interrupting supplies of electricity and water (Section 3.11). Stings from venomous jellyfish such as the box jellyfish *Chironex* and Irukandji are a major occupational health and safety issue for marine industries in tropical countries in the Indo-Pacific and Caribbean. Following the death of two bathers to Irukandji stings in 2002, an estimate of the economic loss to the tourism industry due to negative publicity was A\$65 million (Gershwin *et al.* 2009). Jellyfish are also used in education (see Box 9.1 for special handling of jellyfish).
Human health and pathogens	Zooplankton with chitinous exoskeletons, particularly copepods, are hosts for bacterial pathogens such as *Vibrio cholerae*, which is responsible for ~5 million cases and 120 000 deaths per year. Other *Vibrio* species found in zooplankton include *Vibrio parahaemolyticus* and *Vibrio vulnificus*, which were responsible for significant bacterial pandemics.
Positive human health impacts of plankton	Phytoplankton produce omega-3 fatty acids, which may improve health, neurological development in children, reduce inflammation, or reduce breast cancer. The most common source of omega-3 fatty acids is oily fish and krill, which obtain their fatty acids from phytoplankton. Zooplankton, phytoplankton, bacterioplankton and marine viruses are used in bioprospecting and other commercial products. The antioxidant astaxanthin is a red pigment made from decapod shells and has been used in the treatment of many diseases and conditions (Section 8.3.2).
Toxicity testing	Zooplankton such as *Acartia*, larvae of crabs and prawns and mysids (Section 8.3) are used in acute and sub-lethal pollutant impacts because they are extremely sensitive to toxicants, they can be mass cultured, they have a short life cycle, are easy to handle, and they have distinct stages that provides endpoints to determine toxicity of contaminants. Zooplankton are commonly used in toxicity testing of fresh waters, including cladocerans and calanoid copepods (Section 3.8).
Report cards and ecosystem assessments	Report cards are commonly used to communicate monitoring results in the management of waterways (Box 3.1). They have a simple grading system from 'A' to 'F', reflecting pristine to heavily degraded water quality conditions. Plankton are commonly used as indicators in report cards, including chlorophyll-*a* (phytoplankton biomass, Section 3.4), harmful algal blooms (Sections 3.5, 3.6, 3.7), freshwater zooplankton diversity (Section 3.8) and jellyfish abundance. Report cards are useful for coastal eutrophication and phytoplankton blooms, and there are local management actions that can be taken to reduce nutrient inputs. Ecosystem assessments commonly consider a broader suite of human impacts on marine systems including climate change, ocean acidification, overfishing, species invasion and biodiversity change. Ecosystem assessments thus also use a greater variety of plankton indicators: warm/cold water species and assemblages; calcifying plankton such as coccolithophores, foraminifera, echinoderms and molluscs; metrics of species richness; and changes in timing and distribution of key species and assemblages.
Products from zooplankton	Chitosan is produced from decapod shells and is used in water treatment, cosmetics, food and beverages, agrochemicals and pharmaceuticals (Section 8.3.2.1).
Biomimetics	The elegant and diverse structures of diatoms, coccolithophores, silicoflagellates, tintinnids, radiolarians, foraminiferans and acantharians are being used in a new field called biomimetics (Section 1.2.4). This is the imitation of natural systems and elements to solve human problems such as architectural and engineering designs of towers and buildings. Plankton have also evolved different composite materials to make their structure strong and lightweight, and scientists are studying how plankton lay down composite materials in different orientations to resist stresses in particular directions.

(Continued)

Table 1.1.   (Continued)

Issues-interactions	Section of book
Biomanipulation	Biomanipulation is the intentional manipulation of a freshwater ecosystem to remove excessive phytoplankton blooms (e.g. cyanobacteria) by the addition or removal of species. The idea usually is to reduce the abundance of planktivorous fish, which promotes more zooplankton that eat phytoplankton (Section 3.9).
Biological control	Carnivorous cyclopoid copepods, such as those belonging to *Mesocyclops*, can be used in biological control of mosquito larvae in wells, mines and other breeding habitats where mosquito-eating fish are not effective in controlling them. *Mesocyclops* is effective at controlling the larvae of *Aedes aegypti* mosquito larvae, the vector of dengue fever (Section 3.9).
Invasive species	The global movement of ships, with their uptake and discharge of ballast water, transports many species, particularly plankton, into new regions. Plankton often survive transport in a dark ballast tank; dinoflagellates form protective cysts that can germinate later, and zooplankton continue feeding in the dark (Box 3.2).
Zooplankton fisheries	A total of 1.2 million tonnes of jellyfish are caught per year, mainly in China, and used as food (Box 3.9). Krill is harvested in the Southern Ocean (300 000 tonnes per year) and off Japan (70 000 tonnes per year). Krill is used mainly as fish meal, but also as food for people and in capsules for omega-3 essential fatty acids (Section 8.3). Off Norway, 1000 tonnes per year of the copepod *Calanus finmarchicus* is harvested for omega-3 fatty acids (Section 8.3).
Supporting fisheries	Being at the base of the food web, plankton are important food for fish and prawns, especially as larvae and juveniles. Even in coastal areas, much of the primary production is provided by phytoplankton. Almost all wild-caught fish (80 million tonnes per year) and crustaceans (prawns, shrimps, lobsters and crabs: 3.4 million tonnes in 2009) have larval stages that are planktonic (Section 8.3).
Aquaculture	Plankton are used extensively in aquaculture production for feeding shellfish and fish larvae and juveniles (Box 3.8). Aquaculture farms have large columns of phytoplankton grown under ideal conditions to promote photosynthesis and growth. Some common species include the golden brown flagellate (Prymnesiophyte) *Pavlova lutheri*, the diatom *Chaetoceros calcitrans*, and the green flagellate *Tetraselmis suecica*. Phytoplankton are also used feed zooplankton to rear larvae. For example, larval oysters are reared on the rotifer *Brachionus* (Section 7.2). Zooplankton are also grown to feed prawns, fish larvae and juveniles including the copepods *Acartia*, which has resting eggs, *Oithona* and *Oncaea*, as well as mysids (Section 8.3). Krill are used as fish meal for supporting fish and decapods in marine aquaculture. Zooplankton such as isopods are common parasites on farmed and wild-caught fish and prawns (Section 8.3).
Climate change	Plankton are a key part of taking up CO_2, fixing it via photosynthesis and ultimately having it removed from surface waters and to the abyssal deep by the sinking of their remains and faecal pellets (Sections 1.2.15, 1.2.16). This is the biological pump and removes large amounts of CO_2 from the ocean, and then more CO_2 diffuses from the atmosphere into the ocean, reducing CO_2 levels in the atmosphere. The type of phytoplankton and zooplankton present influences the magnitude of the biological pump. How the biological pump will be stimulated or impeded with climate change is an area of active research (Box 1.1).
Seismic surveys	Marine seismic surveys are widespread throughout the world's oceans for oil and gas exploration. They use intense sound pulses generated by releasing compressed air from airguns. There are concerns about the impact of these sound pulses to marine life, particularly for marine mammals, because sound travels faster, further, and with more energy (lower attenuation) in water than in air. Zooplankton do not have hearing structures and are the same density as sea water so it has been assumed that seismic activity has little impact. However, a recent experimental study by McCauley *et al.* (2017) overturned conventional thinking and suggested that an airgun could cause significant zooplankton mortality (32%). If true, this could have serious consequences for ecosystem function and productivity.
Aquaria	Their lifestyle of living on substrates make harpacticoids the preferred copepod species in marine aquaria, because they clean the substrate and aquarium panels, and their nauplii and copepodites provide food for invertebrates such as corals, clams and sea cucumbers.
Model assessment	The most common models used in water quality assessment are biogeochemical models (Section 3.1). These models typically include the physical oceanography (temperature, salinity, currents), runoff from freshwater sources, nutrient dynamics, primary producers (phytoplankton, benthic microalgae, seagrass, kelp) and zooplankton. Plankton are used in model assessment: a process whereby the model is tested against data. Commonly, chlorophyll-*a* measurements from satellite are used to assess phytoplankton in biogeochemical models. Zooplankton cannot be measured from satellite and thus zooplankton from net samples is usually used.

Particular examples of water quality issues are provided in Chapter 3, which shows how solutions were provided through plankton studies. The plankton issues are structured around a problem for management. These water quality issues should be read before tackling the details of plankton in the subsequent chapters.

Chapter 4 covers how to sample plankton and the advantages and disadvantages of different types of sampling gear. We begin the chapter by giving examples of good and poor sampling designs. We provide an overview of some sampling designs that are necessary to detect environmental impact and change.

The next four chapters are the core of this book – providing a general guide to the major groups of plankton. An overview to identifying larger freshwater and marine phytoplankton is provided in Chapters 5 and 6, respectively. These larger phytoplankton can be observed with a drop of water sandwiched between a microscope slide and coverslip and using a basic compound microscope. We discuss phytoplankton toxicity, but this is a complex topic because there is no simple guide as to whether a particular cell is toxic. Chapters 7 and 8 cover freshwater and marine zooplankton, respectively. We have taken a pragmatic approach to our guide, focusing on what is new for someone working with plankton and drawing on several useful local guides. In Chapter 8, which is the largest chapter in the book, we have classified marine zooplankton into basic body types that are based on the shape, texture, size, and degree of transparency of a specimen, and are only loosely based on taxonomy. For example, one of the body types is 'worm-like', which includes zooplankton from several different phyla that appear worm-like, but only one of which comprises true worms. Once a specimen is placed into a body type, one can then identify its specific taxonomic group based on a few distinguishing features. For each taxonomic group, we describe important information such as its identification, diversity, common species, ecology and human interactions. We illustrate this chapter extensively with comparative images and new line drawings, especially to separate difficult-to-distinguish groups.

In Chapter 9, we show how plankton may be used in the classroom for primary and secondary schools, and for university students studying environmental science. Plankton is an exciting subject and can illustrate key biological and ecological concepts in the curriculum.

1.6 References

Atkinson A, Seigel V, Pakhomov E, Rotherly P (2004) Long-term decline in krill stock and increase in salps within the Southern Ocean. *Nature* **432**, 100–103. doi:10.1038/nature02996

Flombaum P, Gallegos JL, Gordillo RA, Rincon J, Zabala LL, Jiao N (2013) Present and future global distributions of the marine Cyanobacteria *Prochlorococcus* and *Synechococcus*. *Proceedings of the National Academy of Sciences* **110**(24), 9824–9829. doi:10.1073/pnas.1307701110

Fulton EA, Parslow JS, Smith ADM, Johnson CR (2004) Biogeochemical marine ecosystem models II: The effect of physiological detail on model performance. *Ecological Modelling* **173**, 371–406. doi:10.1016/j.ecolmodel.2003.09.024

Gershwin L, De Nardi M, Fenner PJ, Winkel KD (2009) Marine stingers: review of an under-recognized global coastal management issue. *Journal of Coastal Management* **38**, 22–41. doi:10.1080/08920750903345031

Henschke N, Bowden DA, Everett JD, Holmes SP, Kloser RJ, Lee RW, *et al.* (2013) Salp-falls in the Tasman Sea: a major food input to deep sea benthos. *Marine Ecology Progress Series* **491**, 165–175. doi:10.3354/meps10450

Henschke N, Everett JD, Richardson A, Suthers IM (2016) Rethinking the role of salps in the ocean. *Trends in Ecology & Evolution* **31**, 720–733. doi:10.1016/j.tree.2016.06.007

Heron AC (1972) Population ecology of a colonizing species: the pelagic tunicate: *Thalia democratica*. I. population growth rate. *Oecologica* **10**, 294–312.

Hopcroft RR, Roff JC (1995) Zooplankton growth-rates – extraordinary production by the larvacean

Oikopleura dioica in tropical waters. *Journal of Plankton Research* **17**(2), 205–220.

McCauley RD, Day RD, Swadling KM, Fitzgibbon QP, Watson RA (2017) Widely used marine seismic survey air gun operations, negatively impact zooplankton. *Nature Ecology and Evolution* **1**, 1–8. doi:10.1038/s41559-017-0195

McLeod DJ, Hallegraeff GM, Hosie GW, Richardson AJ (2012) Climate-driven range expansion of the red-tide dinoflagellate *Noctiluca scintillans* into the Southern Ocean. *Journal of Plankton Research* **34**, 332–337. doi:10.1093/plankt/fbr112

Neira FJ, Miskiewicz AG, Trnski T (Eds) (1998) *Larvae of Temperate Australian Fishes. Laboratory Guide for Larval Fish Identification.* University of Western Australia Press, Perth, Australia.

Pohl G, Nachtigall W (2015) *Biomimetics for Architecture and Design. Nature – Analogies – Technology.* 1st edn. Springer, Heidelberg, Germany.

Richards WJ (1989) 'Preliminary guide to the identification of the early life history stages of scombrid fishes of the western central Atlantic'. NOAA Technical Memorandum NMFS-SEFC-240, U. S. Department of Commerce, National Oceanic and Atmospheric Administration, National Marine Fisheries Service, Southeast Fisheries Center, Miami FL, USA.

Schminke HK (2007) Entomology for the copepodologist. *Journal of Plankton Research* **29**(1), i149–i162.

Wake LV, Hillen LW (1981) Nature and hydrocarbon content of blooms of the alga *Botryococcus braunii* occurring in Australian freshwater lakes. *Australian Journal of Marine and Freshwater Research* **32**, 353–367. doi:10.1071/MF9810353

Widder EA (2010) Bioluminescence in the Ocean: Origins of Biological, Chemical, and Ecological Diversity. *Science* **328**, 704–708. doi:10.1126/science.1174269

1.7 Further reading

Brotz L, Cheung WWL, Kleisner K, Pakhomov E, Pauly D (2012) Increasing jellyfish populations: trends in Large Marine Ecosystems. *Hydrobiologia* **690**, 3–20. doi:10.1007/s10750-012-1039-7

Champion C, Suthers IM, Smith J (2015) Zooplanktivory is a key process for fish production on a coastal artificial reef. *Marine Ecology Progress Series* **541**, 1–14. doi:10.3354/meps11529

CSIRO Huon Estuary Study Team (2000) 'The Huon Estuary study – the environmental research for integrated catchment management and aquaculture'. Final report to FRDC. Project no 96/284 June 2000. CSIRO Division of Marine Research, Marine Laboratories, Hobart, Australia.

Dennison WC, Abal EG (1999) *Moreton Bay Study: A Scientific Basis for the Healthy Waterways Campaign.* South East Queensland Regional Water Quality Management Strategy, Brisbane, Australia.

Fuiman LA, Werner RG (Eds) (2002) *Fishery Science: The Unique Contributions of Early Life Stages.* Wiley-Blackwell, Oxford, UK.

Ghasemi Y, Rasoul-Aminia S, Naserib AT, Montazeri-Najafabadya N, Mobashera MA, Dabbagha F (2012) Microalgae biofuel potentials. *Applied Biochemistry and Microbiology* **48**, 126–144 [Review]. doi:10.1134/S0003683812020068

Haddock SHD, Moline MA, Case JF (2010) Bioluminescence in the Sea. *Annual Review of Marine Science* **2**(1), 443–493. doi:10.1146/annurev-marine-120308-081028

Hallegraeff GM (2006) *Plankton: A Critical Creation.* University of Tasmania, Hobart, Australia.

Hallegraeff GM (2010) Ocean climate change, phytoplankton community responses, and harmful algal blooms: a formidable predictive challenge. *Journal of Phycology* **46**(2), 220–235. doi:10.1111/j.1529-8817.2010.00815.x

Harris GG, Batley G, Fox D, Hall D, Jernakoff P, Molloy R, *et al.* (1996) 'Port Phillip Bay Environmental Study Final Report'. CSIRO, Canberra, Australia.

Hays GC, Richardson AJ, Robinson C (2005) Climate change and marine plankton. *Trends in Ecology & Evolution* **20**, 337–344. doi:10.1016/j.tree.2005.03.004

Kingsford MJ, Battershill CN (1998) *Studying Temperate Marine Environments.* University of Canterbury Press, Christchurch, New Zealand.

Kingsford MJ, Pitt KA, Gillanders BM (2000) Management of jellyfish fisheries, with special reference to the Order Rhizostomeae. *Oceanography and Marine Biology - an Annual Review* **38**, 85–156.

Kamio M, Furukawa D, Wakabayashi K, Hiei K, Yano H, Sato H, *et al.* (2015) Grooming behavior by elongated third maxillipeds of phyllosoma larvae of the smooth fan lobster riding on jellyfishes. *Journal of Experimental Marine Biology and Ecology* **463**, 115–124. doi:10.1016/j.jembe.2014.11.008

Kunz TJ, Richardson AJ (2006) Impacts of climate change on phytoplankton. In *Impacts of Climate Change on Australian Marine Life: Part C, Literature Review.* (Eds AJ Hobday, TA Okey, ES Poloczanska, TJ Kunz and AJ Richardson) pp. 8–18. Report to the Australian Greenhouse Office, Canberra, Australia.

Kunz TJ, Richardson AJ (2006) Impacts of climate change on zooplankton. In *Impacts of climate change on Australian marine life: Part C, Literature Review.* (Eds AJ Hobday, TA Okey, ES Poloczanska, TJ Kunz and AJ Richardson) pp. 19–26. Report to the Australian Greenhouse Office, Canberra, Australia.

Newell GE, Newell RC (Eds) (1977) *Marine Plankton: A Practical Guide.* Anchor Press, London, UK.

Orr JC, Fabry VJ, Aumont O, Bopp L, Doney SC, Feely RA, *et al.* (2005) Anthropogenic ocean acidification over the twenty-first century and its impact on calcifying organisms. *Nature* **437**, 681–686 doi:10.1038/nature04095.

Parsons TR, Takahashi M, Hargrave B (1984) *Biological Oceanographic Processes.* 3rd edn. Pergamon Press, Oxford, UK.

Richardson AJ (2008) In hot water: zooplankton and climate change. *ICES Journal of Marine Science* **65**, 279–295. doi:10.1093/icesjms/fsn028

Richardson AJ, Bakun A, Hays GC, Gibbons MJ (2009) The jellyfish joyride: causes, consequences and management responses to a more gelatinous future. *Trends in Ecology and Evolution* **24**, 312–322.

Sardet C (2015) *Plankton: Wonders of the Drifting World.* The University of Chicago Press, Chicago IL, USA.

Scott S, Davey M, Dennis J, Horst I, Howe C, Lea-Smith D, Smith A (2010) Biodiesel from algae: challenges and prospects. *Current Opinion in Biotechnology* **21**, 1–10.

Smayda TJ (1997) Harmful algal blooms: their ecophysiology and general relevance to phytoplankton blooms in the sea. *Limnology and Oceanography* **42**, 1137–1153. doi:10.4319/lo.1997.42.5_part_2.1137

2

Plankton processes and the environment

Iain M. Suthers, Anna M. Redden, Lee Bowling,
Tsuyoshi Kobayashi and David Rissik

The diversity and beauty of plankton is completely absorbing: each little plant or animal is exquisite. A few moments at a microscope brings out the curiosity in any of us, and soon the questions come tumbling out. Before your eyes in that teaspoon of water is an ecosystem, from the fixation of carbon dioxide by photosynthetic cells, to herbivores (grazers), carnivores, and carnivores of carnivores. Fresh plankton can be beautifully coloured, or iridescent, or bioluminescent but the majority are often translucent. They are characterised by small size and are surrounded by water that is glutinous or viscous and must therefore work hard to swim. Consequently, the planktonic world can be a lonely place for individuals, even if separated by only a few millimetres because short distances are like a marathon! Virtually all plankton are characterised by spikes and shells to deter predation: the planktonic world is not a peaceful one but a dog-eat-dog world.

Almost all the diversity on land has representatives in plankton, and if one includes bacteria and viruses there is a thousand-fold greater diversity (with all the pharmaceutical implications). It has been estimated that in terms of goods and services the oceans are worth US$21 trillion (Costanza *et al.* 1997), but how can you value a breath of fresh air or a bowl of fresh fish? The following chapter introduces this vital community, their habits and habitats.

2.1 Introduction to plankton ecology

The distribution and abundance of plankton depends on the physical and chemical factors operating in the aquatic environment, which makes plankton so useful for managing natural resources. Plankton habitats are defined largely by temperature, salinity and nutrients (nitrate and phosphate). Within such habitats, the major biological trait that affects plankton ecology is size, from minute viruses and bacteria, to the microscopically visible phytoplankton and small invertebrate larvae, to the large gelatinous zooplankton (jellyfish). In fact, planktonic organisms span seven orders of magnitude in length: from 0.2 μm to ~2 m. A micrometre (μm), or 'micron', is a thousandth of a millimetre (i.e. 1 μm = 0.001 mm). A human hair is about 30 to 100 μm thick (33 to 10 hairs = 1 mm); the standard pin used to package shirts is ~600 μm (0.6 mm) thick; and a dissecting needle used in many science classes as a plankton probe is ~1 mm thick. It will be useful for you to check these dimensions using a microscope and ruler as your microscopic benchmarks, particularly for zooplankton. To appreciate size and scale, another useful measurement is to use copepod size, which are abundant and mostly 1–5 mm (e.g. Chapters 7 and 8).

As there are significant ecological and physiological implications of body size in plankton, we use size as a useful step in classification. The various size categories of plankton are as follows:

- **megaplankton** are those large floating organisms that exceed 20 cm in length. They are represented by very large jellyfish, salps and their relatives.
- **macroplankton** (2–20 cm, Fig. 2.1A) include large visible organisms such as krill, arrow worms, comb jellies and jellyfish.
- **mesoplankton** (0.2–20 mm, Fig. 2.1B) are very common and visible to the naked eye; they are diverse and include copepods, cladocerans, small salps, the larvae of many benthic organisms and fish, and others.
- **microplankton** (20–200 μm, Fig. 2.2A) include large phytoplankton (large single-celled or chain-forming diatoms, dinoflagellates), foraminiferans, ciliates, nauplii (early stages of crustaceans such as copepods and barnacles), and others.
- **nanoplankton** (2–20 μm, Fig. 2.2B) include small phytoplankton (mostly single-celled diatoms), flagellates (both photosynthetic and heterotrophic), small ciliates, radiolarians, coccolithophorids and others.
- **picoplankton** (0.2–2 μm) are mostly bacteria (called bacterioplankton); they require at least 400 × magnification for detection and counting. Marine viruses are even smaller (less than 0.2 μm).

The size categories listed do not reflect particular taxonomic divisions because sizes vary greatly within most taxonomic groups. In addition, size does not reflect any trophic classification. Small plankton may include photosynthetic cells (i.e. autotrophs or 'self-feeders'), herbivores, carnivores or omnivores (i.e. heterotrophs like us). Many phytoplankton cells maintain hundreds of other small symbiotic cells around them, sometimes for their nitrogen fixation (such as by blue-green algae, also known as cyanobacteria). Some organisms even maintain symbiotic relationships with photosynthetically active cells known as zooxanthellae (as in many corals, sea anemones, sponges and clams of tropical coral reefs). Large plankton, such as some jellyfish with symbiotic algae, are akin to carnivorous plants: capturing copepods and small fish for their nitrogen.

Cell size has direct consequences for many physiological processes, including the assimilation of dissolved nutrients from the environment. Picoplankton (cell size: 0.2–2 μm), about the size of bacteria, can dominate the phytoplankton, contributing up to half the chlorophyll-*a* content in coastal waters, up to 90% in nutrient-poor oceanic waters, and producing much of the oxygen we breath.

Low-nutrient (oligotrophic) waters are typically dominated by small phytoplankton cells, which are much more efficient at using small amounts of available nutrients than are large cells. Small phytoplankton have a competitive advantage under low-nutrient conditions because they have a higher cell surface area:volume ratio than large phytoplankton with which to take up available nutrients across their cell membrane. For the most part, large phytoplankton cells appear in abundance primarily in response to periodic nutrient increases (e.g. seasonal rain events) and/or localised inputs. Other features of plankton that are related in some non-linear way with size are growth, carbon content, sinking rates, grazing, swimming, fecundity and longevity (Baird and Suthers 2007).

2.2 Plankton food webs

The most important elements for phytoplankton growth are the macronutrients nitrogen (N) and phosphorus (P) and, for diatoms, silica (Si). Phytoplankton cells take up dissolved forms of N and P across their cell surfaces, normally attaining an atomic ratio of ~106C:16N:1P (i.e. ~42:7:1 by weight) in their tissues (the Redfield ratio). Sometimes the atomic ratios of dissolved nutrients in the water column are different to those required for phytoplankton growth. This provides an important signal to managers and researchers. N:P atomic ratios that are much higher than 16 (say, 25 to 30) suggest that P limitation of algal growth is occurring, which means that the lack of phosphorus is preventing further algal growth. Alternatively, a ratio of less than 10 would imply N limited growth.

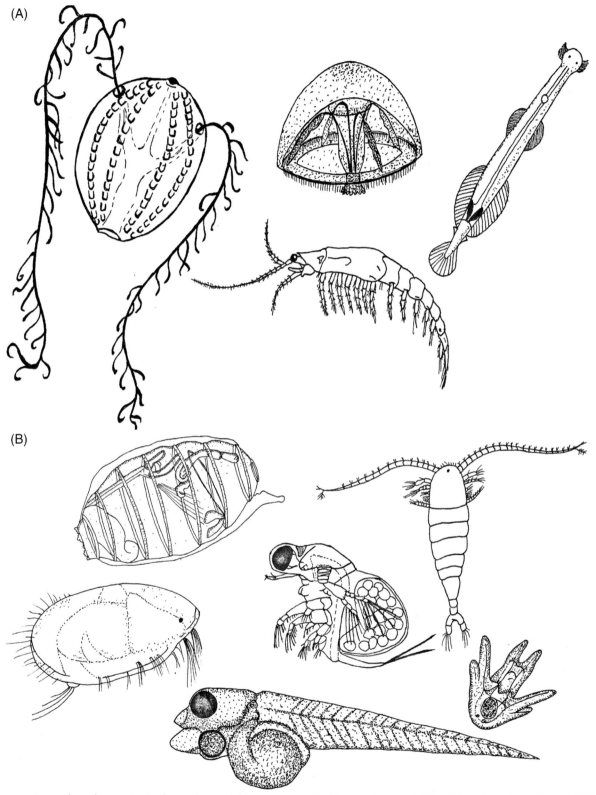

Fig. 2.1. Examples of some typical members of the **(A)** macroplankton (2–20 cm, left to right: ctenophore, krill, jellyfish, arrow worm) and **(B)** mesozooplankton (0.2–20 mm, left to right: ostracod, salp, larval fish, cladoceran, copepod, pluteus larva of a sea urchin).

(A)

(B)

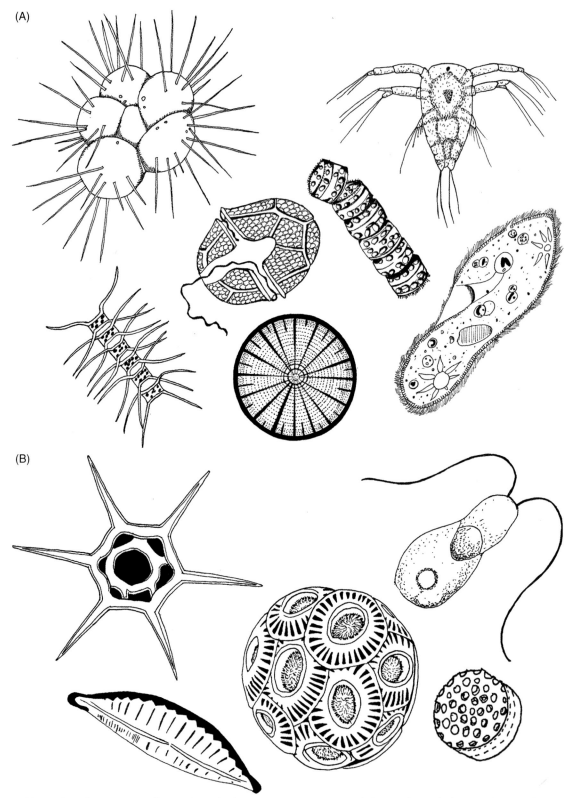

Fig. 2.2. Examples of some typical **(A)** microplankton types (20–200 μm, left to right: radiolarian, diatom chain, armoured dinoflagellate, centric diatom, dinoflagellate chain, nauplius (larval crustacean), ciliate) and **(B)** nanoplankton types (2–20 μm, left to right: silicoflagellate, pennate diatom, coccolithophore, flagellate, diatom).

While phytoplankton growth in freshwater systems is generally P limited, growth in estuarine and oceanic environments is commonly N (and at times also Si) limited. Phytoplankton cells require external sources of other inorganic nutrients, in particular trace metals (Fe, Mg, Zn, Mn and others) and vitamins (thiamine, biotin and B_{12}). These are needed in much lesser quantities and are generally assumed (wrongly at times) to be in sufficient quantities for growth.

In some regions of the world's oceans, phytoplankton cells have access to relatively high levels of N and P yet exhibit low biomass (generally determined by chlorophyll-*a* concentration). A series of elaborate experiments in the equatorial Pacific demonstrated that this 'high-nutrient, low-biomass' phenomenon was due to iron limitation (Behrenfeld *et al.* 1996). This effect is not observed in ocean waters that receive iron via winds that carry iron-rich desert dust (Sigman and Hain 2012).

Field experiments aimed at investigating nutrient deficits and effects on plankton productivity can be effort intensive and expensive to conduct. In many cases, carefully conducted laboratory experiments can be useful for assessing nutrient limitation and for testing the effects of different nutrient combinations (cocktails) on rates of phytoplankton productivity.

In areas of low phytoplankton productivity, most of the phytoplankton growth is sustained through 'regeneration' of nutrients, which serves to maintain biomass levels. This happens when organic matter (e.g. faecal pellets and dead and decaying material) is remineralised to dissolved inorganic nutrients via microbes in the plankton. 'New' production (largely medium to large-sized phytoplankton) occurs in response to external nutrient inputs (from catchments, rivers, atmosphere), or when turbulent diffusion allows deep-water nutrients to cross the thermocline (nutricline) into the surface mixed layer. The ratio of new to regenerated production is referred to as the *f* ratio and has been used as an index of trophic status. The lower the *f* ratio, the greater the dependence on regeneration of nutrients via microbes via planktonic microbes (largely small unicellular cyanobacteria). Conversely, high *f* ratio values are typical of upwelled waters with 'new' nutrients, dominated by diatoms and other large phytoplankton cells on which large zooplankton graze (Dunne *et al.* 2005).

Grazers represent an essential trophic pathway for the transfer of organic carbon from phytoplankton to fish, and they contribute to the nutrient pool by excreting faecal pellets that are either recycled within the water column or used by bottom feeders. Nutrient recycling is also assisted by the 'sloppy feeding' or partial ingestion of cells by herbivorous zooplankters (such as copepods), which results in the release of nutrient-rich cell sap following handling and rupture of captured cells.

Trophic transfer, however, is no longer understood simply as materials and energy passing through producers and a series of consumers in a simple linear chain (the classical food chain). The traditional model of a short marine food chain (phytoplankton → copepod → fish) became obsolete following recognition of the trophic importance of bacterioplankton and protozoans in marine waters (Worden *et al.* 2015). It is now accepted that a significant proportion of phytoplankton production is not consumed directly by zooplankton grazers, but is cycled by the microbial community ('microbial loop') before it becomes available to consumers.

The primary organisms involved in the recycling activities of the microbial loop (Fig. 2.3) are water-column bacteria, heterotrophic flagellates and ciliates. One of the roles of the bacteria is to break down organic molecules contained in non-living particulate organic matter (POM) and dissolved organic matter (DOM) derived from living cells, faecal pellets and dead and decomposing remains. The bacteria convert organic matter to dissolved inorganic nutrients (DIN), such as nitrogen, phosphorus and potassium, which are then available for rapid uptake by phytoplankton. The bacteria are consumed by protozoans (ciliates and nano-flagellates), which are in turn food sources for other zooplankton.

The recycling of POM by the microbial loop also serves to reduce the sedimentation of faecal matter and detritus. This is particularly important in warm, low-nutrient waters, where microbes rapidly and efficiently recycle materials and thus limit the sinking of large amounts of organic matter to the bottom. In cold waters – and during the winter months in many temperate regions – microbial activity is suppressed. The effects are that most of the carbon reaches higher trophic levels directly via the grazing activities of zooplankton, and a large fraction of the carbon fixed during photosynthesis sinks to the bottom, where it is then used by benthic communities.

Numerous feeding strategies are employed by small zooplankton (ciliates and flagellates) including herbivory, carnivory and omnivory. But a strategy commonly used by many is 'mixotrophy' – a feeding strategy that combines characteristics of both autotrophs (which make their own food via photosynthesis) and heterotrophs (which ingest food). Numerous species of ciliates that are known to exhibit mixotrophy contain large numbers of chloroplasts (light-harvesting organelles) sequestered

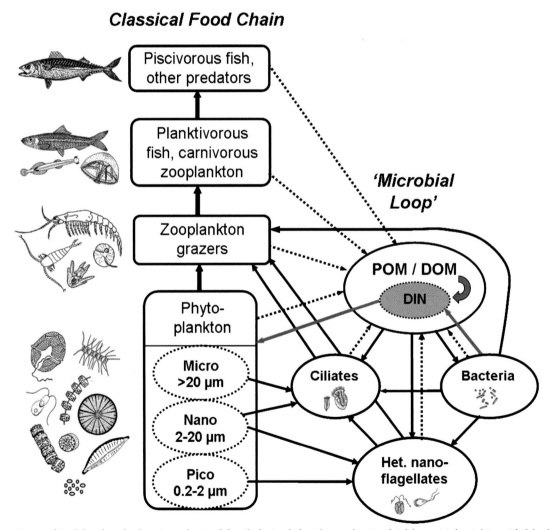

Fig. 2.3. Generalised food web showing classical food chain (left side) and microbial loop (right side), with black arrows showing trophic pathways, flow of particulate and dissolved organic matter (POM, DOM) in excretory products and dead organisms (dashed arrows), and flow of dissolved inorganic nutrients (DIN) to phytoplankton (grey arrows). Het. = heterotrophic (adapted from Beardall and Redden 2007).

from ingested phytoplankton. They derive nutrition from both the direct ingestion of food and by the carbohydrates made by the sequestered photosynthetically active chloroplasts (Stoecker 1987). This nutritional strategy offers great survival and competitive advantages, especially in environments where food resources are highly variable.

2.3 Plankton behaviour, buoyancy and vertical migration

Cell size has a significant impact on the ability of phytoplankton cells to maintain their position at depths with adequate light and nutrients to sustain growth. In general, an increase in cell size results in an increase in sinking rate – with dead cells sinking at faster rates than live cells. Large phytoplankton cells (such as diatoms) are disadvantaged by being highly susceptible to sinking, and may require strong vertical mixing (e.g. caused by upwelling or strong winds) to maintain their position in surface waters.

Sinking of cells can be reduced by morphological structures that increase cell, or colony, resistance to sinking. The flagella of many nanoflagellates serve, in part, to overcome sinking. Adaptations of large and heavy cells (large diatoms and dinoflagellates) to reduce sinking, and to maintain near neutral buoyancy and vertical position in the euphotic zone, include chain formation and cell extensions that provide a high surface area:volume ratio. Cell extensions can be highly numerous and include protuberances, spines, horns, wings and hair-like structures. They increase frictional drag and also increase the effective size of phytoplankton cells, which makes them more difficult for zooplankton grazers to capture and ingest. Another advantage of cell extensions – particularly diatom spines – is that they can house large numbers of chloroplasts and thus increase the ability of cells to harvest light for photosynthesis.

Cell density, and thus rate of sinking, is also affected by the composition of cells. Silica-laden diatoms are particularly heavy. Mechanisms to control cell density, and thus location within the water column, may include production of gas vacuoles and the accumulation of fats and oils, which are lighter than water. Cell ageing and nutritional state of phytoplankton cells are physiological conditions that affect cell density. Post-bloom nutrient-starved diatoms tend to sink significantly faster than nutrient-rich diatoms (Sarthou *et al.* 2005). This effect is frequently demonstrated in temperate and polar waters, where mass sinking of phytoplankton blooms occurs following nutrient exhaustion (Thompson *et al.* 2008). A large proportion of bloom material may settle to the bottom as diatom flocs or aggregates (>0.5 mm) composed of algal cells, zooplankton remains, faecal pellets and other forms of detritus. These highly visible settling flocs are commonly referred to as 'marine snow'.

Zooplankton features that increase drag, and thus reduce sinking, include long, thin or flattened body shapes, and projections such as hairs, long spines and wings. Buoyancy may also be assisted by small droplets of oil. Many planktonic animals can swim reasonably well, or are able to control their position by selecting different depths and currents, or by adjusting buoyancy. Many species of crustacean zooplankton – especially the adult forms – are strong swimmers and conduct diel (over 24 hours) vertical migrations through the water column (Fig. 2.4). This involves rising to surface waters at dusk and grazing heavily on phytoplankton cells throughout the night, before descending to deeper waters well before dawn (although some interesting cases of reverse migrations are known: that is, rising up in the day, and dropping back down at night; Bayly 1986). The distance travelled during diel vertical migration can range from a very short distance (less than 2 m in coastal lagoons) to hundreds of metres up and down in 24 hours in oceanic waters).

Diel vertical migration (DVM) is triggered by changes in light intensity, and is largely an adaptation to avoid visually feeding predators, particularly fish. Migratory patterns can be variable, and are known to differ with the sex and age of the species, habitat type and season (Hays 2003; Brierley 2014). Many gelatinous plankton (such as jellyfish)

Day	Night

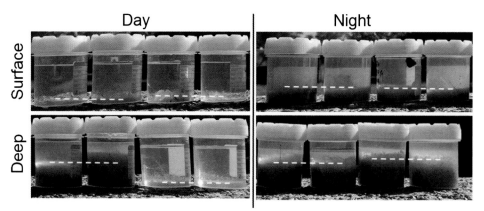

Fig. 2.4.   Representative catches of zooplankton during the day and during the night. Both top panels were 5-minute surface plankton tows (see Fig. 4.5A for the method); both bottom panels were from near-bottom tows with a depressor. A white dashed line indicates the top of the settled biovolume; note how there is virtually no zooplankton in the surface daytime tows compared with the night. Trends at depth are a little more complicated. In some years there may be no difference between day and night zooplankton abundance (photo: Simon Gorta).

and larval crustaceans (such as prawns) exhibit tidal-driven vertical migrations into estuaries. They move up into the flood tide waters – especially at night – and are transported into the estuary, and move lower in the water column during ebb tides to avoid being carried out. Such migrations are entrained into the circadian rhythm of many organisms, such that some diel and tidal activities continue to be observed even after the organisms are removed from their natural environment (e.g. when maintained in a laboratory).

2.4 Life cycles of zooplankton

In general, the smallest plankton have the shortest life cycles (generation time): bacteria and flagellates generally multiply within a few hours to 1 day. Most mesozooplankton have life cycles of a few weeks, while the macro- and megaplankton usually have life cycles spanning many months and longer.

Many zooplankton spend their entire life cycle as plankton (e.g. copepods, salps and some jellyfish) and are called holoplankton. The meroplankton, which are seasonally abundant, especially in coastal waters, are only planktonic for part of their lives (usually at the larval stage). Most bear little, if any, resemblance to the adult form and drift for days to weeks before they metamorphose and

assume benthic or nektonic lifestyles. Examples of meroplankton include the larvae of sea urchins, starfish, crustaceans, marine worms and most fish (Fig. 2.5).

The general copepod life cycle includes six naupliar stages (larvae) and five copepodid stages (juveniles) before becoming an adult. Each stage is separated by a moult and, as the stages progress, the trunk of the copepod develops segmentation. Sexes are separate, sperm is transferred in a spermatophore from the male to the female, and eggs are either enclosed in a sac until ready to hatch or released as they are produced. Development times from egg to adult are typically 2 to 6 weeks, and are significantly affected by temperature and food availability. The lifespan of adults may be from one to several months.

Barnacles also have free-swimming nauplius stages, followed by a carapace-covered cyprid stage after the final naupliar moult. Cyprid larvae are attracted to settle on hard substrates by the presence of other barnacles, ensuring settlement in areas suitable for barnacle survival and for obtaining future mates. After settling, the cyprid releases a substance to permanently cement itself to the substrate. Calcareous plates then grow and surround the body. The appendages face upwards to form cirri that sweep food particles towards the mouth. The adults are hermaphroditic (each with

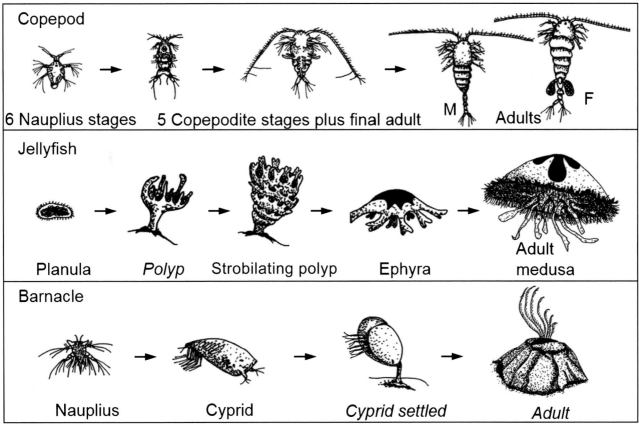

Fig. 2.5. Life stages (larval to adult form) of a typical copepod, barnacle and jellyfish. Names in italics refer to those life stages that are not planktonic, when the animal becomes attached to hard surfaces.

both male and female gonads) and reproduce sexually by cross fertilisation. The adult broods the fertilised eggs within the shell until they develop into nauplius larvae. Over 10 000 larvae may be released by a single adult.

Life cycles of jellyfish are meroplanktonic, with generally two adult morphologies: the sessile polyp and the planktonic medusa (the typical jellyfish morphology). The sexes are separate and mature adult medusae release eggs and sperm, which, upon fertilisation, form free-swimming larvae known as planulae. After a few days to weeks, the planulae settle on hard substrates and metamorphose into tiny sessile polyps (which look like upside-down jellyfish), which clone themselves and bud (strobilate). Juvenile jellyfish (ephyrae) peel off from the stack, float into the plankton as young jellies and grow into adult medusae. This transformation can take a few weeks up to a few

years, depending on the temperature and species of jellyfish.

2.5 Freshwater habitats of plankton

There is a wide variety of inland aquatic systems – ranging from rivers and streams to lakes and reservoirs, farm dams and ponds, billabongs and wetlands (Fig. 2.6). Due to low rainfall and high evaporation, there is often a scarcity of permanent water bodies. Rivers and streams are often ephemeral – containing flowing water only after rainfall. Natural lakes are rare; reservoirs built to conserve water for town water supply and for irrigation are more common.

Inland waters – as distinct from estuarine or marine environments – are often fresh, with low concentrations of dissolved salts. However, some inland waters can be salt lakes with salinities

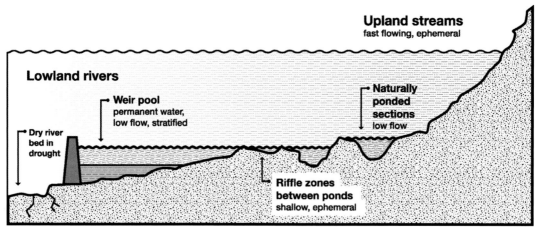

Fig. 2.6. Diagram of a stream network and pool formation as plankton habitat. Upland streams provide an input of nutrients, but the naturally ponded sections have a reduced flow rate, which allows water residence times to match phytoplankton cell doubling times. Riffle zones can provide habitat for benthic forms. The weir provides permanent water, but can become stratified and without atmospheric mixing or photosynthetic production, the lower layer becomes de-oxygenated.

greater than that of sea water. Williams (1980), in arbitrary terms, defined fresh water as that with a salinity of less than 3 g/L of dissolved salts. In lowland areas with low rainfall and high evaporation, salts are often dominated by sodium and chloride, rather like sea water. In upland headwater streams and reservoirs, calcium and magnesium bicarbonates may be the predominant salts present (contributing to water hardness).

During flood events, rivers may break out of the confines of their river channel, with their waters then spreading out over the floodplain. On these occasions, they can also transport large quantities of sediment and nutrients downstream from the catchment. During droughts, stream flow in permanent rivers is sustained by drainage from adjacent groundwater systems, while many others cease flowing, with only isolated pools remaining. The characteristically shallow nature, steep gradients and high flow velocities of upland rivers and streams keeps their waters well mixed (Fig. 2.6). In lowland country, gradients are small and channels become broad and meandering, or split into anabranches and distributary channels – with many terminating in extensive wetland areas. Lowland rivers may be impounded in natural ponds or by constructed weirs where flow velocities decrease and the resultant ponds become deeper and more

lake-like, including stratification of the water column during summer. Nutrient and light availability, rate of flow, and stratification will all affect plankton community composition and abundance (Mitrovic *et al.* 2003).

Flowing river systems are generally not good habitats for plankton, because the organisms entrained within the water column are continually displaced downstream. However, some of the larger lowland rivers may develop their own riverine phytoplankton communities that develop within parcels of water as these traverse the length of the river. Most algal growth in smaller, shallower, faster flowing streams, however, is confined to clumps of filamentous algae attached to a secure substrate to prevent themselves from being washed away, and to films of microscopic algae coating the surfaces of rocks, mud, sticks and aquatic macrophytes. These algae obtain the substances they require to sustain their growth as the water flows over them. The weir pools and ponded sections of lowland rivers and streams may, however, become suitable habitats for phytoplankton to form blooms. Some rivers also have small embayments, inlets or backwater areas where water movement may be minimal. These areas – known as 'dead zones' – are also areas where phytoplankton can develop (Mitrovic *et al.* 2001).

Lakes, reservoirs, farm dams, ponds, billabongs and wetlands are characterised by prolonged residence times of the water they contain and the limited mixing of water within them – apart from that caused by wind-driven currents and internal-heat-transfer processes. Deeper lakes and reservoirs undergo strong thermal stratification during the warmer months of the year, caused by the preferential solar heating of the surface waters. Water density decreases as temperature increases, so warm water overlies colder water and creates horizontal density gradients that resist vertical mixing and enhance the stability of the water column. Chemical and biological demand for oxygen in deeper regions, accompanied by limited replenishment from the surface due to the lack of vertical mixing, can lead to very low oxygen levels in deep lake waters (Smith *et al.* 2011). Deoxygenation of the deeper waters has major effects on the chemistry of other substances, especially nutrients, which can be mobilised from the lake sediments under such conditions. The thermal stratification and mixing regimes of lakes and reservoirs influences water column stability, nutrient and light availability at different times of the year and, consequently, the plankton community.

Farm dams are often very turbid environments, so lack of light within the water column may limit phytoplankton growth. These, and other small ponds, are often typified by high amounts of organic substances in the water, which is often thought to favour certain kinds of motile unicellular algae known as euglenoids (Section 5.6). Wetlands and billabongs are shallow and much of the submerged area is occupied by aquatic macrophytes and large macroalgae, known as charophytes, that grow from the sediments. These macrophytes, and algae that grow attached to them (termed epiphytes) may compete with phytoplankton for light and nutrients, so that wetlands may not be good habitats for phytoplankton and zooplankton. Shallow water bodies may be clear water, macrophyte-dominated systems, or turbid, nutrient-enriched, phytoplankton-dominated systems (Scheffer 1998). In very dry regions, suitable habitats for plankton may only be present for short periods following brief periods of rainfall, and ponds may dry up rapidly often for long periods. The plankton associated with these environment have evolved strategies that still enable them to thrive in such harsh environments.

2.6 Estuarine and coastal habitats of plankton

Estuaries are the tidal portions of river mouths, bays and coastal lagoons, and can be dominated by hypersaline, marine or freshwater conditions. This includes inter-tidal wetlands, where water levels can vary in response to the tidal levels of the adjacent waterway, perched freshwater swamps, and coastal lagoons which are intermittently connected to the ocean.

The tidal ranges of estuaries undergo a regular fortnightly cycle in response to the monthly orbit of the moon around the Earth, where gravity affects the tidal range. Tidal range varies considerably, increasing to a maximum range over a week (spring tides) and then decreasing to a minimum range over the following week (neap tides). Solstice tides, or king tides, occur in June and December of each year, when the sun is directly over the Tropics of Cancer and Capricorn, respectively.

The characteristics of tides vary across spatial scales. For example, on the south-east coast of Australia, tides are generally semi-diurnal with high and low tides occurring almost twice a day. These tides have diurnal inequality where the heights of two consecutive tides varies (Fig. 2.7). Tides elsewhere have different characteristics: for example, many regions in Western Australia experience one tidal cycle each day (a diurnal tide), due to the ocean basin and topography.

Inside estuaries, the timing and dynamics of tidal currents become more complicated. Meanders around topography can slow tidal movement upstream, such that peak tides upstream occur hours after peak tides on the coast. The tidal limit of an estuary is the region of an estuary where there are no discernible changes to water levels as a

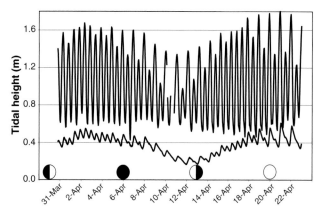

Fig. 2.7. Progression of the tides within a day, and over a lunar month. The upper line shows tidal fluctuation on the open coast, while the lower line shows the damped tide inside a nearby coastal lagoon.

result of tidal movement. The salinity limit is where there are no measurable changes to salinity over tidal cycles. The tidal limits and saline limits are often different, with tidal limits generally being further upstream.

Flood and ebb tides have different velocities, which can result in more water moving upstream into estuaries at flood tides than leaving at low tides. This can change the flow regimes of these systems. The shapes of estuaries can also influence the behaviour of tidal movement. In some estuaries with long thin channels upstream of a wide embayment near the ocean, the change of shape can force the upstream tidal range to be greater than that downstream. Alternatively, tidal movement becomes attenuated rapidly in estuaries with thin channels connecting them to the ocean, but which have wide reaches upstream. Influencing the depth or width of estuaries through dredging activities or by seawall construction can affect their hydrology.

Run-off from the land can result in estuaries becoming vertically stratified, with less dense, brackish, turbid water on top and denser, salty, clear, oceanic water beneath. This salty layer is sometimes termed 'the salt wedge' and can penetrate many kilometres upstream, along the bottom.

Even a wind- and tidally mixed estuary is remarkably structured into different planktonic habitats. The most obvious is where the 'estuarine plume' of brown brackish water meets the clear blue ocean water. Within a matter of minutes, or metres, you could be sampling completely different water. If you are not aware of this change, then your 'replicate' samples will be very different – making any comparisons very difficult. The estuarine plume is usually less dense by nature of lower salinity (even fractionally less), and is also identified by colour, and by being warmer in summer and cooler in winter than the ocean. An estuarine plume is usually quite shallow – less than a few metres deep (Fig. 2.8).

Where the brown estuarine water meets the blue oceanic water, there is a convergence where the denser ocean water wedges underneath the estuarine plume (Kingsford and Suthers 1996), leaving any buoyant material trapped at the surface as an oily looking line of water, mixed with flotsam. This convergence line is known as a slick, or a 'linear oceanographic feature' which influences the distribution of plankton (Kingsford 1990). Not only are these slicks evident near the estuary mouth on the ebb tide, they are evident on the flood tide, often as a 'V-shaped' front (Fig. 2.8A). This is because the ocean water is retarded by the shoreline, while the ocean water in the central channel can push further upstream.

In the coastal ocean, the sun warms the surface waters and, along with wind mixing and some fresh water create a surface mixed layer that may be 2–50 m deep. The layer may completely disappear during the winter storms, or have a series of layers during hot calm days. The thermocline is the temperature boundary between the two layers. Other similar boundaries include haloclines (by salinity), pycnoclines (by density), or nutriclines (by nutrients). At the temperature boundary, phytoplankton find the best of light and nutrient conditions and frequently bloom – forming a sub-surface chlorophyll maximum.

Other convergence lines are evident behind islands and headlands (e.g. Kingsford and Suthers 1996; Suthers et al. 2004). Pre-settlement fish and invertebrates may be concentrated in these slicks, which are often moved onto reefs or seagrass beds as the tide turns. In this oceanographic way, some areas characteristically receive more young prawns and fish than other parts of estuaries and deserve to be protected (or rehabilitated). The tidal wakes and eddies may only exist for up to 6 hours of a sinusoidal varying current, while the wake of an oceanic island can last for weeks.

The wakes of islands in shallow water may bring deep or benthic plankton near to the surface by eddy pumping, or by the tidal current scouring around the sides of an island and bringing material to the surface (Suthers *et al.* 2004). Whatever the mechanism, while often complex, the

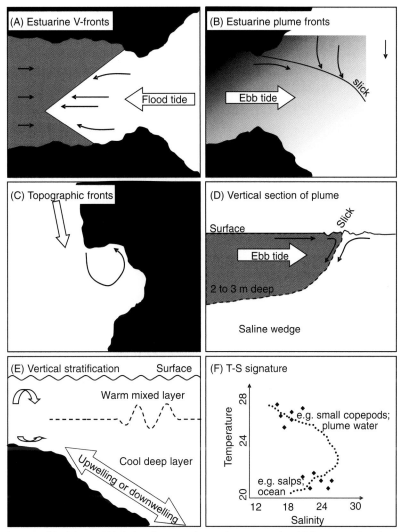

Fig. 2.8.   Estuarine and coastal habitats. **(A)** A landscape view of an estuarine V-front, as the channel walls retard the flood tide and the saltier (denser) coastal water wedges beneath the estuarine water. **(B)** An estuarine plume front showing the ebb tide flow of brackish (less-dense) water flowing on top of coastal water, which has a coastal flow deflecting the plume. **(C)** A topographic front generated in the lee of a headland or island. **(D)** A vertical section of an estuarine plume front, showing the convergence and sinking along the thermocline or halocline (dashed line) front creating a slick of buoyant material (foam, flotsam). **(E)** Vertical stratification showing a thermocline (dashed line), an internal wave, the breakdown of stratification in shallow water and the potential for upwelling or downwelling. **(F)** Temperature–salinity (T–S) signature of a water mass determined from a series of temperature and salinity measurements (line of dots). The depth or distance down-estuary are implied from the least dense (top left) to most dense (bottom right). The dominant types of plankton and water mass associated with particular T–S characteristics are indicated.

wakes are often obvious from the slightly turbid plumes shown in remote sensing, or from an aircraft.

On a calm sunny morning in coastal waters, one may see rows of slicks, 100–200 m apart and parallel to the shore. These are generated by internal waves, which are waves moving along the thermocline (similar to the familiar air–water waves). These waves are created by sudden tide changes or currents at particular submarine cliffs, when they propagate towards the coast. At the leading edge of each wave is a slight downwelling, which traps any buoyant particles such as oils and, possibly, plankton.

The key to sampling such a variable environment as an estuary is to always record temperature and salinity with a calibrated electronic meter. Talk to fishers about the local tides and typical currents. Spend some time looking at the waterway with drift objects, such as oranges, to appreciate the individual traits and the appropriate spatial and temporal scales before making any comparisons.

2.7 References

Baird ME, Suthers IM (2007) A size-resolved pelagic ecosystem model. *Ecological Modelling* **203**, 185–203. doi:10.1016/j.ecolmodel.2006.11.025

Bayly IAE (1986) Aspects of diel vertical migration in zooplankton, and its enigma variations. In *Limnology in Australia*. (Eds P De Deckker and WD Williams) pp. 349–368. CSIRO, Melbourne, Australia.

Beardall J, Redden AM (2007) Ecology of marine microalgae: a physiological perspective. In *Algae of Australia: Introduction*. (Eds PM McCarthy and AE Orchard) pp. 405–433. ABRS, Canberra; CSIRO Publishing, Melbourne, Australia.

Behrenfeld MJ, Bale AT, Kolber ZS, Aiken J, Falkowski PG (1996) Confirmation of iron limitation of phytoplankton photosynthesis in the equatorial Pacific Ocean. *Nature* **383**, 508–511. doi:10.1038/383508a0

Brierley AS (2014) Diel vertical migration. *Current Biology* **24**, R1074–R1076. doi:10.1016/j.cub.2014.08.054

Costanza R, D'Argee R, De Groot R (1997) The value of the world's ecosystem services and natural capital. *Nature* **387**, 253–260. doi:10.1038/387253a0

Dunne JP, Armstrong RA, Gnanadesikan A, Sarmiento JL (2005) Empirical and mechanistic models for the particle export ratio. *Global Biogeochemical Cycles* **19**, GB4026. doi:10.1029/2004GB002390

Hays GC (2003) A review of the adaptive significance and ecosystem consequences of zooplankton diel vertical migrations. *Hydrobiologia* **503**, 163–170. doi:10.1023/B:HYDR.0000008476.23617.b0

Kingsford MJ (1990) Linear oceanographic features: a focus for research on recruitment processes. *Australian Journal of Ecology* **15**, 391–401.

Kingsford MJ, Suthers IM (1996) The influence of the tide on patterns of ichthyoplankton abundance in the vicinity of an estuarine front, Botany Bay, Australia. *Estuarine, Coastal and Shelf Science* **43**, 33–54. doi:10.1006/ecss.1996.0056

Mitrovic SM, Bowling LC, Buckney RT (2001) Quantifying potential benefits to *Microcystis aeruginosa* through disentrainment by buoyancy within an embayment of a freshwater river. *Journal of Freshwater Ecology* **16**, 151–157. doi:10.1080/02705060.2001.9663800

Mitrovic SM, Oliver RL, Rees C, Bowling LC, Buckney RT (2003) Critical flow velocities for the growth and dominance of *Anabaena circinalis* in some turbid freshwater rivers. *Freshwater Biology* **48**, 164–174. doi:10.1046/j.1365-2427.2003.00957.x

Sarthou G, Timmermans KR, Blain S, Tréguer P (2005) Growth physiology and fate of diatoms in the ocean: a review. *Journal of Sea Research* **53**, 25–42. doi:10.1016/j.seares.2004.01.007

Scheffer M (1998) *Ecology of Shallow Lakes*. Chapman and Hall, London, UK.

Sigman DM, Hain MP (2012) The biological productivity of the ocean. *Nature Education Knowledge* **3**(10), 21. <https://www.nature.com/scitable/

knowledge/library/the-biological-productivity-of-the-ocean-section-70631104>.

Smith J, Baumgartner LJ, Suthers IM, Taylor MD (2011) Distribution and movement of a stocked freshwater fish: implications of a variable habitat volume for stocking programs. *Marine and Freshwater Research* **62**, 1342–1353. doi:10.1071/MF11120

Stoecker DK (1987) Photosynthesis found in some single-cell marine animals. *Oceanus* **30**, 49–53.

Suthers IM, Taggart CT, Kelley D, Rissik D, Middleton JH (2004) Entrainment and advection in an island's tidal wake, as revealed by light attenuance, zooplankton and ichthyoplankton. *Limnology and Oceanography* **49**, 283–296. doi:10.4319/lo.2004.49.1.0283

Thompson RJ, Deibel D, Redden AM, McKenzie CH (2008) Vertical flux and fate of particulate matter in a Newfoundland fjord at sub-zero water temperatures during spring. *Marine Ecology Progress Series* **357**, 33–49. doi:10.3354/meps07277

Worden AZ, Follows MJ, Giovannoni SJ, Wilken S, Zimmerman AE, Keeling PJ (2015) Rethinking the marine carbon cycle: factoring in the multifarious lifestyles of microbes. *Science* **347**, 1257594. doi:10.1126/science.1257594

Williams WD (1980) *Australian Freshwater Life*. Macmillan, Melbourne, Australia.

2.8 Further reading

Castellani C, Edwards M (2000) *Marine Plankton: A Practical Guide to Ecology, Methodology, and Taxonomy*. Academic Press, New York, USA.

Daborn GR, Redden AM (2018) Estuaries. In *The Wetland Book: II: Distribution, Description and Conservation*. (Eds CM Finlayson, R Milton, C Prentice and NC Davidson) pp. 37–54. Springer eBook, Dordrecht, The Netherlands.

Miller CB, Wheeler PA (2012) *Biological Oceanography*. John Wiley & Sons, New York, USA.

3

Use of plankton for management

David Rissik, Penelope Ajani, Lee Bowling, Mark Gibbs, Tsuyoshi Kobayashi, Kylie Pitt, Anthony J. Richardson and Iain M. Suthers

Plankton ecology is integral to the management of most aquatic systems and their catchments. As primary producers and secondary grazers, plankton responds rapidly to changes in nutrients, which are affected by runoff from catchments and discharge of point source pollutants (Fig. 1.2; Cloern *et al.* 2014; Capuzzo *et al.* 2018). Plankton responds to changes in light availability, which is influenced by the turbidity (muddiness) of water. Plankton also has a role in carbon uptake in aquatic systems and is affected by changes in climate, ocean currents and temperatures. Some phytoplankton contain toxins that can bioaccumulate in grazers such as oysters, prawns and fish and are harmful to humans. Others can cause direct impacts on humans through contact or through aerosols. By integrating environmental conditions, plankton are sensitive indicators of changes in ecosystem health over time (McQuatters-Gollop *et al.* 2017). Therefore, it is necessary to monitor changes in planktonic systems and to understand their causes, so that appropriate management actions can be taken.

Plankton are difficult to identify, which has led to several innovative approaches to help understand changes to communities, their functions and biomass, including using pigments or size classes to simplify functional groups, and by models. This chapter covers the role of plankton in management and builds on Section 1.4 that summarises the human interactions with plankton and highlights some of the management importance of plankton.

3.1 Plankton models and management

Models are simplified representations of the real world, used to help understand the major processes taking place or as tools for prediction. The use of models can help guide management decisions by indicating the potential outcome and helping isolate the causes of an effect. Models help adaptive management by allowing trial and error of potential solutions. Models can vary significantly in their degree of sophistication, ranging from cartoons or flow diagrams to highly sophisticated, data hungry numerical representations of systems.

Simple conceptual models or flow diagrams outline the processes considered important, without focusing on details. More sophisticated models, such as hydrodynamic or biogeochemical studies, quantify processes in a simple, time-averaged manner, or build on a collection of quantitative descriptions of the rates of processes. Sometimes, models of differing sophistication are used in a project, each providing an alternative view.

Managers frequently use models for a simplified understanding of what processes are driving a particular water-quality issue such as an algal bloom. For example, information on the residence times of water in different parts of an estuary indicate where phytoplankton are concentrated over time (Fig. 3.1), or where nutrient discharge may cause a phytoplankton bloom (a population explosion). Often the process of developing models draws a diversity of people into discussions and workshops. This is a valuable aspect of model development because it supports communication, generates knowledge, and creates a better understanding of links between science, environmental management and planning.

Models can be data hungry and expensive to develop or run. There is a wide range of free or commercial models available. Selecting the right models to investigate management concerns can be difficult for non-specialist users. To aid model selection and to ensure models are fit for purpose, it is essential that consideration be given to several issues associated with models and their use.

Users should clearly define the model requirements and ensure they have sufficient knowledge to manage model development. Effective pre-model planning should include a clear articulation of the problem – an understanding of what managers expect from the model, including how they will use the outputs – and knowledge of the spatial and temporal scales that need to be covered by the model. Unless the needs of the user are clearly defined, it is difficult for modellers to suggest an appropriate approach. It is important to be aware of the level of uncertainty with the results or predictions of the model, and therefore the risk of acting on the results. The more complex the problem, the greater will be the uncertainty associated with the model's output.

Consider how long the model should be useable. If a model is used on a regular basis, you may need training or advice to update it with new or better information. This may be expensive and may require a business case explaining the value

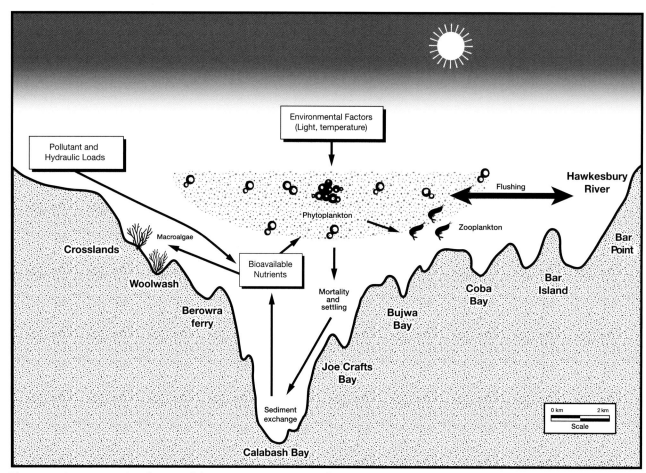

Fig. 3.1. A cross-sectional view of the Berowra estuary (northern Sydney, flowing into the Hawkesbury River estuary) showing the relationship of depth (water volume) and thus water retention time, matching doubling time, which results in phytoplankton blooms.

proposition and including being aware of issues associated with the model licence or the computational cost. Consider the data and process understanding that are available for the system that will underpin the model. Quality of data is determined by methods (Chapter 4) and by the spatial and temporal scales over which the data were collected. Longer periods enable trends to be incorporated into models and provide greater certainty in their predictions.

Model assessment measures the accuracy of the model. Calibration can be used to assess the accuracy. Calibration is when the underlying parameters are adjusted to ensure that there is strong correlation between the observed data at a specific field site and the results from model simulations. Good correlations signify that the model results are meaningful, which can increase the certainty that users have in the model output. Model validation is a more broad assessment to examine the underlying principles of the model, for a broader application.

Users should understand the assumptions underlying the production of the model. By undertaking a reality check of model output, managers can reduce the distrust and cynicism about model predictions that often exists in the broader community. In some cases, such as hydrodynamic models, reality checks may be simply assessing the correlation of collected versus predicted data. If the correlation is sufficiently strong and based on a large enough dataset (many points), then the model could be considered to be adequate. In more complex decision support models with a variety of variables and based on poor quality data or expert opinion, models can be assessed to determine which variables resulted in certain outcomes. Reality checking can provide other useful information to managers. If outcomes are counter-intuitive, and are shown to be based on poor quality data or information with a high level of uncertainty, then managers are able to focus data collection or knowledge generation on those specific variables and improve the model quality.

Acting on modelling results can be challenging, particularly when results may differ from community or stakeholder perceptions. Regular engagement with stakeholders throughout the process helps their awareness of why a particular model is being developed, how it will be used, and what data will be gathered to underpin the model. If stakeholders are confident in the processes, they will be more likely to act on that prediction and be satisfied with the management response.

3.2 Coastal and natural resource management

Management plans need to respond to changes in conditions, requirements and available knowledge. The key to adaptive management is scientific knowledge supported by a well-designed monitoring program. It is difficult to maximise environmental outcomes in relation to remediation costs if outcomes are not monitored. To ensure good science that supports management, the following hallmarks should be considered (see Table 3.1):

1. Objectives should be clearly stated in the management plan to ensure they are transferrable to specific monitoring programs. Management objectives should be measurable and linked to timeframes for assessment. Any monitoring not linked to clear objectives is a waste of resources.
2. Sampling designs should be articulated and peer reviewed to ensure that they meet the management objectives. Testable hypotheses (Table 3.1) help to ensure that appropriate statistics can assess the objectives.
3. What indicators will be used to investigate each issue and over what time period? Good management plans include suitable performance indicators for each proposed action that can be used to indicate change within a year or a within a decade. Management plans should be adaptive, and their accompanying performance indicators should be reviewed and changed if necessary.

Conceptual models of the system and the various functions taking place within the system can be useful when determining appropriate indicators for specific actions (Dennison and Abal 1999).

4. To determine whether any changes are the result of management intervention or natural variation, it is necessary to collect reference information from several control sites (Chapter 4). The selection of appropriate control locations is a critical component of monitoring and needs expert advice. An increase in the frequency of algal bloom reports may be a natural phenomenon affecting several regions simultaneously, rather than the result of a local impact. External reference sites (e.g. in collaboration with another local council) are essential.

5. It is important to peer review reports that result from monitoring activities to provide advice about the degree of success of management actions in achieving objectives. There are several actions and outcomes of monitoring, such as changed management actions or finding new methods. Report cards are an excellent way of presenting monitoring results to decision makers, stakeholders and the community, because they describe a variety of aspects of monitoring or to pull evidence together to present a narrative. (Box 3.1).

The steps listed earlier may appear elaborate and may be expensive to undertake. There are, however, ways in which costs can be reduced. These include integrating monitoring programs among councils and government agencies, sharing control sites, and working together with university groups. By working with experts, it is also possible

Table 3.1. Hallmarks of adaptive management plans

Hallmark	Examples of good practice	Examples of poor practice
1. Set clear and unambiguous objectives, linked to a timeframe such that their performance can be assessed.	Within 10 years we wish to have fencing and establish 10 m wide riparian vegetation belts between this tributary and a particular bridge.	Monitoring to satisfy public expectation that the water quality is good enough, but not linking to action.
2. Establish testable hypotheses, within the context of your sampling design and known variability.	By investing in fencing, the frequency of algal blooms (or percentage seagrass fouling, or water clarity) should significantly improve from 1990 levels.	The dairies and piggeries are responsible for the eutrophication in the estuary [an ambit claim].
3. Select suitable indicators that will respond to your management plan	The ratio of diatoms:dinoflagellates will increase to reference site levels after we implement this management plan. Or, we propose that a 20% increase in suspended sediment is a trigger for action (based on some credible study).	The plankton and biodiversity will improve after we implement this management plan [no quantitative measure of improvement].
4. Select sampling locations that include reference or control locations	To test the environmental worth of the new sewage treatment plant, we will sample before, during and after construction, as well as in neighbouring embayments and an adjacent estuary	We will sample before, during and after we build artificial wetlands at the stormwater discharge [no study of neighbouring sites or estuaries to provide a context for the comparison].
5. Use appropriate data interpretation and reporting	The number of replicate samples was in proportion to the sample-to-sample variability, and in proportion to the magnitude of change [see Section 4.2 for what constitutes a replicate sample].	There was a significant change in water-quality parameters before versus after construction [what level of change is required?].

Box 3.1 Report cards and ecosystem assessments

A key part of any monitoring program is the communication of results. Effective communication of complex data and information can help to increase understanding, build capacity and importantly can lead to action. Once decision makers are clear about what monitoring results are showing, they are more likely to commit to decisions, particularly when other stakeholders and the community are also aware of what the monitoring results are showing.

Report cards are commonly used to communicate monitoring results in the management of coastal waterways. The first report card used in Australia described the health of Moreton Bay in Queensland (http://hlw.org.au/report-card). Report cards are easily understandable because they grade ecosystem health from 'A' meaning pristine to 'F' being highly degraded. As communication tools, they provide a narrative about what the results are showing. Many report cards show information about the pressures on waterways, the condition of receiving water (both physico-chemical and ecological) and provide an indication of the management responses that have been undertaken. Increasingly report cards are also providing socio-economic information.

Report cards are generally designed together with stakeholders, ensuring that they provide information that is targeted to their area of concern. Report cards provide information simply and succinctly, but are underpinned by well-designed scientific programs, based on a good understanding of the systems of interest. Report cards have a robust and transparent process for integrating data from different sources, assessing the data, and presenting report card grades for different areas, different themes and different times.

Plankton, particularly phytoplankton, is a key component of many report cards. The use of chlorophyll-a as a proxy for phytoplankton biomass provides useful information about the biological response of a waterway to catchment inputs (nutrients and sediment). The exact indicators selected reflect the pressures that affect the system of interest and the scale at which change is likely to occur.

Report cards are useful for coastal situations, such as waterways where there is a fairly direct link between eutrophication and phytoplankton biomass, and there are local management actions that can be taken to reduce nutrient inputs. In such situations, an 'A' to 'F' grading system makes sense. However, this concept does not always apply to monitoring data over larger space scales or further offshore in the ocean. Ecosystem assessments report the trend of indicators to summarise changes in an ecosystem, but without giving a grade 'A' to 'F' as the concept of 'good' or 'bad' is not always applicable. For example, with climate change, species are moving poleward as the ocean warms. Ecosystem assessments such as the IPCC's Oceans chapter (Hoegh-Guldberg et al. 2014) have used phytoplankton and zooplankton species movement to describe the spatial rearrangement of marine systems. This is a natural adaptive response to warmer oceans, and will have ecosystem consequences such as changing productivity regimes (increasing in some regions and decreasing in others) and flow-on effects to higher trophic levels such as fish, which will benefit some species and negatively impact others.

to design studies that may not be overly expensive, but that can fulfil the objectives of the monitoring study. If monitoring is deemed too expensive to undertake appropriately, perhaps the management action should not be carried out because its success cannot be quantified.

3.3 Management of geographically persistent algal blooms in an estuary

Some estuaries have persistent blooms in particular areas. Harmful algal blooms also occur intermittently, which result in closure of the oyster aquaculture facilities situated in the downstream reach of the estuary. Closure of the estuary following algal blooms has a significant impact on the local community due to the importance of the area for boating and swimming. Berowra Estuary (near Sydney) is a tidally dominated drowned river valley estuary, which joins the Hawkesbury River estuary ~24 km from the Pacific Ocean (Rissik et al. 2006). The estuary has a waterway area of ~13 km^2 and drains a catchment of ~310 km^2. The estuary flushing time – the time taken for water in a specified region of the estuary to be replaced (diluted) by incoming fresh water or by tidal dynamics – is a key variable to help understand the location of phytoplankton blooms. Upstream, the estuary is narrow and shallow; mid-stream the estuary is wide and deep; and downstream the estuary is wide and shallow. These factors translate to flushing times of 1.5 days for the upstream site, 7 days for the middle region site, and 1 day for the downstream site (Fig. 3.1). Flushing times in the middle region were sufficiently long for both high phytoplankton and zooplankton growth to take place in warm summer temperatures (Rissik *et al.* 2006).

Most primary production occurs in the middle, low-flushing, bloom area. Zooplankton responds in the areas with the highest phytoplankton biomass, especially by small zooplankton such as copepod nauplii, suggesting that when more food was available, there was higher zooplankton production. However, persistently high phytoplankton concentrations in the mid-reaches of the estuary indicated that phytoplankton production was at a rate at which biomass could not be controlled by zooplankton grazing. Only when other factors that reduced phytoplankton growth, such as reduced light intensity, could the zooplankton control the bloom.

From a manager's perspective, flushing times in various reaches of an estuary are an important determinant of phytoplankton biomass. The most effective options to reduce blooms are those which result in fewer nutrients being discharged to the estuary from run-off, sewage discharge from homes and boats. The Berowra estuary receives tertiary treated discharge from two sewage treatment plants and also receives stormwater from several drains. Solutions to the problem involved working with sewage-treatment managers, repairing pipes, educational campaigns, and building nutrient-reduction devices, such as constructed wetlands and gross pollution traps. To assist management, a strategically placed algal bloom monitoring buoy was moored in the deep waters near Calabash Point (Fig. 3.1), which automatically sends an e-mail to the local council when the chlorophyll-*a* level exceeds 20 µg/L (or 20 µg L^{-1}).

3.4 Coastal water discolouration and harmful algal blooms and their management

Phytoplankton are able to reproduce rapidly in favourable conditions and a bloom can occur. Blooms can be red, green, purple, yellow, brown, blue, milky or even colourless. They may be natural or the result of human activities. Some blooms are beneficial to the ecosystem, while others can be harmful, so it is important to know what species (it is often only one) make up the bloom and what conditions caused the bloom. Some water discolourations are unrelated to phytoplankton and are a result of silty water (reddish) or drainage from acid sulphate soils (greenish).

Natural phytoplankton blooms in coastal waters may be due to fluctuations in the essential nutrients (such as nitrate, phosphate and silicate) from run-off from the land, or from cold, nutrient-rich bottom waters upwelled to the surface by winds or ocean currents (see Chapter 2). Such blooms may be simply harmless transient pulses in response to episodic nutrient enrichment. Sometimes (generally through human activities), excessive nutrients entering a waterway can lead to excessive growth of phytoplankton, in turn shading important plants such as seagrass, or leading to excessive oxygen uptake in the waterway at night and causing fish kills – this is known as eutrophication (see Box 3.3). Eutrophication can affect human health, ecosystem function and fish resources, as well as the recreational amenity of beaches and embayments. Whatever factors affect their formation, the incidence of algal blooms may be increasing, as evident in the increased global distribution of paralytic shellfish poisoning (Hallegraeff *et al.* 2003). Increased connectivity of waterways across the globe as a result of ship-based transport has contributed to the spread of harmful algae and is a focus of management action (Box 3.2).

Phytoplankton blooms have different effects depending on the species and their toxicity. Some may cause harmless water discolouration. Some may be non-toxic, but may be harmful to marine ecosystems (by either rotting and decreasing oxygen concentrations or by shading seagrass). Others may contain toxins that are transferred to filter-feeding bivalves (shellfish) such as mussels, oysters, scallops and clams, and thereby to fish, marine mammals, birds and humans. Phytoplankton blooms that have the potential to cause harm are commonly referred to as harmful algal blooms (HABs).

Most blooms are simply harmless water discolourations. If algal blooms are sufficiently extensive, especially in enclosed or partially enclosed areas (such as coastal lagoons and estuaries), they can cause fish kills. This may be due to changes in dissolved oxygen availability (hypoxia) or by mechanical damage to fish gills. As blooms in estuaries and enclosed systems decompose, they can also cause sulphur-based odours, which can be unpleasant for nearby residents or waterway users. Phytoplankton spines, such as those observed in the diatom genus *Chaeotceros*, may lodge in fish gills and cause an inflammatory response, making them susceptible to infection.

Other blooms can produce biotoxins that can kill fish. The dinoflagellate *Karlodinium* produces karlotoxins, which are highly active against blood cells, causing cell lysis in the gills of fish and other marine life. Another type of dinoflagellate,

Box 3.2 Invasive species from ballast water

Shipping movements across the globe may transport certain phytoplankton to new regions. Plankton are transported in the ballast tanks of ships, having been pumped into the ballast tanks in a port and then pumped out of the tanks once they reach their destination. Harbour environments are an excellent habitat for plankton – often having long residence times and high nutrient supplies either from the sediment or from the surrounding, generally urbanised, catchments (Hallegraeff 1998).

Ballast water is more likely to transport taxa that are able to survive in conditions where there is no available light, such as dinoflagellates. Once light becomes limiting in a ballast tank, dinoflagellates form protective coats around their cells (cysts) and sink – almost like seeds. Once the ship reaches its destination, the ballast water is pumped out and the dinoflagellate cysts sink to the bottom of the waterway. When nutrients and light become sufficient, the cysts germinate and the resultant cells undergo a reproductive process and the cells begin to grow and multiply.

Studies have identified a large number of species in ballast water. Many of these are cosmopolitan species and do not contain toxins; others, however, contain toxins and have the potential to cause major problems in areas to where they are transported and released. Preventing the transport of species in ballast water is difficult and requires global cooperation. Strategies include:

- exchanging ballast water at sea, or flushing ballast water at sea (200 nautical miles from land) when conditions are suitable
- treating ballast water, either while taking up or discharging ballast water
- discharge to a ballast water reception facility
- retention (sometimes known as tank-to-tank transfers).

Vessels with sealed tanks, which only use potable water, or which only take up and discharge in the same place (within port limits), will not need to install specialised treatment facilities. Vessels do not need to use an approved method of ballast water management if all ballast water in a tank is taken up and discharged in the same place. The discharge of ballast water is governed globally under the Ballast Water Management Convention; there are also additional national regulations. Under the Convention, ships are required, according to a timetable of implementation, to comply with standards. One standard requires ships to carry out a ballast water exchange, and specifies the volume of water that must be replaced. This standard involves exchanging the discharge water taken up from the last port, with new sea water; it must occur at a minimum of 200 nautical miles from shore. A more stringent standard requires the use of an approved ballast water treatment system.

Amphidinium, produces a group of biotoxins known as amphidinolides, which may similarly damage the epithelial and gill epithelial cells of fish.

Human illness associated with HABs is due to naturally occurring toxins that are transferred to humans through the consumption of shellfish or fish. Typically, shellfish simply filter toxic phytoplankton and remain unaffected, while the toxins are retained. The most significant public health problems caused by HABs are amnesic shellfish poisoning (ASP, see Box 6.3), ciguatera fish poisoning (CFP), diarrhetic shellfish poisoning (DSP), neurotoxic shellfish poisoning (NSP) and paralytic shellfish poisoning (PSP). Each of these syndromes is the result of different phytoplankton species that produce a range of toxins and risks to humans. All these syndromes are caused by toxins synthesised by dinoflagellates except for ASP, which is caused by diatoms (Hallegraeff *et al.* 2003).

Ciguatera fish poisoning (CFP) is a severe illness causing immediate vomiting and diarrhoea, but in ensuing years causes tingling in the fingertips, where hot feels cold, and vice versa. It typically occurs when people eat certain fish from near coral reefs, such as some snapper, mackerel and surgeon fish. CFP is caused by the benthic dinoflagellate *Gambierdiscus*. The food chain leading back to this toxic dinoflagellate can be complex – including copepods, shellfish and other prey species – but the fish species that become toxic from *Gambierdiscus* are usually known by regular fishers and avoided at certain times of the year.

Tasmanian paralytic shellfish toxin (PST) events have been caused in the past by the dinoflagellate *Gymnodium* (but PST can also be caused by *Gonyaulax*) and more recently by species belonging to the *Alexandrium* group. Scientists have determined that *Gymnodium* was introduced into Tasmania by ballast water in the early 1980s, based on cysts in layered, dated sediments. The cysts can remain viable in the mud for many years and germinate to form blooms after a sequence of events, involving water temperatures warmer than 14°C, a rainfall trigger, followed by calm conditions for 14 days (Hallegraeff *et al.* 2003). A large PST event on the east coast of Tasmania resulted in widespread harvest closures of mussels, oysters, scallops, rock lobster and abalone over a period of 6 months along 350 km of coastline and total economic losses of ~A\$23 million (Ajani *et al.* 2016). Subsequently, once again, species of *Alexandrium* bloomed along the east coast of Tasmania, with concomitant high levels of PSTs in shellfish. These more recent years also had the first reported cases of human illness in Australia in 30 years, suggesting these types of blooms are increasing (Ajani *et al.* 2016).

Making prediction of harmful algal blooms is difficult because toxic species are not always toxic, depending on the situation. It is also anticipated that other phytoplankton species may prove to be toxic in the future under certain conditions. Only ~40 of the more than 1200 species of dinoflagellates are known to be toxic, and many of the non-toxic varieties are beneficial to the environment and to aquaculture. *Symbiodinium microadriaticum* is one of the symbiotic algae ('zooxanthellae') that live with corals in our tropical reefs: they provide corals with essential sugars and beautiful colours.

The impact of HABs on the shellfish aquaculture industry has resulted in farmers, scientists and natural resource managers working together to develop effective HAB management programs. These can include quality assurance programs, biotoxin management programs and algal contingency plans to prevent any harm to the public. Management of blooms requires providing information to the public and waterway users about the causes of blooms and relevant issues such as toxicity. One example of this collaborative management is the development of a decision support tool to help coastal managers monitor and respond to HABs in New South Wales (NSW), Australia. This tool is called Algalert (https://sites.google.com/site/algalert/home), which is a centralised system that gives timely, consistent information about HABs found in coastal waters with the aim of reducing public health incidents.

Preventing or reducing the discharge of excess nutrients into estuaries and the coastal zone is the most effective means of managing eutrophication and minimising HABs. Understanding the pathways of nutrient enrichment in each system is essential. In urban areas, possible strategies include education programs, controls on nutrient sources, removing pollutants, upgrading sewerage systems, replanting riparian zones, and even maintaining high abundances of natural filter feeders such as mussels and oysters. Natural oyster and mussel reefs have been lost in many systems around the world and restoration of these reefs is becoming a major initiative to improve estuarine water quality. In many rural areas, land clearing and poor land-management practices have contributed to poor water quality. Clearing of vegetation is a major cause of land degradation and poor-quality run-off.

Two major challenges for managers are increasing population growth in coastal areas, and the direct and indirect impacts of climate change. Climate change is causing warmer water temperatures, increasing ocean acidity, higher sea levels, and changes to episodic events such as rainfall and cyclones. Although the effects vary throughout the world, it is important for managers to consider how and when these changes may influence their systems, and to plan for change.

3.5 Monitoring phytoplankton over the long-term

Red tides have become a common sight in urbanised coastal waters, often during spring and summer. Frequently mistaken for a pollution event (such as dumped paint), phytoplankton blooms may be highly visible and raise public concern. About 60% of Sydney's reported red tides are formed by surface concentrations of the dinoflagellate *Noctiluca scintillans* (Ajani *et al.* 2001a; Fig. 3.2). Fortunately, this species is non-toxic. This species is distributed worldwide and is often present in pristine waters. Red tides of *Noctiluca* can cause some irritation to the skin and eyes. Fish and other marine organisms may avoid the bloom area due to the concentrations of ammonia associated with the bloom. Ammonium chloride is stored in vacuoles

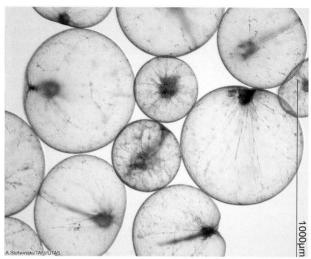

Fig. 3.2.   Magnified image of *Noctiluca scintillans* (photo: Anita Slotwinski, Tasmanian Aquaculture and Fisheries Institute).

of *Noctiluca* cells increasing their buoyancy, especially towards the end (senescent) stages of blooms.

Regional oceanographic processes are the main mechanisms for driving seasonal variability of plankton communities for the NSW coast. Increased flow of the East Australian Current (EAC) and upwelling-favourable northerly winds during the spring–summer stimulates slope water intrusion events that bring cold nutrient-rich water into the coastal zone and encourage phytoplankton growth (Lee *et al.* 2001b). Waters of the EAC originate in the Coral Sea and are characteristically oligotrophic (nutrient poor). Slope-water intrusions deliver episodic influxes of nitrogen (as nitrate) up onto the shelf and towards the coastal zone. Research has shown that algal blooms appear in response to slope water intrusion events, irrespective of the proximity to other major nutrient sources such as major riverine discharges or Sydney's deep ocean outfalls. Blooms begin with small chain-forming diatoms (*Skeletonema, Thalassiosira, Leptocylindrus, Asterionella*), followed by large diatoms (*Eucampia, Detonula, Lauderia*) and finally by large dinoflagellates (*Protoperidinium, Ceratium*).

In some years, the dominance of the small diatom *Thalassiosira partheneia* and an increased presence of *Noctiluca scintillans* may be related to climate. Comparatively lower concentrations of nutrients and overall warmer water temperatures occurred relative to previous years when eastern Australia was experiencing effects of an El Nino–Southern Oscillation (ENSO) event (Lee *et al.* 2001a). Warmer water temperatures and strong southward flow of the EAC were also reflected in the increased presence of tropical phytoplankton indicator species (such as *Bacteriastrum, Ceratium gravidum* and *Trichodesmium erythraeum*) compared with four decades ago. More recent work at a long-term monitoring station has confirmed an increase in tropical taxa and a shift in the community composition at this station, with the continued year round presence of *Noctiluca.*

Spring and summer blooms of *Noctiluca* most often occurred off Sydney during, or soon after, diatom blooms dominated by *Thalassiosira* and examination of the cell contents of *Noctiluca* has confirmed *Thalassiosira* as the dominant prey item (i.e. the single celled dinoflagellate is carnivorous, preying on diatoms; Dela-Cruz *et al.* 2002). Additionally, laboratory studies have found *Thalassiosira* to be an optimal food source for *Noctiluca.* The shift towards *Thalassiosira* as the dominant diatom bloom species at a long-term coastal station may also be the contributing factor towards the increased and year-round prevalence of *Noctiluca* in NSW coastal waters (Ajani *et al.* 2001b). It appears that diatom blooms are not occurring with greater intensity or frequency than before, but the red-tide forming dinoflagellate *Noctiluca scintillans* appears more prevalent. However, most blooms in the coastal ocean are natural phenomenon. Long-term monitoring is required to resolve the effects of climatic variability on phytoplankton populations compared with increasing anthropogenic nutrient loads and chronic impacts.

Of greater concern is the potential for a shift in prey species. That is, there is the potential for an increase in occurrence of a phytoplankton species that is the preferred food source of a harmful algal species. Blooms of harmful algal species such as *Alexandrium, Gymnodinium, Karenia, Dinophysis* and *Pseudonitzschia* occur in south-eastern Australian waters. Toxic algal blooms are a significant potential threat to our coastal environment, local economies and a risk to human health. Modern research methods using remote sensing techniques and on-ground implementation of a state-wide network of moored long-term ocean reference stations would provide an opportunity to monitor physio-chemical and biological oceanography on better spatial and temporal scales.

3.6 Managing blooms of freshwater cyanobacteria

Algal blooms cause several problems for managers of fresh waters. Surface scums may occur during blooms of cyanobacteria (blue-green algae), flagellated green algae and euglenids, because these organisms can float or swim to the surface and accumulate. The presence of these scums, and other algal growth, can lower the aesthetic and recreational amenity of water bodies. Blooms of cyanobacteria can also impart musty, earthy tastes and odours, and/or toxins into the water, while blooms of green algae can impart grassy tastes and odours, and blooms of some chrysophytes and other flagellated algae can create fishy tastes and odours. The presence of high concentrations of algal cells in the water can also cause problems for water treatment plants by blocking filters and fine nozzles in irrigation systems (Box 3.3).

Similar to marine waters, several environmental factors drive the formation of algal blooms in freshwater environments. Because of the diversity of freshwater algae, different species and strains of species have considerably differing environmental tolerances and requirements, so that one set of water-quality characteristics may suit some species of phytoplankton, while another set may suit completely different species. For example, blooms of cyanobacteria may be enhanced by nutrient-enriched, warm waters that are slightly alkaline, while chrysophytes predominate in cold, soft (low ion concentrations), oligotrophic waters that are slightly acidic. This section will concentrate on the factors causing cyanobacterial blooms in fresh water, because of their importance in public health and their risk to livestock and wildlife.

Box 3.3 Effects of eutrophication

Eutrophication is the process that increases biological productivity within an ecosystem and in particular algal blooms. The causes are many, but are usually associated with an increase in nutrients from agricultural run-off, sewage discharge, or from the sediment of waterways. Algal blooms can cause large daily variations in pH and dissolved oxygen. By day, algal photosynthesis removes carbon dioxide from the water, allowing the pH to increase, and produces oxygen, which can lead to supersaturation of dissolved oxygen. At night, cellular respiration by the algae, and other organisms in the water, increases the amount of carbon dioxide dissolved in the water, and causes pH to fall, while dissolved oxygen can fall to quite low concentrations. Large daily changes in pH in raw waters used for town water supply are not desirable, as water-treatment processes work best at a constant pH. Low dissolved oxygen concentrations at night may stress, or even kill, fish and other aquatic organisms. Decomposing cells and the absence of oxygen can also lead to the production of noxious gases, such as hydrogen sulphide and methane, and to high concentrations of ammonia, which may be toxic to aquatic organisms. Anoxic conditions also lead to reducing chemical conditions at the sediment–water interface, and the mobilisation of soluble forms of nutrients, especially phosphorus, from the sediments, which can lead to future algal blooms. Metals – in particular iron and manganese – are also mobilised under anoxic conditions, and their presence in a town water supply can cause discolouration, taste and staining of laundry.

Cyanobacterial blooms tend to occur when nutrient loads are high. Depending on the system, phosphorus (Box 3.4) and/or nitrogen (Box 3.5) can stimulate blooms. Cyanobacteria and eukaryotic algae also require other nutrients for growth, such as iron, but these are generally available in most fresh waters at concentrations that do not limit growth. Many temperate species of cyanobacteria that form noxious blooms have optimal growth rates above 20°C (Robarts and Zohary 1987), so typically occur during spring, summer and autumn. Of recent concern are several species of potentially toxic cyanobacteria thought to be of pantropical origin that have become invasive species into warm temperate regions, in particular *Chrysosporum ovalisporum* and *Cylindrospermopsis racaborski* (Sukenik *et al.* 2012, 2015). There was a major bloom of *Chrysosporum ovalisporum* over many hundreds of kilometres of the Murray River, central Australia at high biomass during the

Box 3.4 Key nutrient: phosphorus

Phosphorus can be measured in two ways – as soluble reactive phosphorus or as total phosphorus. Soluble reactive phosphorus represents the phosphorus that is immediately available for algal growth within the water column. Total phosphorus includes not only the soluble forms, but also that bound up in the cells of existing phytoplankton and other microscopic aquatic organisms, in organic detritus, and in part of the suspended particulate mineral material. Much of the total phosphorus is thus not immediately available for phytoplankton growth, but may become available in the near future. In many Australian inland waters, soluble reactive phosphorus represents only 10–30% of the total phosphorus. Although cyanobacteria can grow at lower concentrations, they tend to become more prevalent as total phosphorus concentrations rise, especially above 10 µg/L. Various algal and cyanobacterial species respond to different total phosphorus concentrations. For example, very tiny celled cyanobacteria from the Order Chroococcales are better able to scavenge available phosphorus at low concentrations than some of the larger celled species, such as *Dolichospermum circinale*, which require higher concentrations. In terms of the number of cells present per millilitre of water, the Chroococcales may bloom at low total phosphorus concentrations, although, because of their tiny size, these large cell numbers still represent very little biomass. However, total phosphorus concentrations above 20 µg/L – and especially above 30 µg/L – favour most cyanobacteria. Dissolved phosphorus can be precipitated from the water by adding a type of bentonite clay ('Phoslock').

Box 3.5 Key nutrient: nitrogen

Nitrogen availability can be measured in terms of readily bioavailable forms, such as oxidised nitrogen ions (nitrate NO_3^- and nitrite NO_2^-) and ammonia (NH_3), and also as total nitrogen, which also includes the organic and bound forms of nitrogen. Algal presence increases as nitrogen becomes more readily available at higher concentrations, especially once total nitrogen exceeds 1000 µg/L, provided other factors are not limiting the growth. The form of nitrogen may also influence the type of phytoplankton present. Cyanobacteria from the Order Chroococcales prefer nitrogen to be present in the form of ammonia (or the ionic form ammonium NH_4^+), while other cyanobacteria and eukaryotic algae more readily use nitrate. Some heterotrophic flagellated algae (such as non-photosynthetic dinoflagellates) may use organic sources of nitrogen. Some cyanobacteria are, however, less reliant on ambient nitrogen concentrations, because they can fix atmospheric nitrogen to obtain their needs if concentrations in the water are low. Nitrogen fixation is especially common in the Order Nostacales, although species from other orders can also do this. Most phytoplankton (diatoms, dinoflagellates), including many cyanobacteria, cannot fix atmospheric nitrogen.

summer and autumn of 2016 when it had previously only been reported from that water body occasionally and at low abundance (Crawford *et al.* 2017). Ongoing global warming is likely to result in a greater occurrence of harmful cyanobacterial blooms (O'Neil *et al.* 2012; Paerl and Paul 2012).

Cyanobacteria generally prefer calm, non-turbulent conditions, as this allows them to maintain their buoyancy and float towards the surface and light, or to sink into deeper waters as required. Deeper lakes, weir pools, and reaches of rivers become thermally layered (stratified) in summer when their surface waters are warmed up by the sun (Fig. 2.6). This stratification of the water column creates stability and reduces turbulence. Such conditions are ideal for cyanobacterial blooms, but are unsuitable environments for many of the larger, heavier non-flagellated eukaryotic algae, such as green algae and diatoms, which require turbulence to keep them suspended within the water column and to prevent them from sinking.

Algal bloom development is also facilitated by water retention times. Retention times (the period of time required for the water in a lake, reservoir or weir pool to be replaced by new water) longer than 2 weeks tends to favour cyanobacterial growth (Mitrovic *et al.* 2003). Conversely, high flow rates in rivers are not conducive for bloom formation, as the algal cells are displaced downstream (although certain algal species are distinctively riverine and

continue to live in discrete packages of water as these move downstream).

The availability of light is another factor that can promote blooms during spring and summer, particularly in temperate regions. Phytoplankton cells also need to be close enough to the surface (the euphotic zone) to obtain sufficient light for photosynthesis, so that food production equals or exceeds loss by respiration. The maximum depth for photosynthesis is usually the depth at which only 1% of the light penetrating the surface of the water remains. Light penetration is limited by dissolved organic substances, which often stain the water a yellow to brown colour, and suspended particulate matter. These substances in the water also change the spectral distribution of the light away from the blue wavelengths that are most useful to algae, towards a predominance of yellow to red wavelengths. This is outside the main range of wavelengths absorbed by chlorophylls, but many algae have additional pigments, such as carotenes and xanthophylls – and in cyanobacteria phycocyanin and phycoerythrins – so that they are still able to harvest light within these wavelengths.

Turbidity or suspended particulate matter is a major factor influencing the underwater light availability of inland waters. Turbidity is a measure of the amount of light scattered by these particles, and is often used as a surrogate measure of the amount of suspended particulate matter. Depending on the species, cyanobacteria can bloom in both high and low turbidity. Blooms occur in low turbidity water, where light is plentiful for photosynthesis, and in some relatively turbid weir pools it has been demonstrated that once turbidity falls below a certain level and the water becomes clearer, then the chance of cyanobacterial blooms increases considerably (Mitrovic *et al.* 2003). Blooms also occur in highly turbid water. As well as having ancillary pigments for light harvesting in light-restricted waters, cyanobacteria can use their positive buoyancy in non-turbulent turbid waters to rise to the surface, where there is sufficient light for their needs. Cyanobacteria also have low light requirements in comparison with many eukaryotic algae, enabling them to grow in light-restricted turbid environments and, in fact, prolonged exposure to high light intensities is detrimental – resulting in cell death.

Salinity, and the ionic composition of these salts, together with pH can also affect algal presence in fresh waters. Little is known of the salinity tolerances of most freshwater species of phytoplankton. Two potentially toxic species of cyanobacteria, *Dolichospermum circinale* and *Microcystis aeruginosa*, have salt tolerances of up to 5–6 g of salt per litre (~15% seawater) before they die (Winder and Cheng 1995), which is well above the salinity of water considered to be 'fresh' (~2% seawater or <3 g/L). Therefore, salinity may select for a particular species of cyanobacterium. For example, there have been changes in species composition from species of *Dolichospermum* to the more salt-tolerant *Anabaenopsis* in some parts of the Darling River in NSW where there is saline groundwater inflow under low flow conditions. In South Australia, *Dolichospermum circinale* in the Murray River tends to be replaced by the brackish water species *Nodularia spumigena* in Lake Alexandrina, where salinities are higher. The pH tolerance varies from species to species. For example, many chrysophyte algae prefer slightly acidic, soft water environments, while cyanobacteria in general grow better in slightly alkaline (pH 8.0–8.5) waters. However, blooms of phytoplankton also increase the pH, as they use and replace the carbon dioxide in the water through photosynthesis and respiration on a daily basis.

The main concern about algal blooms is the ability of some, but not all, to produce potent toxins that create a public health hazard and can lead to the deaths of domestic animals and wildlife. In fresh waters, only some species of cyanobacteria are known to produce toxins, although all likely produce contact irritants. Cyanobacterial contact irritants cause skin and eye irritations and digestive tract upsets in recreational water users who come into contact with them, or swallow water containing them. The potency of these contact irritants varies from species to species, while the response of people coming into contact with them also varies greatly, with epidemiological studies suggesting that ~15% of people being susceptible to them, while others are not (Pilotto *et al.* 2004; Stewart *et al.* 2006).

There are two main types of toxins produced by cyanobacteria: hepatotoxins and the neurotoxins. Hepatotoxins cause the breakdown of cells within the liver and other internal organs of the poisoned victim, and may lead to death by internal haemorrhage. Neurotoxins attack the nervous system of the poisoned victim and may lead to death from respiratory failure. In addition, some of these substances have been identified as cancer-promoting substances. Each year in Australia, cyanobacterial blooms cause the deaths of agricultural livestock drinking contaminated water. Deaths of humans at a renal dialysis clinic in Brazil have also been attributed to cyanobacterial toxins in the water used in their treatment. To date, only about 12 species of cyanobacteria have been shown to produce toxins in Australia. Research has indicated that ~40% of blooms within the Murray–Darling Basin are toxic (Baker and Humpage 1994), with neurotoxic *Dolichospermum circinale* predominating. Hepatotoxic species include *Microcystis aeruginosa*, *Nodularia spumigena*, *Chrysosporum ovalisporum* and *Cylindrospermopsis raciborskii* (see Box 3.6).

Also of concern is a slow-acting neurotoxin considered to be produced by many species of cyanobacteria (and also by diatoms) known as β-methylamino-L-alanine (BMAA). BMAA causes

Box 3.6 Analysis of cyanobacterial toxins

There is a range of methods by which the toxicity of cyanobacterial blooms can be assessed.

High-pressure liquid chromatography (HPLC)

This is used to determine the concentration of common hepatotoxins in water samples. HPLC can also be used for the determination of saxitoxin (a neurotoxin) concentrations in water, although different analytical and detection methods are required. There is no single HPLC analysis that will test for all toxins simultaneously.

Liquid chromatography-mass spectrometry (LC-MS)

Also used for hepatotoxin analysis, especially for the toxins produced by *Cylindrospermopsis raciborskii*.

Tandem mass spectrometry (MS-MS) coupled with some form of chromatography (e.g. HPLC)

This is an increasingly used detection method for many cyanotoxins. Fluorescence detection (for saxitoxins) and diode array UV detection (for microcystin) are also used.

Enzyme linked immunosorbent assay (ELISA)

These employ antibodies raised to react with certain hepatotoxins. Differences in the cross-reactivities of the antibodies used in different ELISA test kits to the range of hepatotoxins possible in environmental samples may influence their relative performance, and produce over or underestimates of toxin concentration. They therefore cannot be relied on as quantitative assays, unless the bloom is ongoing with a known and consistent toxin profile.

Protein phosphatase inhibition assay (PPI)

The hepatotoxin microcystin is a potent inhibitor of protein phosphatases, and a colourimetric test is used to detect this enzyme inhibition. The test can provide overestimations of toxin content as cyanobacterial cellular compounds other than the toxins may also cause inhibition.

Polymerase chain reaction (PCR)

This method amplifies the DNA within cyanobacterial cells, and detects the presence of gene sequences that code for toxin biosynthesis. As such, it provides a rapid screening test of the potential for the cyanobacteria within a bloom to produce toxins (if the genes responsible for toxin production are present, the bloom can produce toxins – if the genes are absent, the bloom will not be toxic). The test does not provide a quantitative measurement of any toxins present. PCR is currently used mainly as a research tool, and is not yet commercially available for routine sample analysis.

Mouse bioassay

This used to be the traditional method of toxicity assessment. Concentrated samples of cyanobacteria are required. Known concentrations of sterile cyanobacterial cellular extracts are administered to test mice by intra-peritoneal injection. From these tests, the concentration that will kill 50% of mice (the LD_{50}) can be calculated. The time to death indicates whether the sample is hepatotoxic or neurotoxic – the latter being most rapid. Autopsy also indicates any internal organ damage due to hepatotoxins. Because of animal ethics considerations, mouse bioassays are rarely used these days.

disruption to motor neurone cells and is hypothesised to be a possible cause of neurodegenerative disease (Bradley *et al.* 2013).

3.7 Phytoplankton monitoring in New Zealand for toxic shellfish poisoning

Shellfish are an important resource in New Zealand and have great cultural importance for Maori

and, more recently, for New Zealanders of European descent. Over the past three decades, mussels have formed the basis of a large aquaculture industry (with an annual revenue of more than NZ$200 million). Mussels, oysters and other important bivalves are filter feeders of phytoplankton (see Box 3.7) and thus can be an efficient vector for transferring biotoxins from phytoplankton to humans via the consumption of shellfish. While

Box 3.7 Depletion of phytoplankton around New Zealand mussel farms

Shellfish growers are farmers: they sow the seed, tend the crop and then harvest the product. Hence, there are many similarities between shellfish aquaculture and horticulture, but there are major differences. Most terrestrial farmers have property rights in the form of land tenure or leases and hence they have control over the land and soil. For example, terrestrial farmers have the ability to manipulate, in part, the growing conditions through the use of irrigation and fertilisers. By contrast, shellfish farming involves placing the crop in the water and allowing it to grow under the influence of a natural food supply. The farmers have little control over the food availability.

The shellfish industry in New Zealand is expanding into new growing areas or further developing existing areas. How many shellfish farms can be established without having an undue adverse effect on the environment? It may seem surprising that bivalves can overgraze the phytoplankton production of an estuary, and exceed the ecological carrying capacity. Shellfish farming applicants must provide predictions of the likely extraction of phytoplankton that will result if a farm is established, and some guide to the possible impacts of this extraction to the greater ecosystem. Predictions are derived from simple analytical models to complex coupled hydrodynamic-ecosystem models. However, the level of uncertainty often increases with the complexity of the models. These models are generally nutrient–phytoplankton–mussel growth models that typically ignore all other plants and animals in the system. The other principal weakness of these types of models is that bottom-up drivers of phytoplankton production are nutrient inputs. In inshore areas, nutrients are derived from run-off and, in some cases, from local oceanographic events.

The development of the shellfish aquaculture industry in New Zealand has also led to a renewed interest in the abundance and distribution of phytoplankton in coastal waters. In particular, farmers have an interest in understanding the availability of phytoplankton for farm planning and management – and other stakeholders and regulators have an interest in understanding how the establishment of shellfish farms may influence other marine animals and communities that rely on phytoplankton.

these naturally occurring toxins are not harmful to the shellfish, they can be fatal to humans. Several large-scale monitoring programs are in place in New Zealand to minimise these threats.

Prior to 1992, toxic shellfish poisoning (TSP) resulting from the consumption of filter-feeding shellfish grazing on phytoplankton had not been officially reported in New Zealand. However, awareness of the risk of toxic phytoplankton was raised in the summer of 1992–93 when 180 cases of illnesses fitting the case definition for neurotoxic shellfish poisoning were reported. Although this event was relatively localised to a section of the coastline, a blanket closure of commercial and recreational shellfish harvesting was enforced nationwide. This seemingly extreme response enabled management structures to be developed, and provided a coordinated approach to contend with the TSP event and future HABs events.

In this context, New Zealand's National Marine Biotoxin Management Plan (NMBMP) was established. An independent phytoplankton laboratory constitutes the first tier of monitoring for toxic microalgae, which is divided into the commercial (industry) and non-commercial (public health) sectors. The laboratory gives an early warning of potential blooms at up to 250 representational sites around the coast. Risks associated with toxic species are defined by the New Zealand Food Safety Authority and a conservative approach is taken to trigger flesh testing, with regulatory decisions being made based on flesh test results. This introduces the second tier of HAB monitoring: biotoxin testing. In conjunction with water sampling sites, shellfish are collected weekly and tested for marine biotoxins. If potentially toxic phytoplankton are identified in the water samples, a search for the toxin group is made in the flesh sample. These two complimentary monitoring systems optimise sampling effort, cost and reporting time constraints. For example, where phytoplankton testing represents a spot sample in time, flesh testing resolves these spatial and temporal issues to a degree, because shellfish act as bioaccumulators, concentrating toxins in their flesh. Conversely, a lag period is often observed between the detection of toxic phytoplankton in the water column and when shellfish accumulate the toxin to a level where it is detectable. This lead time provides early warning to managers if further action is required. Therefore, combining phytoplankton and biotoxin monitoring provides a comprehensive, efficient and cost-effective system for detecting HABs and their biotoxins.

For example, the system was used to identify a particular species of *Pseudo-nitzschia* that produces a novel form of domoic acid (iso-DA). Not all species of *Pseudo-nitzschia* produce toxins, but differentiating *Pseudo-nitzschia* species with light microscopy is almost impossible. As a solution to this problem, a suite of DNA probes have been developed and are offered as a routine test with compliance to ISO 17025 standard. At one stage, *Pseudo-nitzschia* cells (3.6×10^4/L) were present at the same site and time in the Marlborough Sounds as shellfish were found containing iso-DA. Because the phytoplankton monitoring requires both live and preserved water samples, *Pseudo-nitzschia* species from sites where iso-DA was detected were able to be isolated and cultured from the live water sample. Cultures of each isolate were identified to the species level using DNA probes and stressed to enhance DA production. Analysis of the different forms of was carried out using liquid chromatography mass spectrometry (LC-MS) and *Pseudo-nitzschia australis* was identified as the producer of the novel iso-DA.

Another example was a bloom of *Gymnodinium catenatum* that was tracked as it extended along the coast of the North Island using phytoplankton and biotoxin monitoring. Low levels of PSP toxins were detected in some routine flesh samples and reactive sampling of the water around these areas resulted in the detection of *G. catenatum*. Routine sampling for phytoplankton monitoring was limited in this area by high surf and the exposed nature of the coastline. From the original point of detection, it soon became clear that the bloom was intensifying and expanding – both in terms of cell numbers and shellfish toxicity levels. Within 1 month, *G.*

catenatum had spread into the far north of the North Island, with resting cysts of this species detected in high numbers. Resting cysts can germinate later into the usual form of the species when environmental conditions are favourable – sometimes many years later. This posed a major problem to the industry, as contaminated drift weed that naturally washes ashore on this beach supplies around 80% of mussel spat required for seeding out mussel farms around New Zealand. With the detection of *G. catenatum*, a voluntary ban was imposed to prevent transport of contaminated weed to unaffected areas around New Zealand. The future production of the mussel industry was in serious jeopardy as it faced spat shortages for the crop next season. Compounding this problem was the timing of the bloom, which coincided with the prime collecting time for spat and for re-seeding marine farms. In response to the dilemma marine farmers and the industry were facing, several methods were developed to eradicate cysts from the weed to which the spat were attached. Decontamination of spat has been successful and there are several cleansing plants in operation, allowing 'clean' spat to be transferred into unaffected aquaculture areas.

Although this was the first recorded presence of this species, sediment cores taken from around highly affected areas suggest that resting cysts have been dormant in the sediments since at least 1981, and even as far back as 1921 in some areas. This inferred that *G. catenatum* was not a recently introduced species, as first speculated, but had in fact been in New Zealand waters in recent history. There will always be new species discovered, new toxins detected, new regulatory demands and the need for new technologies to be developed. Any monitoring program must be adaptive and amenable to evolve to best mediate effects of HABs and marine biotoxins.

3.8 Managing water quality using freshwater zooplankton

Monitoring and assessment of the freshwater environment are often based on turbidity, pH, dissolved oxygen, biological oxygen demand and nutrients.

Point measurements of these physio-chemical traits vary over hours to weeks, and from metres to kilometres, whereas we need traits that integrate this small-scale variation (Fig. 1.2). Zooplankton have been used widely as indicators to monitor and assess various forms of pollution including acidification, eutrophication, pesticide pollution, algal toxins and climate warming (see Table 1.1 for a summary). In addition, zooplankton have been used to improve water quality, particularly using knowledge of their feeding behaviours. We provide two examples here: biomanipulation and mosquito control.

In the Northern Hemisphere, acidification (i.e. the lowering of pH) due to acid rain from airborne pollutants such as sulphur dioxide and nitrous oxides had adverse effects on a broad range of organisms in freshwater ecosystems. Zooplankton species richness is reduced with increasing acidification. The cladoceran or water flea, *Daphnia*, is eliminated, while smaller crustaceans (especially *Bosmina* and some calanoid copepods) and rotifers become dominant. With the concomitant loss of fish, cyclopoid copepods may become the top predators in the lake, together with macroinvertebrates such as corixid beetles and phantom-midge larvae.

The relative abundance of the rotifer *Keratella taurocephala* is a good indicator of low pH in North American lakes, while the littoral cladocerans *Alona rustica* and *Cantholeberis curvirostris* are associated with acidic lakes in Norway. Zooplankton have been used to assess natural and artificial recoveries of lakes from acidification by the addition of lime. With recovery of acidified lakes, the increase in species richness and return of acid-sensitive species of zooplankton have been reported (Locke and Sprules 1994; Walseng and Karlsen 2001).

Eutrophication of lakes and ponds also changes the size structure, species composition, and biomass of zooplankton. Typically, total zooplankton biomass increases with increasing eutrophication and is accompanied both by replacement of particular species and functional groups by others. For example, eutrophication leads to increased importance of rotifers, ciliated protozoans, cyclopoid copepods and cladocerans, particularly small cladocerans. However, there are reduced numbers of calanoid copepods and large cladocerans. Some zooplankton species are specific indicators of either eutrophy or oligotrophy in temperate lakes in the Northern Hemisphere (Table 3.2). The rotifer *Asplanchna brightwelli* is listed as an indicator of eutrophy in an Australian river.

In addition, the process of lake eutrophication in the past can be studied by means of the examination of exoskeletons (exuviae) of cladocerans in sediments. By checking abundances and changes in species compositions of the cladoceran remains collected in sediment core samples, the timing and trajectory of eutrophication and loss of littoral habitats can be inferred and used to support other paleolimnological evidence of lake eutrophication (Jeppesen *et al.* 2001).

Discharge of pesticides such as herbicides and insecticides from agricultural and pastoral lands into rivers and dams has adverse effects on the freshwater environment and human health. Zooplankton have been used as test or monitoring organisms to assess the acute and chronic toxicity, bioconcentration and biomagnification of these chemicals. In normal agricultural practice, protection of crops from pest organisms is achieved with the application of more than one chemical for different target organisms. The effects of combinations of pesticides on freshwater ecosystems may be synergetic, resulting in greater harm than expected.

Table 3.2. Indicators of trophic status in lakes in the Northern Hemisphere (Gannon and Stemberger 1978; Gulati 1983)

Trophic status	Animal group	Species
Eutrophy	Rotifers	*Anuraeopsis fissa, Brachionus angularis, Filinia longiseta, Keratella tecta, Polyarthra euryptera, Pompholyx sulcate, Trichocerca cylindrica,* and *Trichocerca pusilla*
Oligotrophy	Calanoid copepods	*Limnocalanus macrurus* and *Senecella calanoides*

Large cladocerans and calanoid copepods in general are more sensitive to pesticide toxicity than microzooplankton, such as *Bosmina*, *Ceriodaphnia*, rotifers and cyclopoid copepods. Therefore, an increase in microzooplankton could occur following pesticide applications, which may lead to an increase in certain groups of phytoplankton due to decreased zooplankton grazing pressure (Hanazato 2001). The feeding performance of zooplankton such as *Daphnia* is inhibited by concentrations of the pesticide endosulfan (DeLorenzo *et al.* 2002).

The cyanobacteria *Microcystis* and *Dolichospermum* may produce intracellular toxins and release them into surrounding waters, especially when they are in a senescent growth phase or when an algicide has been applied. Zooplankton such as *Daphnia*, copepods and rotifers are used ecotoxicologically as test organisms to assess the direct and indirect effects of cyanotoxins. High concentrations of cyanotoxins kill zooplankton, including *Daphnia*, while low concentrations of cyanotoxins reduce the growth and reproduction of various zooplankton (DeMott *et al.* 1991). Even filtered water that had been used to grow toxic cyanobacteria (such as *Dolichospermum*) is reported to have a negative effect on the feeding activity of *Daphnia*. Warmer temperatures may exacerbate effects of cyanotoxins on zooplankton. Zooplankton such as *Daphnia* can accumulate cyanotoxins in their bodies and may transfer them to higher trophic levels, such as fish.

3.9 Remediation of phytoplankton blooms and biomanipulation

Freshwater phytoplankton are eaten by herbivorous zooplankton such as *Daphnia*, and zooplankton are eaten by fish. The removal or reduction of zooplanktivorous fish stimulates the growth of zooplankton, which will then eat more phytoplankton. The reduced phytoplankton abundance results in more light for photosynthetic macrophytes, so the excess nutrients that were taken up by phytoplankton are now taken up by macrophytes on the bottom. The reduced phytoplankton abundance

and greater macrophyte cover leads to an improvement of water quality and clearer water.

Intentional manipulation of an aquatic ecosystem such as this is known as biomanipulation (Shapiro 1990). The biomanipulation approach includes the introduction of phytoplankton-eating fish and use of macrophytes (large plants). It focuses on the manipulation of zooplankton-eating fish and zooplankton to increase grazing pressure on phytoplankton. Biomanipulation has been used in ponds, lakes and reservoirs, particularly in the Northern Hemisphere. Because biological interactions are often complex in aquatic ecosystems, biomanipulation trials can meet with both success and failure. The average success rate of biomanipulations is ~60% (Mehner *et al.* 2002). Biomanipulation is most likely to be successful in shallow eutrophic lakes.

Zooplankton can also be used in the biological control of mosquitos. Studies have been carried out on the use of carnivorous copepods (especially cyclopoids belonging to the genus *Mesocyclops*) as biological agents for control of mosquito larvae in wells, mines and other breeding habitats especially where mosquito-eating fish are ineffective in controlling them. Such studies are important, because certain mosquitoes are a vector of viruses that cause fatal diseases to humans (such as dengue and Ross River fevers). Carnivorous copepods may be used as an environmentally acceptable and persistent agent for the control of such mosquitoes if operationally feasible procedures for the rearing and field introduction of carnivorous copepods are established, together with other environmentally friendly control measures. An advantage of using cyclopoid copepods for biological control of mosquitos is that it is easy to train locals and school children to collect, rear and distribute them.

3.10 Managing phytoplankton blooms through grazing by zooplankton

The assimilation of high algal biomass by mussels is an under-appreciated management consideration for maintaining water quality. The invasion of the Great Lakes in the north-eastern US by the zebra mussel *Dreissena polymorpha* has fundamentally altered the ecology of those lakes. By filtering out the phytoplankton, zooplankton populations have collapsed and so have the zooplanktivorous fish, including the 'alewife' *Alosa pseudoharengus*, which was introduced. Zebra mussels have also decimated the phytoplankton concentration in Hudson River and San Francisco Bay. The pygmy mussel *Xenostrobus securis* was implicated in the rapid demise of phytoplankton blooms in an estuary on the central coast of New South Wales, Australia (Moore *et al.* 2006). *Xenostrobus* aggregates on the mangrove aerial roots in brackish waters. Up to 25% of the decline in phytoplankton blooms was attributed to the pygmy mussel, but the remaining 75% could have been caused by zooplankton or population decay by salinity stress (Moore *et al.* 2006).

Zooplankton can reduce the frequency of harmful algal blooms by grazing bloom species down to low concentrations and the zooplankton biomass can increase. Analysing sufficient zooplankton samples to understand the interactions taking place in estuaries can be time consuming and answers can be achieved more rapidly by using a particle counting and sizing device. The abundance of various size categories of zooplankton can yield a useful estimate of grazing and production rates, because metabolic rate is predictably related to body size (Section 2.1). Biomass is passed from smaller to larger particles via predation (Fig. 3.3). Particle size is measured by an optical plankton counter or image analysis as area, and converted to biomass assuming a density of water and the volume of a sphere (see Section 4.9).

To assess the effect of catchments on zooplankton, the zooplankton size frequency distribution was compared in three contrasting rural estuaries (Moore and Suthers 2006). One estuary had a forested and less-developed catchment (the Wallingat River), while the other two estuaries had catchments dominated by dairy farming and hence had enhanced nutrient flows. The monthly variation was related to rainfall and nutrient supply to the estuaries. There were significant differences in the

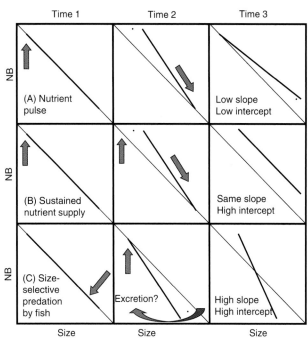

Fig. 3.3. Sketch of possible bottom-up and top-down processes altering the biomass size frequency distribution ('spectrum') of zooplankton during three time periods. **(A)** A nutrient pulse stimulates phytoplankton and increasing the (normalised) biomass concentration of small zooplankton particles, which is passed by predation to larger particles. **(B)** A sustained nutrient supply increases the biomass and intercept. **(C)** Size-selective predation by larval and juvenile fish could steepen the slope, and their excreted nutrients could increase the production of smaller particles (reproduced from Suthers *et al.* 2006).

size distribution of zooplankton between large estuaries with rural catchments and nutrient enrichment, versus the small estuary with a forested catchment (Fig. 3.4). The more pristine estuary often had a steeper slope of the zooplankton size spectrum and lower overall biomass, possibly due to the greater water clarity allowing visual-feeders such as fish to predate the larger zooplankton and thus steepen the slope (Fig. 3.3).

The role that zooplankton play in assimilating algal biomass was shown clearly in work conducted in a small coastal lake in the northern beaches area of Sydney (Rissik *et al.* 2009). The lake is closed off from the ocean for long periods of time, which removes the influence of tidal flushing and enables biological responses after a large rainfall event, after a prolonged summer dry period. Nutrients

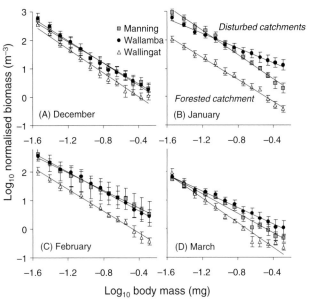

Fig. 3.4. The average normalised biomass size spectrum (NBSS) for zooplankton caught in a 100 μm mesh net in three temperate estuaries, during four summer months (reproduced from Moore and Suthers 2006).

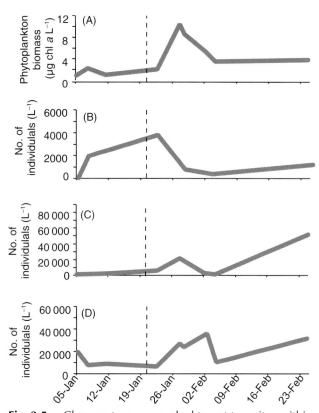

Fig. 3.5. Changes to average plankton at two sites within Dee Why lagoon over the study period. Vertical dashed lines indicate the main, initial rain event. **(A)** phytoplankton biomass (μg Chl-*a*/L). **(B)** *Oithona*, an adult copepod, which doubled in abundance within 48 hours. **(C)** Copepod nauplii. **(D)** adult *Acartia bispinosa* (adapted from Rissik *et al.* 2009).

(ammonia and oxidised nitrogen) significantly increased the day after initial rainfall, before returning to pre-rainfall conditions within 5 days. In response, phytoplankton biomass grew 10-fold within a week after the initial rainfall and declined to near original levels 2 weeks later (Fig. 3.5A). Blooms of diatoms followed the rainfall within a week, which returned to pre-rainfall levels within 2 weeks.

Zooplankton increased in response to the higher phytoplankton concentrations and were responsible for the rapid decline in phytoplankton. Some zooplankton responded within 24 hours, with two-fold increase in adult stages of the calanoid copepod *Oithona* sp. (Fig. 3.5B), followed a week later by nauplii (Fig. 3.5C) and adult *Acartia bispinosa* (Fig. 3.5D). The influx of adult copepods was presumably from resting populations that were previously under sampled by the plankton net. The zooplankton community returned to the initial state by 2 weeks and then matured to a centric diatom and *Acartia* dominated population after 5 weeks.

This section illustrates the remarkable ecosystem role of zooplankton in estuaries, in connecting primary production with fish and fisheries. The trophic connection between phytoplankton and zooplankton in lakes and estuaries is often exploited by the aquaculture industry (Box 3.8), because these species are robust to changes in temperature and salinity. Sometimes, however, jellyfish and ctenophores can bloom in estuaries and interfere with this connection.

3.11 Management of jellyfish blooms and fisheries

Dense concentrations of jellyfish are usually referred to as 'blooms'. 'True blooms' arise due to the proliferation and growth of individuals, which results in a rapid increase in the biomass of the population (Graham *et al.* 2001). Sometimes, however, 'apparent blooms' (Graham *et al.* 2001) form because

Box 3.8 Plankton in aquaculture

Some key species of phytoplankton and zooplankton are essential for the aquaculture industry, mainly as prey for larval forms of commercially important species (and sometimes as prey, to rear the prey!). Many aquaculture facilities are dominated by large columns of phytoplankton, with bubbling CO_2 and under bright lights to promote photosynthesis and growth.

Some useful phytoplankton, such as the golden brown flagellate (Prymnesiophyte) *Pavlova lutheri*, a diatom *Chaetoceros calcitrans* and the green flagellate *Tetraselmis suecica*, are small (<10 μm across) and thus difficult to view under a microscope. These phytoplankton are used to for rearing larval oysters, and are also the prey for zooplankton that are then used to rear fish larvae. For example, *Chaetoceros calcitrans* and *Tetraselmis suecica* are used to rear the rotifer *Brachionus*: an estuarine species and one of the few commercially important rotifers (see Chapter 7). These phytoplankton are also used to feed the brine shrimp *Artemia* for rearing fish larvae. Brine shrimp are found worldwide in highly saline ponds where they are safe from predators. They are used in aquaculture because their eggs are conveniently stored as dry cysts and when transferred to water hatch into the larval nauplius stage within a day, ready for larval fish. Brine shrimp are also well-known fish as food in aquaria and are kept as pets – 'sea monkeys' (Fig. 1.7).

Aquaculture industries are constantly seeking ways to reduce operating costs when rearing commercially important species. One successful method is to fertilise the production of local copepod species in aquaculture ponds. A useful copepod in aquaculture in Australia is *Gladioferens imparipes*, which thrives in estuaries and therefore has many natural attributes as food for larval fish.

oceanographic features, such as tidal currents, concentrate sparsely scattered individuals within a small area (Graham *et al.* 2001). Apparent blooms, therefore, represent a change in the distribution rather than the size of the population (see Fig. 2.5 for the life cycle of a typical scyphozoan jellyfish).

Blooms can impart major socio-economic impacts on coastal infrastructure and commercial fishing, aquaculture and tourism industries (Lucas *et al.* 2014). For example, jellyfish often clog the cooling water intakes of coastal power plants and the seawater intakes of desalination plants, interrupting supplies of electricity and water. Jellyfish can also clog the condensers and bow-thrusters of ships, which interfere with their ability to manoeuvre in port. Indeed in 2005, the USS Ronald Regan was forced to depart the Port of Brisbane ahead of schedule because blooms of the blue blubber jellyfish *Catostylus mosaicus* kept blocking its condensers. Fisheries and aquaculture are also affected when jellyfish clog and damage nets and their stings damage and devalue, or even kill, the catch. Fishers in Peru are estimated to lose approximately US$160 million in revenue per ton of jellyfish caught in their nets (Quiñones *et al.* 2013) and Irish

and Scottish salmon aquaculture ventures have lost millions of dollars when blooms of jellyfish have drifted amongst the fish cages and killed the stock (Lucas *et al.* 2014). Tourism is also affected when blooms of stinging jellyfish close beaches or simply deter people from swimming. This is particularly true in tropical regions of the world where the box jellyfish *Chironex* and a suite of species known as Irukandji can cause serious problems or even death of bathers (see Box 9.1 for treatment of jellyfish stings).

Concern that jellyfish blooms might be increasing globally has intensified efforts to document jellyfish population trends and identify the potential causes of jellyfish blooms. We now know that while some regions of the world have experienced sustained increases in jellyfish, other regions have experienced decreases (Condon *et al.* 2013), hence increased jellyfish blooms are a regional, rather than a global, phenomenon. The definitive causes of problematic blooms have rarely been identified, although human activities such as overfishing, excessive inputs of nutrients into coastal waters and climate change are speculated to facilitate blooms (Richardson *et al.* 2009). It is important to

Box 3.9 Jellyfish fisheries in China

Most of the world's jellyfish are harvested from Liaodong Bay in China. The fishery is managed by regulating the amount of time that jellyfish can be harvested each year. In the early 1980s, ~2000 boats harvested jellyfish in Liaodong Bay and the fishery was open for 2 weeks each year (You *et al.* 2016). Since then the number of boats in the fishery has grown to more than 10 000 and the fishery is now open for only a few hours each year! Overfishing, coupled with substantial illegal fishing (i.e. fishing outside the designated time) resulted in stocks of jellyfish declining in Liaodong Bay in the early 2000s (You *et al.* 2016). To compensate for the decline in jellyfish, juvenile jellyfish are now produced via aquaculture and used to restock Liaodong Bay. Each year ~0.1–0.3 billion juvenile jellyfish are released, with an estimated recapture rate of 3% (You *et al.* 2016).

Upon capture, jellyfish are semi-dried using a combination of alum and salt and processing can take 20–40 days. The dried product is initially prepared by soaking it in cold water to remove the salt. The jellyfish is then shredded into strips, blanched in boiling water, and mixed with sauces and other ingredients such as chicken or seafood and served cold as a salad.

note, however, that a relatively small number of the ~200 species of 'true' jellyfish (Scyphozoa) cause problematic blooms and management of these blooms may need to be species-specific.

Dried jellyfish is eaten in many Asian nations and is used in traditional Chinese medicine to treat a range of health disorders including arthritis, ulcers and asthma (You *et al.* 2016). Jellyfish have been harvested in China for over 1700 years (Omori and Nakano, 2001) but large-scale commercial harvesting of jellyfish commenced in Asia around 1970 (Box 3.9). The annual global harvest is around a million tonnes (Brotz 2016). Although the major fishing grounds are concentrated around Asia, increased demand for jellyfish has seen small-scale jellyfish fisheries emerge in other regions, including Australia, the Middle East, Europe and the USA. Approximately 17 species of jellyfish are harvested for food around the world (You *et al.* 2016).

3.12 Conclusions

It is important to understand the role of plankton in any aquatic system that is being managed. Their link to virtually every aspect of the waterway demands that we know more about them, and place more effort into understanding their drivers, responses, trophic roles and increasingly the differences between different taxa. There are great opportunities that can be derived with a better knowledge of plankton. Monitoring is time consuming and costly, but modern technology is helping to make it easier to work with these creatures, and their utility will continue to increase.

3.13 References

Ajani PA, Hallegraeff GM, Allen D, Coughlan A, Richardson AJ, Armand LK, *et al.* (2016) Establishing baselines: a review of eighty years of phytoplankton diversity and biomass in southeastern Australia. *Oceanography and Marine Biology - an Annual Review* **54**, 387–412. doi:10.1201/9781315368597-9

Ajani P, Hallegraeff G, Pritchard T (2001a) Historic overview of algal blooms in marine and estuarine waters of New South Wales, Australia. *Proceedings of the Linnean Society of New South Wales* **123**, 1–22.

Ajani P, Lee RS, Pritchard TR, Krogh M (2001b) Phytoplankton dynamics at a long-term coastal station off Sydney, Australia. *Journal of Coastal Research* **34**, 60–73.

Baker PD, Humpage AR (1994) Toxicity associated with commonly occurring cyanobacteria in surface waters of the Murray-Darling basin, Australia. *Australian Journal of Freshwater Research* **45**, 773–786.

Bradley WG, Borenstein AR, Nelson M, Codd GA, Rosen BH, Stommel EW, *et al.* (2013) Is exposure to cyanobacteria an environmental risk factor for amyotrophic lateral sclerosis and other neurodegenerative disease? *Amyotrophic Lateral Sclerosis & Frontotemporal Degeneration* **14**, 325–333. doi:10.3109/21678421.2012.750364

Brotz L (2016) Jellyfish fisheries – a global assessment. In *Global Atlas of Marine Fisheries: A Critical Appraisal of Catches and Ecosystem Impacts.* (Eds D Pauly and D Zeller) pp. 110–124. Island Press, Washington DC, USA.

Capuzzo E, Lynam CP, Barry J, *et al.* (2018) A decline in primary production in the North Sea over 25 years, associated with reductions in zooplankton abundance and fish stock recruitment. *Global Change Biology* **8**(24), e352–e364. doi.org/10.1111/gcb.13916e364

Cloern JE, Foster SQ, Kleckner AE (2014) Phytoplankton primary production in the world's estuarine-coastal ecosystems. *Biogeosciences* **11**, 2477–2501. doi:10.5194/bg-11-2477-2014

Condon RH, Duarte CM, Pitt KA, Robinson KL, Lucas CH, Sutherland KR *et al.* (2013) Recurrent jellyfish blooms are a consequence of global oscillations. *Proceedings of the National Academy of Sciences of the United States of America* **110**, 1000–1005. doi:10.1073/pnas.1210920110

Crawford A, Holliday J, Merrick C, Brayan J, van Asten M, Bowling L (2017) Use of three monitoring approaches to manage a major *Chrysosporum ovalisporum* bloom in the Murray River, Australia, 2016. *Environmental Monitoring and Assessment* **189**, 202. doi:10.1007/s10661-017-5916-4

Dela-Cruz J, Ajani P, Lee R, Pritchard TR, Suthers IM (2002) Temporal abundance patterns of the red tide dinoflagellate *Noctiluca scintillans* along the southeast coast of Australia. *Marine Ecology Progress Series* **236**, 75–88. doi:10.3354/meps236075

DeLorenzo ME, Taylor LA, Lund SA, Pennington PL, Strozier ED, Fulton MH (2002) Toxicity and bioconcentration potential of the agricultural pesticide endosulfan in phytoplankton and zooplankton. *Archives of Environmental Contamination and Toxicology* **42**, 173–181. doi:10.1007/s00244-001-0008-3

DeMott WR, Zhang Q, Carmichael WW (1991) Effects of toxic cyanobacteria and purified toxins on the survival and feeding of a copepod and three species of *Daphnia. Limnology and Oceanography* **36**, 1346–1357. doi:10.4319/lo.1991.36.7.1346

Dennison WC, Abal EG (1999) *Moreton Bay Study: A Scientific Basis for the Healthy Waterways Campaign.* South East Queensland Regional Water Quality Management Strategy Team, Brisbane, Australia.

Gannon JE, Stemberger RS (1978) Zooplankton (especially crustaceans and rotifers) as indicators of water quality. *Transactions of the American Microscopical Society* **97**, 16–35.

Graham WM, Pages F, Hamner WM (2001) A physical context for gelatinous zooplankton aggregations: a review. *Hydrobiologia* **451**, 199–212. doi:10.1023/A:1011876004427

Gulati RD (1983) Zooplankton and its grazing as indicators of trophic status in Dutch lakes. *Environmental Monitoring and Assessment* **3**, 343–354. doi:10.1007/BF00396229

Hallegraeff GM (1998) Transport of toxic dinoflagellates via ships' ballast water: bioeconomic risk assessment and efficacy of possible ballast water management strategies. *Marine Ecology Progress Series* **168**, 297–309.

Hallegraeff GM, Anderson DM, Cembella AD (2003) *Manual on Harmful Marine Microalgae.* Monographs on Oceanographic Methodology 11. UNESCO Publishing, Paris, France.

Hanazato T (2001) Pesticide effects on freshwater zooplankton: an ecological perspective. *Environmental Pollution* **112**, 1–10. doi:10.1016/S0269-7491(00)00110-X

Hoegh-Guldberg O, Cai R, Poloczanska ES, Brewer PG Sundby S, Hilmi K, *et al.* (2014) The ocean. In *Climate Change 2014: Impacts, Adaptation, and Vulnerability. Part B: Regional Aspects. Contribution of Working Group II to the Fifth Assessment Report of the Intergovernmental Panel on Climate Change.* (Eds VR Barros, CB Field, DJ Dokken, MD Mastrandrea, KJ Mach, TE Bilir, *et al.*) pp. 1655–1731. Cambridge University Press, Cambridge, UK.

Jeppesen E, Leavitt P, De Meester L, Jensen JP (2001) Functional ecology and palaeolimnology: using cladoceran remains to reconstruct anthropogenic impact. *Trends in Ecology & Evolution* **16**, 191–198. doi:10.1016/S0169-5347(01)02100-0

Lee R, Ajani P, Wallace S, Pritchard TR, Black KP (2001a) Anomalous upwelling along Australia's East Coast. *Journal of Coastal Research* **34**, 87–95.

Lee RS, Ajani P, Krogh M, Pritchard TR (2001b) Resolving climatic variance in the context of retrospective phytoplankton pattern investigations off the east coast of Australia. *Journal of Coastal Research* **34**, 96–109.

Locke A, Sprules WG (1994) Effects of lake acidification and recovery on the stability of zooplankton food webs. *Ecology* **75**, 498–506. doi:10.2307/1939553

Lucas CH, Gelcich S, Uye S-I (2014) Living with jellyfish: management and adaptation strategies. In *Jellyfish Blooms*. (Eds KA Pitt and CH Lucas) pp. 129–150. Springer, New York, USA. doi:10.1007/978-94-007-7015-7_6

McQuatters-Gollop A, Johns DG, Bresnan E, Skinner J, Rombouts I, Stern R, *et al.* (2017) From microscope to management: the critical value of plankton taxonomy to marine policy and biodiversity conservation. *Marine Policy* **83**, 1–10. doi:10.1016/j.marpol.2017.05.022

Mehner T, Benndorf J, Kasprzak P, Koschel R (2002) Biomanipulation of lake ecosystems: successful applications and expanding complexity in the underlying science. *Freshwater Biology* **47**, 2453–2465. doi:10.1046/j.1365-2427.2002.01003.x

Mitrovic SM, Oliver RL, Rees C, Bowling LC, Buckney RT (2003) Critical flow velocities for the growth and dominance of *Anabaena circinalis* in some turbid freshwater rivers. *Freshwater Biology* **48**, 164–174. doi:10.1046/j.1365-2427.2003.00957.x

Moore SK, Suthers IM (2006) Evaluation and correction of subresolved particles by the optical plankton counter in three Australian estuaries with pristine to highly modified catchments. *Journal of Geophysical Research* **111**, C05S04. doi:10.1029/2005JC002920

Moore SK, Baird ME, Suthers IM (2006) Relative impacts of physical and biological processes on nutrient and phytoplankton dynamics in a shallow estuary after a storm event. *Estuaries and Coasts* **29**, 81–95.

Omori M, Nakano E (2001) Jellyfish fisheries in southeast Asia. *Hydrobiologia* **451**, 19–26. doi:10.1023/A:1011879821323

O'Neil JM, Davis TW, Burford MA, Gobler CJ (2012) The rise of harmful cyanobacterial blooms: the potential roles of eutrophication and climate change. *Harmful Algae* **14**, 313–334. doi:10.1016/j.hal.2011.10.027

Paerl HW, Paul VJ (2012) Climate change: links to global expansion of harmful cyanobacteria. *Water Research* **46**, 1349–1363. doi:10.1016/j.watres.2011.08.002

Pilotto LS, Hobson P, Burch MD, Ranmuthugala G, Attewell R, Weightman W (2004) Acute skin irritant effects of cyanobacteria (blue-green algae) in healthy volunteers. *Australian and New Zealand Journal of Public Health* **28**, 220–224. doi:10.1111/j.1467-842X.2004.tb00699.x

Quiñones J, Monroy A, Acha EME, Mianzan HW (2013) Jellyfish bycatch diminishes profit in an anchovy fishery off Peru. *Fisheries Research* **139**, 47–50.

Richardson AJ, Bakun A, Hays GC, Gibbons MJ (2009) The jellyfish joyride: causes, consequences and management responses to a more gelatinous future. *Trends in Ecology & Evolution* **24**, 312–322. doi:10.1016/j.tree.2009.01.010

Rissik D, Doherty M, van Senden D (2006) A management focussed investigation into phytoplankton blooms in a sub-tropical Australian estuary. *Journal of Ecosystem Health and Management*, **9**, 365–378.

Rissik D, Ho Shon E, Newell B, Baird ME, Suthers IM (2009) Plankton dynamics due to rainfall, eutrophication, dilution, grazing and assimilation in an urbanized coastal lagoon. *Estuarine, Coastal and Shelf Science* **84**, 99–107. doi:10.1016/j.ecss.2009.06.009

Robarts RD, Zohary T (1987) Temperature effects on photosynthetic capacity, respiration, and growth-rates of bloom-forming cyanobacteria. *New Zealand Journal of Marine and Freshwater Research* **21**, 391–399. doi:10.1080/00288330.1987.9 516235 .

Shapiro J (1990) Biomanipulation: the next phase making it stable. *Hydrobiologia* **200/201**, 13–27.

Stewart I, Robertson IM, Webb PM, Schluter PJ, Shaw GR (2006) Cutaneous hypersensitivity reactions to freshwater cyanobacteria – human volunteer studies. *BMC Dermatology* **6**, 6. doi:10.1186/1471-5945-6-6

Sukenik A, Hadas O, Kaplan A, Quesada A (2012) Invasion of Nostocales (cyanobacteria) to subtropical and temperate freshwater lakes – physiological, regional, and global driving forces. *Frontiers in Microbiology* **3**, 1–9. doi:10.3389/fmicb.2012.00086

Sukenik A, Quesada A, Salmaso N (2015) Global expansion of toxic and non-toxic cyanobacteria: effect of ecosystem functioning. *Biodiversity and Conservation* **24**, 889–908. doi:10.1007/s10531-015-0905-9

Suthers IM, Taggart CT, Rissik D, Baird ME (2006) Day and night ichthyoplankton assemblages and the zooplankton biomass size spectrum in a deep ocean island wake. *Marine Ecology Progress Series* **322**, 225–238. doi:10.3354/meps322225

Walseng B, Karlsen LR (2001) Planktonic and littoral microcrustaceans as indices of recovery in limed lakes in SE Norway. *Water, Air, and Soil Pollution* **130**, 1313–1318. doi:10.1023/A:1013953806377

Winder JA, Cheng DMH (1995) 'Quantification of factors controlling the development of *Anabaena circinalis* blooms'. Research Report No. 88. Urban Research Association of Australia, Melbourne, Australia.

You K, Bian Y, Ma C, Chi X, Liu Z, Zhang Y (2016) Study on the carry capacity of edible jellyfish fishery in Liaodong Bay. *Journal of Ocean University of China* **15**, 471–479. doi:10.1007/s11802-016-2924-x

3.14 Further reading

Dakin WJ, Colefax AN (1933) The marine plankton of the coastal waters of New South Wales. I. Chief planktonic forms and their seasonal distribution. *Proceedings of the Linnean Society of New South Wales* **58**, 186–222.

Harris RP, Wiebe PH, Lenz J, Skjoldal HR, Huntley M (Eds) (2000) *ICES Zooplankton Methodology Manual*. Academic Press, London, Uk.

4

Sampling methods for plankton

Iain M. Suthers, Lee Bowling, Tsuyoshi Kobayashi and David Rissik

A plankton study requires careful thought before field work starts. You should consider what your organisation needs to know about water quality, and how can this be achieved cost effectively. Perhaps there are some trouble spots such as phytoplankton blooms after rainfall. A pilot study is a good start. What about the future – should a monitoring program be in place, and for how long? Perhaps you could consider previous sampling and if those sites could be re-visited. And, most importantly, how will you determine if change is due to natural variability or changes in catchment management? This chapter will examine the issues necessary to launch such a study.

Many issues need to be resolved before collecting samples. The degree of sorting the sample needs to be understood because the cost of collecting the samples is possibly a third of the total budget. Long-term data provides an important baseline against which future changes can be assessed. A general monitoring program should be conservative and easily interpreted by all, without relying on a single individual. Good monitoring should also discriminate natural changes at control sites, as well as changes caused by humans. Otherwise, potential developers may use this natural variation to hide any environmental impact that they may have caused.

4.1 Introduction to sampling methods

When preparing for sampling, time invested in formulating unambiguous questions and appropriate methods and analyses, is time well spent. You must decide to what degree are the samples to be analysed (e.g. just biomass, or by size, or to phylum level, or right down to species?). Many issues can be investigated by using biomass, size or classifying plankton into broad taxonomic groups. Try to imagine the data and even the graph that you seek and then plan a program that will put data onto that graph. A day's pilot study will fine-tune the planning.

There is no single, generic sampling method – the method chosen must suit the question. Plankton has a patchy distribution in both space (vertical and horizontal) and time (between day and night, winter and summer). This means, for example, that sampling with a particular size of mesh, or during the night, or during the ebb tide will influence the results and the interpretation. This chapter is about:

- determining a robust sampling design in relation to the objectives
- the observation, preparation and quantification of plankton samples
- the preservation of plankton samples.

Defining your question is perhaps the most important and difficult part of plankton studies because it requires you to consider exactly what information your organisation requires in the short and long term. With a defined question, the proposed statistical analysis that answers this question must be considered before data collection even begins. If in doubt get advice, because sampling effort has been wasted in the past by not considering the final analysis.

Conflicting advice on statistics is typical, and it is up to you to rationalise differing views. One typical outcome of an environmental impact study is 'no significant effect' due to insufficient replication and statistical power. The long-term implications of wrongly concluding no significant impact are more severe (such as permanent loss of biodiversity) than wrongly concluding a significant impact and changing the development proposal.

There is great value in attempting to integrate old data or data collected in previous studies by you or others. This is because there is a gradual, declining standard of our environment, which is not easily noticed in the short term (between years), but changes are evident by comparing over decades. Some data should not be used where sampling and analytical methods and other influencing factors are not documented or have changed considerably.

Salinity and temperature are key environmental traits that place the plankton into a context. Therefore, measure the temperature and salinity at every station, because the water mass or vertical stratification can influence plankton communities (Fig. 2.8). For example, a vertical profile of salinity and temperature at several sites can enable you to assess whether the waterway is stratified, horizontally or vertically. Estuarine plankton communities vary according to the physical and chemical water properties, which in turn vary seasonally and yearly and daily due to tides.

At the most upstream reaches with salinities between 0–3, the plankton community consists mainly of a relatively few freshwater taxa. At salinities 5–7, the communities are true brackish-estuarine species (euryhaline species), with an increasing dominance of stenohaline marine taxa as the salinity approaches 30–35. A useful conceptual model of estuarine species abundance with salinity may be summarised with a Remane diagram (Whitfield *et al.* 2012), revealing a modest abundance of freshwater species, a paucity of true brackish species and a predominance of estuarine and marine species.

4.2 Sampling design and environmental variability

4.2.1 Independent samples

Good sampling design requires that each level of a sampling program (e.g. estuaries or bays), and each replicate sample, is independent of each other. Dividing a plankton sample in half is not a replicate. For example, samples should not be collected simultaneously (e.g. the paired nets of a bongo net (Section 4.7) are not independent replicates) and replicates sampled immediately downstream of another sample are not independent.

To assess water quality you may need an additional level of analysis – that of an independent estuary or lake. This is because the water body of interest may be changing throughout all bays and coves due to changes in its catchment (of major concern to a manager), or due to global or regional changes (also of concern, but not within a manager's mandate). Consequently, a parallel sampling program should be conducted in a related or similar water body (Fig. 4.1). It may be difficult to convince managers to invest funds outside their constituency, despite the need to benchmark your own investigation. One approach is to do the regional comparison during a particular summer month only – in a separate analysis – or by collaborating with other groups.

4.2.2 Spatial and temporal scale

Many ecological processes may be relevant at the small scale (minutes or metres) or at the large scale (years or tens of kilometres). Water-quality managers generally operate within a 1- to 20-year timeframe and a 1- to 20-km spatial scale. Consequently, you will find in this section reference to a sampling hierarchy: from the level of sub-sample to site, to

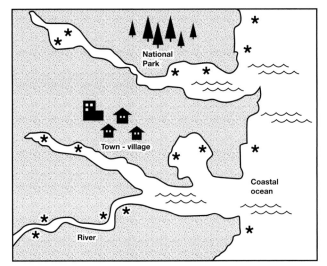

Fig. 4.1. A possible monitoring design of an estuary or river, illustrating the importance of replicating sites in each region (with a replicated sample from each), and possibly an external reference estuary. The external site is to ensure that changes in your estuary are not due to some global or climate phenomenon. The external data could be another municipality's monitoring program – it does not have to be pristine. Perhaps the town or village was wishing to install an artificial wetland or perhaps fencing along one of the rivers. This design could assess the environmental cost–benefit of rehabilitation, for example. Your sampling needs to have replicate sites to ensure that the changes you are observing are not just peculiar to one site and unrelated to the wetland or fence.

embayment, to water body; and from the level of day, month and perhaps year. This traditional sampling strategy is particularly appropriate in marine ecology for the analysis of variance (ANOVA), which partitions the variability among the factors and their levels (Box 4.1). Although there may be some logistical constraints, this approach explicitly lays out your sampling proposal, defining a hierarchical sampling design (e.g. estuaries, months, days, sites and replicate tows), even if you choose not to use this particular analysis (see Kingsford and Battershill 1998). Regression and correlation are useful methods to further test your findings.

4.2.3 Variance, sample size and replication

Natural variability underscores all biological sciences, from evolution to ecology. Variability is what determines the importance of an average value. For example, finding that the average summer chlorophyll values in your pond increase over 10 years from 0.5 to 2.0 µg/L may not be as important if the range during each summer was 0.5 to 10 µg/L (expressed as µg L^{-1} in scientific publications). For this reason, ecologists are often concerned with the degree of variability, as well as the average value (Box 4.1).

Box 4.1 Variance, patchiness and statistical power

The sum of the squared differences between each observation and the overall mean value ($\bar{x}$) is known as the variance (s^2), and the square root of the variance is termed the standard deviation (s). The standard deviation may be compared between ponds or days or species by a coefficient of variation (CV), expressed as (100 × s/$\bar{x}$). It is the variance that determines confidence limits and significant differences. An estimate of the number of replicates (n) needed to be within 15% of the average value may be calculated from a pilot study by $n = (s/0.15 \times \bar{x})^2$ (Kingsford and Battershill 1998, p. 53).

Biological and physical processes can promote patchiness or clumping, such as cell division, or cell buoyancy. A random distribution of cells is a combination of clumped and uniform distribution of animals. The degree of clumped distribution is described as patchiness, which can vary among species or in space (such as at the centimetre, metre and kilometre scales, such that there are patches of patches, rather like suburbs, towns and cities).

The concept of statistical power pervades any environmental impact assessment. The cheapest approach to an environmental impact is to take just two replicates in an impacted and non-impacted area. Inevitably, the natural variability will swamp any difference between the two areas, and you would wrongly conclude no significant impact (a type II error). Such a flawed comparison would be an example of low statistical power, because of low replication or high variability. Managers need to be wary of quick and cheap assessments – and also be aware of any attempts to avoid environmental responsibility for an impact, under the guise of natural variability.

Variation associated with the natural patchiness of plants and animals can be 10- or 100-fold greater than the variation in physical characteristics, such as sediment type or water temperature. The degree of variation is often in proportion to the average value (i.e. a large value has a greater capacity for variation than a small value) and, in part, to the spatial and temporal range (samples taken at metres or seconds apart vary less than those taken at kilometres or months apart, Box 4.2). Variability may occur at temporal scales of less than a few hours and at spatial scales less than hundreds of metres and most questions for water-quality management occur at scales greater than this.

As a first step, we often sample large volumes of water with a plankton net or pole sampler that integrates, or mixes, this small-scale variability. Plankton net tows are different to many other kinds of ecological sampling (such as benthic cores, quadrats, fish counts or bottle samples of water), because a tow over 5–10 minutes integrates many fine-scale patches. Therefore the variance among replicate tows is often small (Box 4.2) and while we may generally collect three or four replicate tows, you may need to only sort and identify two. Nevertheless, we do need to know the degree of variability at the scale of our sampling device, and at each and every level above. To further pool all the samples of a particular region would become pointless – because the value and statistical power of just two replicates exceeds one pooled sample.

Your pilot study may indicate the need to consider additional factors, such as replicate days or months, to partition the variability. These could otherwise overwhelm your variable of interest (such as an estuary with, or without, sewage treatment). Normally every factor of your analysis should be replicated (i.e. 2 or 3 replicate days). By inserting additional factors, samples numbers and costs can quickly escalate. However, without partitioning the variability, you would have to take many more replicates at the level of your sampling unit, and the statistical power would be lower.

We may also need to quantify variability at the level of our sampling device by taking replicate

Box 4.2 Where plankton variance may be expected

Relative coefficient of variation (CV) of plankton within an estuary (if all other factors are constant). The table is based on our experience with towed nets (100–500 μm mesh) of at least 3 min duration (i.e. integrating many fine-scale patches) and should be used as a guide only. The number of dots represents the approximate variability in plankton that could be expected by sampling, for example, at one site before and after rain. Patchiness (variability) in (a) time is generally greater than (b) spatial patchiness, but sampling over time takes more organisation and effort.

Factor	Relative CV
(a) Temporal	
Before/after rainfall	•••••
Day/night	••••
Morning/afternoon	•
Between flood/ebb tides	••••
Among days	•
Among weeks	••
Among seasons	•••••
Between two years	•
(b) Spatial	
Among estuaries	•••
Among habitats within an estuary	••
Among sites within a habitat	•
Within a site (i.e. among replicates)	•
Among sub-samples	•
Surface/depth (between 0 and 5 m)	••

samples. The number of replicates needed is in proportion to the variability (Box 4.1), which is frequently determined by a pilot study. A pilot study provides a useful snapshot, but the variance due to seasons or rainfall may change months after the pilot study. Instead, many scientists estimate by 'taking two, three or four samples', which, for many plankton studies, may actually be appropriate, providing there is a suitable hierarchy of sampling levels. In summary, your final design will depend on whether you are planning a baseline study, an impact study, a monitoring study or to determine patterns and processes (Kingsford and Battershill 1998). Variability in plankton samples can be dealt with by integration (taking larger samples) or stratification (recognise regions or days to block your data) and replication (increase your base sample size in proportion to the variability).

You may consider a cost–benefit analysis, whereby you balance the competing needs of a limited budget, increased replication and/or inserting levels into your sampling design, or integrating variability with larger samples. There are formal ways of balancing these competing needs explicitly in terms of dollars to variance (Kingsford and Battershill 1998). For plankton studies involving identification, the major costs are the sorting and analysis, rather than the collection costs.

4.3 Typical sampling designs: where and when to sample

Anticipate that your established monitoring program of water quality may be incorporated into an unexpected impact assessment, such as a chemical spill. A robust monitoring program can account for the intrinsic, natural variation, and statistical or graphical methods can partition natural and artificial variability among your sampling sites. Data collected over long periods of time can be used to explore the response by plankton in time and space and to infer a process. It is also possible to manipulate the environment experimentally in a way that specifically adds or subtracts a component that you believe to be important in influencing water quality.

Temporal variation must be accounted for – despite the factors of interest often being spatial (impacted versus control sites). Day/night effects incorporate a large proportion of zooplankton variability, due to diel vertical migration, selective tidal stream transport and net avoidance. Significantly more and larger zooplankton is caught at night, but this community may have a significant benthic component, which may be useful if you are interested in the small animals that inhabit the sediment. Whether you sample during the day or at night is not crucial, so long as you are consistent and avoid the effects of vertical migration around dawn and dusk. Similarly, you should consistently sample the ebb or flood tide, depending on your question. If you sample only on the ebb tide, you will be sampling water that has spent at least the past 6 hours in the estuary, and thus reflects estuarine conditions. Because plankton can rapidly increase over days and weeks, a robust plankton sampling design should include daily to weekly variation. If the seasonal component is not important to you, then you could just sample the midsummer months, on an annual basis.

Choose sites on the basis of logistics and safety, and avoid areas with conspicuous fronts and foam lines. Pay careful attention to tidal characteristics, estuarine flow, wind strength and direction, which can influence plankton abundance. Some sites may characteristically support a bloom (e.g. Fig. 3.1).

The bathymetry where sampling occurs may have a large effect on plankton composition in lakes and estuaries. Such areas are often well mixed from top to bottom and an oblique – or near-surface tow – is adequate. Vertical phytoplankton hauls or pole samplers will also mix or ignore any vertical structure. In general, the effect of depth is ignored in sampling areas <5 m depth, provided the sampling protocol is consistent. Ensure that you record at least the temperature, salinity and turbidity at every station.

You may sample plankton at point stations, or along transects, or a grid of stations. The survey method used will depend primarily on tidal and current features of the inshore sub-tidal habitat, and on study objectives. A transect of stations is appropriate if an alongshore or across-shore gradient in plankton is suspected.

How often should you sample? If plankton monitoring is your goal, then sampling every 2 to 3 days (during a similar phase of tide), on each of 2 to 3 midsummer months is a good start. Representative regions should be sampled with at least two stations in each, with two to four replicate, depth-integrated samples at each station. To monitor the effect of rainfall and run-off does require a degree of opportunistic sampling to coincide with a downpour (see Fig. 3.5, Moore *et al.* 2006).

4.4 Measurement of water quality

Estuarine water quality is dependent on several factors, such as loads of nutrients and sediments to the system, recycling of nutrients within the system, reworking of sediment and other integrating factors within the system (such as assimilation, flushing and light penetration). Water-quality parameters can be separated into those that are toxic to organisms at certain levels and those that have indirect effects on organisms by changing the nature of the system. Water quality can be determined using a variety of means, including direct measurement of specific variables, such as nutrients, or by measuring other variables, such as phytoplankton biomass or biodiversity. Phytoplankton biomass is a useful indicator because phytoplankton integrate many water-quality attributes over a variety of timescales and, although temporally and

spatially variable, are less so than factors such as nutrients (Fig. 1.2).

Water temperature (T), along with salinity (S), characterises the 'T–S signature' of water habitats (Box 4.3). The actual differences in T and S are small, yet tiny changes of just 0.1°C in temperature or 0.01 in salinity can be the planktonic equivalent of moving from a desert to a rainforest (see Fig. 2.8).

Some probes may have a chlorophyll fluorescence sensor (Fig. 4.2). This instrument shines a blue light into the water, which, in turn, causes the chlorophyll to fluorescence (i.e. the chlorophyll molecule emits a photon). Once calibrated with actual extractions in the laboratory, the fluorescence is roughly proportional to the actual biomass of chlorophyll. The advantage of fluorescence over absorbance is that it only needs *in situ* concentrations – no extraction into solvents is necessary. The disadvantage is that many factors such as tannins or cell physiology influence fluorescence, and the signal is at best ± 50% precise. Other commercial fluorescence sensors make *in-situ* measurements of other pigments. Examples include sensors that measure phycocyanin presence (that indicate the

Box 4.3 Electronic determination of salinity

Salinity used to be determined chemically, such as from the concentration of chlorine ions – which uniformly account for 55.0% of total ions. A kilogram, or nearly a litre, of seawater typically contains ~35 g of salts (or 3.5% weight for volume), and therefore has been expressed as 35 parts per thousand (ppt). Today, one of the most common methods of estimating salinity is by its electrical conductivity. This modern method of salinity is a ratio of two electronic signals, so today there are no units for salinity (e.g. 'the salinity was 35 last week'). For a given temperature, conductivity of water varies linearly with ion concentration – making measurement of electrical current between two submerged electrodes a convenient measurement (Fig. 4.2A,B). Alternatively, salinity can be measured by inducing an electric field around the sensor, which is linearly proportional to the concentration of ions. Particular attention should be paid to this type of sensor as spuriously low readings will be recorded if it touches the side of the bucket, or even seagrass.

A simple measurement of salinity is the refractive index of water, which is measured with a portable refractometer using just a few drops of water (Fig. 4.2C,D). The refractometer is calibrated for a direct read-out of S at 20°C. Salinity may be expressed in parts per thousand (ppt), or practical salinity units (psu), or usually without units (because the electrical method is actually a ratio). Unlike temperature, salinity is ecologically conservative parameter, and so it is an excellent indicator of circulation in an estuary. Together with water pressure, temperature and salinity determine the density of sea water. The density of pure water at 15°C is 1000 kg/m^3 (i.e. 1 kg/L), while warm sea water at 25°C and a typical oceanic salinity of 35 is ~1023.3 kg/m^3 (i.e. 1.023 kg/L). The density is therefore expressed as rho (ρ = 1.0233). Oceanographers abbreviate this to sigma (in this case, σ = 23.3; same units by convention).

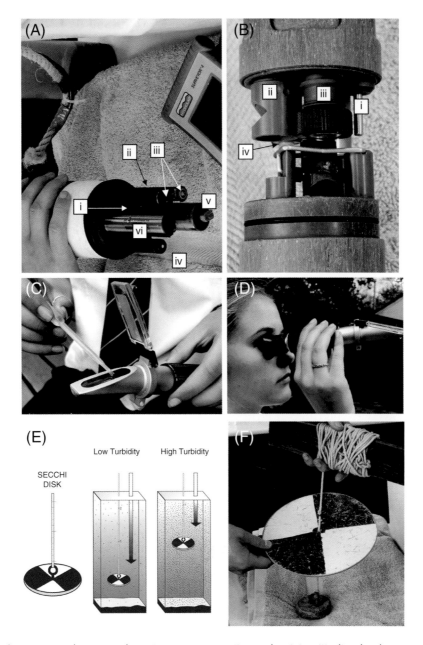

Fig. 4.2. (A, B) Typical commercial CTD probes: (i) temperature (ii) conductivity (iii) dissolved oxygen and associated stirrer (iv) pH and reference electrode (partially hidden) (v) turbidity (vi) chlorophyll; **(C, D)** using a refractometer; **(E, F)** a Secchi disc and its deployment for measuring turbidity.

amount of cyanobacteria (blue-green algae) present in freshwater environments) or phycoerythrin (to determine cyanobacterial and cryptophyte presence in marine waters).

Turbidity refers to the interference of light by suspended matter, soluble coloured organic compounds or plankton in the water. The measurement of turbidity is used as an indirect indicator of the concentration of suspended matter, and is important for evaluating the available light for photosynthetic use by aquatic plants and algae. One method of measuring turbidity is with an electronic transmissometer, which measures light attenuation in water optically, yielding a percentage transmittance. A much simpler, traditional method is to use a Secchi disc (Fig. 4.2E,F): a black and white disc that is lowered in water to the point where it is just barely visible in order to measure the depth of light penetration. If you can see the bottom of the water body then it is not possible to measure a Secchi depth, nor can it be used in low light.

Total suspended solids (TSS) refer to the concentration of suspended solid matter in water. TSS is measured by weighing the undissolved material trapped on a 0.45 μm filter after filtration. The constituents that pass through the filter are designated total dissolved solids (TDS) and are comprised mainly of ions such as sodium, potassium, calcium, magnesium, chloride, sulphate, and (bi-) carbonate. It should be noted that there is a direct proportional relationship between suspended solids and turbidity. The solids in suspension may include sediment or detrital particles and plankton.

Dissolved oxygen (DO) is the traditional and ubiquitous indicator of aquatic health. It determines the ability of aerobic organisms to survive and, in most cases, higher dissolved oxygen is better. The concentration of dissolved oxygen depends upon temperature (an inverse relationship), salinity, wind and water turbulence, atmospheric pressure, the presence of oxygen-demanding compounds and organisms, and photosynthesis. Of these, DO is introduced into the water column principally through re-aeration, (simple mechanical agitation by wind) and through photosynthesis. DO is typically around 4–8 mg/L, or reported as percentage saturation, when 100% is in equilibrium with the air. Therefore high percentage saturation occurs during the day due to algal photosynthesis, and low (hypoxic, less than 1.5 mg/L DO) or anoxic water (around 0 mg/L) occurs late at night due to respiration and decomposition. Even at 100% saturation, warm salty water holds less DO than cool fresh water. Dissolved oxygen deficit is the difference between the capacity of the water to hold oxygen and the actual amount of DO in the water (the converse of percentage saturation). A large deficit is an indicator of some oxygen demanding stress on natural waters, while a low deficit is an indicator of generally unstressed conditions (DO gives no indication of possible toxic contamination).

pH is a measure of acidity or alkalinity of the water. High pH indicates that the water is alkaline and low pH indicates that the water is acidic. Generally, pH exhibits low variability in coastal situations due to the high buffering capacity of sea water. Departures from the normal range of 7–9 are therefore especially significant because the pH scale is logarithmic (i.e. departures from 10^7 and 10^9 moles of hydrogen ions per litre are significant). Therefore, if you average several pH values, the correct procedure is to average the anti-log. Low pH occurs following rainfall events on areas with exposed acid sulphate soils. The sulphuric acid run-off from these exposed soils can cause direct mortality of biota, as well as a variety of sub-lethal effects. Acid run-off also influences the chemistry of estuaries and can also damage infrastructure.

Biochemical oxygen demand (BOD) is an indirect measure of biodegradable organic compounds in water, and is determined by measuring the dissolved oxygen decrease in a controlled water sample over a 5-day period. During this period, aerobic (oxygen-consuming) bacteria decompose organic matter in the sample and consume dissolved oxygen in proportion to the amount of organic material that is present. In general, a high BOD reflects high concentrations of substances that can be biologically degraded, thereby consuming oxygen and potentially resulting in low dissolved oxygen in the receiving water. The BOD test was developed for samples dominated by oxygen-demanding pollutants such as sewage. While its merit as a pollution parameter continues to be debated, BOD has the advantage of being used over a long period.

4.5 Sampling methods for phytoplankton

You should choose a method based on your question, the precision required and your budget. If your purpose was to collect a sample to determine what species were present in an algal bloom, and not for any comparative purposes, it is possible to collect three samples, mix them together in a bucket and then take a sub-sample for counting. This sub-sample will provide an indication of the average counts, but will give no indication of the variation between the samples.

Visual assessment is the least expensive way to monitor phytoplankton – by estimating phytoplankton abundance based on water colour, Secchi depth, area of bloom or from a satellite image. There is a smart phone application, which provides an estimate of chlorophyll (it is worth checking the app for your waters by making some traditional extractions). Satellite images of water colour (chlorophyll) can be very precise to just a few square metres for your particular area, but these can be expensive and use a generic calibration. Local calibration of these images to your particular area in terms of sediment load or pigment diversity is an ongoing activity for most organisations.

You can also make fine-mesh net collections (20 µm mesh; see Fig. 4.3, Section 4.7), to concentrate rare species. Net collections of phytoplankton are suitable for larger cells, such as some diatoms, but the bulk of phytoplankton in the sea and inland waters is much less than 20 µm. Consequently, a plankton tow is regarded as a qualitative measure due to avoidance and particularly extrusion of particles through the mesh.

Quantitative samplers include surface water samples, which are collected by dipping a well-rinsed bucket over the side of the boat. A sample may be collected from the shore or bank with an empty sample jar on a pole. Integrated samples are usually taken from the surface to 3 m depth or more. These samplers can be made from a 2–5 cm diameter PVC or hosepipe, based on the principle that an integrated sample is taken through the photic zone of the water column (note that a sample taken from a bloom or surface scum is useful to identify the culprit, but is not a representative sample of the algal population). The entire sample is then released into a clean bucket – repeated up to three times – and a 100 mL sub-sample is then removed from the bucket and preserved for phytoplankton identification.

Water samples from specific depths can be collected using diaphragm pumps or water bottles, such as Niskin bottles. Water bottle casts ('hydro-cast') can be conducted using a rope over the side of a boat, and a heavy metal 'messenger' then slides down the rope to close the bottle.

At least two replicate water samples should be collected at each station or depth, and their unique numbers recorded on the field data sheet. An extra water sample from each hydro-cast should be retained in case of laboratory mistakes. Label each bottle with a unique identifying number for the laboratory. A pad of self-adhesive labels is useful, such that the same number can be used on the various samples for nutrients, chlorophyll, phytoplankton and zooplankton and the data sheet.

4.6 Analysis of phytoplankton samples

Phytoplankton samples collected using appropriate quantitative sampling methods can be analysed using a fluorescence probe on the salinity meter (Section 4.4), or in the laboratory by the measurement of the chlorophyll-*a* concentrations within the samples (Box 4.4).

The chlorophyll-*a* concentration provides an estimate of the biomass of phytoplankton presence in a water sample, but it will not provide any information on the composition of the phytoplankton present. To do this, you will need to identify and count each taxon (i.e. each species or 'type') present using a microscope and a counting chamber. The data obtained by these means will provide an estimate of the number of cells per mL (cells/mL) of each taxon and can be used to describe the composition of the entire phytoplankton community, the dominance of each taxon within that community, and changes in community abundance and composition over time. However, because different species of phytoplankton have cell sizes that differ greatly from each other, total cell counts are often unreliable for describing these changes. For example, a large cell count of a very small-sized algal species may be replaced over time by a smaller number of cells of a much larger size. Using just the cell count data, you may deduce that algal presence has decreased, whereas, in fact, algal biomass may

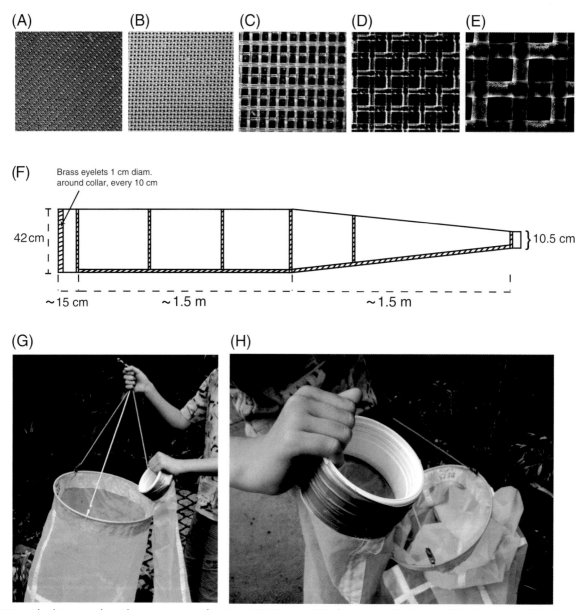

Fig. 4.3. Plankton mesh at the same magnification: **(A)** 15 μm, 10% free area; **(B)** 48 μm, 31% free area; **(C)** 150 μm, 51% free area; **(D)** 250 μm, 44% free area; **(E)** 500 μm, 39% free area. **(F)** Typical design for a 40 cm diameter ring net. **(G)** Mouth of a plankton net showing the bridle and attachments. **(H)** The cod-end of the net, showing the thread made to suit the sampling jars.

have increased. It may therefore be important, depending on the objectives of your study, to also determine the biomass present of each algal taxon identified and counted within the sample. Biomass is usually initially calculated as a biovolume (mm³/L), which is converted to biomass by assuming that algal cells have a density similar to that of water (therefore a biovolume of 1 mm³/L equals a biomass of 1 mg/L). Most correctly, biovolume estimates should be done by:

1. measuring the size of the cells of each species
2. converting this to an average cell volume for this species using a standard geometric formula best representing the shape of the cell
3. multiplying the cell count by this average cell volume to obtain a total volume for all of the cells for that species (Hillebrand *et al.* 1999).

This is often laborious as it needs to be repeated for each species present in the sample. Sometimes

Box 4.4 Extraction and quantification of chlorophyll

Chlorophyll-*a* is an indirect measure of phytoplankton standing stock and represents the biomass of the main photosynthetic pigment per unit volume or area of water (the biomass of chlorophyll represents ~0.45% of the actual cell biomass). Chlorophyll biomass is reported as micrograms per litre (μg/L) or milligrams per cubic metre (mg/m^3) or per square metre (mg/m^2). The areal measurement (mg/m^2) is calculated from the concentration, multiplied by the average depth of the lake or estuary. The chlorophyll-*a* content is measured in the laboratory using either the fluorescence or absorbance technique (Strickland and Parsons 1972; Ritchie 2008). Water samples are filtered onto a filter paper with a gentle vacuum. The actual sample volume can range from 100 mL to 4 L, as long as you can see that the filter paper is distinctively green or brown. You should work in a shaded room because, in this state, chlorophyll can degrade in bright light. The sample may be wrapped in foil and frozen for up to 3 months for later analysis. The filter paper is extracted into 99% ethanol and the light absorbance at four wavelengths is recorded with a calibrated spectrophotometer (Ritchie 2008), and the absorbance is entered into a spreadsheet containing the necessary coefficients. Calibration is done with a sample of pure chlorophyll-*a*. There are also additions to stabilise the chlorophyll extract, and an additional step to acidify the extract to remove the absorbance of phaeopigments (e.g. tannins). Alternatively, the fluorescence of the extracted chlorophyll can be determined – this is a more sensitive method. Research laboratories may examine chlorophyll-*a* and all the other accessory pigments using high-performance liquid chromatography (HPLC) of the extract. It is worth noting that all these methods quantify the biomass (mg/m^3) and not the production or rate (mg/m^3/year). Phytoplankton biomass and production are usually correlated, but not always. Measurement of phytoplankton biomass and production is complex, but for many of us using a calibrated water quality meter with a fluorescence sensor is an excellent start.

published tables of standard cell sizes for various species are used instead, if the error involved is considered acceptable in comparison with the costs of using actual measurements.

Samples are best preserved using Lugol's iodine solution for both freshwater and marine samples (although it may damage some of the small flagellates). Some laboratories will not analyse samples preserved with substances such as formaldehyde, because these are carcinogenic and represent an occupational health and safety hazard. Samples collected from a dense algal bloom can be analysed directly, but they usually need to be concentrated before analysis. This is usually done using a 100 mL aliquot of the sample that has already been well mixed by shaking the sample bottle before subsampling. The aliquot is poured into a 100 mL measuring cylinder and left to stand for a minimum of 24 hours. If small nanoplankton are present, a longer sedimentation time may be necessary. The Lugol's iodine preservative helps the cells sink more rapidly. After the required sedimentation period, most of the phytoplankton cells will have settled to the bottom of the measuring cylinder. The top 90 mL can then be drawn off using a suction pipette, taking care not to disturb the algal cells at the bottom of the cylinder. This gives a 10× concentration.

The identification and counting of phytoplankton cells is something that cannot be learned rapidly or easily, but takes much patience, practice and experience to do correctly. There are several taxonomic guides and keys that have been published to assist in the identification of both freshwater and marine algae (see Chapters 5 and 6).

There are several methods available for counting algal cells in samples. The easiest method is using a Sedgwick-Rafter cell. However, other methods (such as a Lund cell or an inverted microscope) are useful providing they can be used with at least as good an accuracy and precision as counts using a Sedgwick-Rafter cell. The Sedgwick-Rafter cell is a four-sided counting chamber that is 50 mm long by 20 mm wide by 1 mm deep, giving a bottom area of 1000 mm^2, and an internal volume of 1 mL. The cell has a grid engraved on the bottom, with lines 1 mm apart. If correctly calibrated and filled, the volume of sample covering each grid square is 1 mm^3. Both glass and plastic versions are available, with the glass cells being better, but more expensive. The cells are used on the stage of a normal compound microscope – preferably one with binocular eye-pieces. Counting is done a 100× magnification, with higher power being used to identify small sized algal cells. A very thin microscope cover slip (No. 1 thickness) is required to cover the cell.

Immediately before commencing a count, the phytoplankton cells in the bottom of the measuring cylinder are resuspended into the remaining 10 mL of sample left in the measuring cylinder by swirling, and a further sub-sample of ~1 mL of this collected with a Pasteur pipette. This is then decanted carefully into the counting chamber of the Sedgwick-Rafter cell. The cell is full once the cover slip, which should be placed obliquely over the cell before filling with one corner open, just begins to float and can be rotated to completely cover the chamber. This avoids introducing air bubbles into the sample. The cell should not be overfilled. Once filled, the counting cell should be left to stand on the stage of the microscope for 15 minutes, to allow the algal cells to settle to the bottom. It is not necessary to count all the cells in the entire area of the Sedgwick-Rafter cell. However, a minimum of 30 grid squares should be counted. These should be selected randomly, as there is differential sedimentation of algal cells within the counting cell, with more algae sedimenting closer to the walls than in the centre ('edge effects'). Counting traverses across the width of the cell helps to overcome these edge effects and will cover 40 grid squares. A second requirement is that a sufficient number of algae are counted to provide a counting precision of ±30%. This involves counting at least 23 'units' for all of the most dominant algal taxa present. A 'unit' is either an algal cell, filament or colony, depending whether the species being counted is unicellular, filamentous or colonial. If counting 30 grid squares or two traverses does not yield a sufficient number of units (i.e. more than 23), then additional grid squares or traverses will need to be counted. Record

the number of grid squares counted as well as the number of algal units counted. If an algal unit lies across the line engraved in the base of the Sedgwick-Rafter cell to delineate a grid square, so that it falls within two squares, the simple rule is that if it lies on the right side of the grid square, include it in the count, but if it lies on the left side, exclude it. Similarly, if it falls across the top line of the square, include it, but exclude any algal units falling across the bottom line. Algal units are often smaller than the width of the lines engraved in the Sedgwick-Rafter cell, so the same applies for any algal units lying within the grid lines delineating a square.

The number of algal units present per mL within the actual water body is calculated as:

$$\text{No. of units/mL} = \frac{(\text{units counted} \times 1000 \text{ mm}^3)}{\begin{array}{l}(\text{no. of grid squares counted} \times \\ \text{concentration factor, which is} \\ \text{typically 10})\end{array}}$$

For filamentous and colonial algae, it is then necessary to convert the count in units/mL to cells/mL. Many green algae have a set number of cells per colony (e.g. 4, 8, 16, or 32), so, when this is known, it is easy to multiply the units by the cell number per colony to obtain cells/mL. However, many other phytoplankton species, especially cyanobacteria, have a variable number of cells per filament or colony. In this instance, it is necessary to count the number of cells in 20 to 30 randomly selected filaments or colonies, and then obtain an average number of cells per colony from these counts.

Further problems arise when samples contain large-sized colonies or tangled aggregations of filaments containing thousands of cells, where it is impossible to count all the cells in each colony or aggregation. In these situations, it is necessary to estimate a portion of the colony or aggregation – say 5% or 10% of the total colony size – and count or estimate the number of cells within that portion. Remember that the colonies or aggregations are three dimensional, with cells overlying cells, and outside of the focal plane at which you are viewing the colony. Once you have an estimate of the number of cells in 5% or 10% of the colony, multiply this by 20 or by 10, respectively, to obtain an estimate of the total cells per colony.

When you do these estimates of average cell numbers per filament or colony to obtain a count in terms of cells/mL, the errors can be quite large and are in addition to any statistical counting error. The need to make these estimates arises only during blooms and becomes acceptable because of immediate management needs. Methods to break up large colonies into smaller units to make counting easier (homogenisation, addition of chemicals or sonification) are often inadequate and may destroy a large proportion of the cells present.

Counting algal cells by microscopy is usually time consuming and laborious, although necessary to identify the species present in the water bodies, and the data may not be available for management purposes without a considerable delay. New technologies are being developed that aim to provide near real-time data for management purposes. These include *in vivo* fluorometry, where fluorometric probes measure particular pigments such as chlorophyll-*a* and phycocyanin within the actual water body (Zamyadi *et al.* 2016).

Remote sensing is also increasingly being developed to monitor algal blooms, and in particular cyanobacterial blooms. Satellite remote sensing is used to monitor large-scale blooms in the Baltic Sea (Hansson and Håkansson 2007) and Lake Erie (Wynne *et al.* 2013; Wynne and Stumpf 2015). Aerial remote sensing platforms have also been trialled, and Bowling *et al.* (2017) has reported on the use of a hand-held 'point-and-shoot' proximal remote sensing instrument for monitoring cyanobacteria. These instruments collect spectral information reflected from the surface of water bodies. Algorithms then use data from certain sections of the spectra such as absorbance between 610–640 nm (due to phycocyanin) and 670–680 nm (due to chlorophyll-*a*) and convert them into a concentration value for the pigment.

When sampling freshwater cyanobacterial blooms, it may be important to determine whether the bloom is producing toxins or otherwise. One litre water samples should be collected and chilled for subsequent toxin analysis. Several methods are available, in particular chromatography with various forms of detection, depending on the toxin. Tandem mass spectrometry coupled with high-performance liquid chromatography is often used. Other methods for toxin analysis include enzyme-linked immunosorbent assays and protein phosphatase inhibition assays (Kaushik and Balasubramanian 2013). Genetic methods are also used, especially those that target specific genes involved in the biosynthesis of particular toxins, in particular polymerase chain reaction (PCR) (Pacheco *et al.* 2016). However, these can determine only the potential of a bloom to produce toxins (if high numbers of gene copy numbers are present) but cannot inform if toxins are actually being produced and present or not. The combined use of microscopy, PCR and toxin analysis helped manage a cyanobacterial bloom in the Murray River (Crawford *et al.* 2017).

4.7 Sampling methods for zooplankton

4.7.1 Mesh size, extrusion and avoidance

Zooplankton is typically collected with a fine mesh net, but using buckets or dip nets around bright lights is also possible. The appropriateness of mesh size can be determined through the trade-off between the net avoidance of zooplankton and net extrusion of zooplankton. With towed plankton nets, the smallest mesh size will never sample all the zooplankton, because larger and better swimming zooplankton will sense the pressure wave in front of a small mesh net and dodge it (this is known as net avoidance). If you use larger mesh, then the smaller zooplankton will be extruded through the mesh. We must accept that our sample is a selective view of plankton, but it will be a consistent view. The standard UNESCO mesh size for sampling zooplankton is 200 μm mesh (Harris *et al.* 2000) (Fig. 4.3D), but a 100 μm mesh is useful in estuaries because small zooplankton respond to environmental variability more rapidly than larger zooplankton. Many larval fish biologists in Australia use 500 µm mesh, knowing full well that fish eggs and small, unidentifiable larvae will be extruded through the mesh. Ultimately, net size should be determined in accordance with the objectives of your study.

Vertical hauls provide a depth-integrated plankton sample (Fig. 4.4D,E), and are useful for broad-scale spatial surveys of microplankton (<200 µm, small zooplankton and phytoplankton). The vessel must be stationary, and the net is either hauled up from a specified depth (an up-cast, Fig. 4.4E), or a heavy metal ring (10–20 kg) carries the

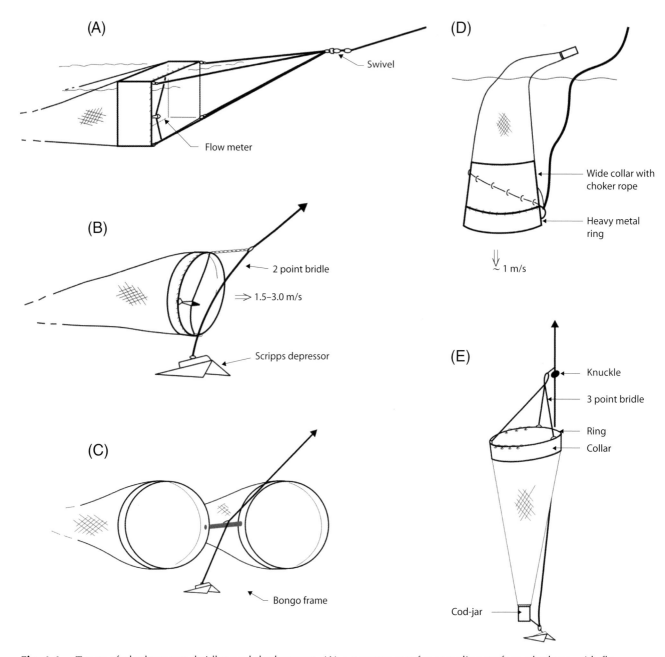

Fig. 4.4. Types of plankton net, bridles and deployment: **(A)** a neuston net for sampling surface plankton with flow meter and 4-point bridle; **(B)** a standard ring net configuration (design in Fig. 4.3F), with a two-point bridle, a Scripps depressor, and flow meter; **(C)** a bongo net sampler with a single bridle and no tow rope in front of the net; and **(D)** a drop net or **(E)** a lift net (i.e. down-cast and up-cast).

net down to a specified depth (a drop net or down-cast; see Fig. 4.4D).

Zooplankton is collected horizontally by slowly towing the net at a constant speed – around 1–2 m/s (Fig. 4.5A). Any faster will increase the extent of extrusion, and any slower may increase the incidence of avoidance. Nets may be fitted with a flow meter to determine the volume of water filtered

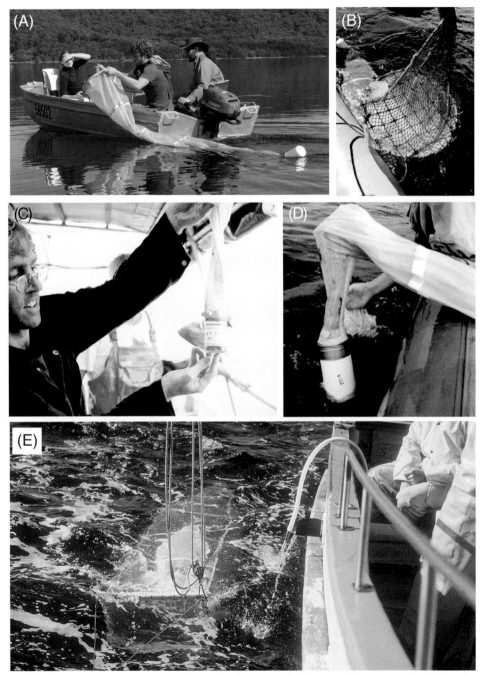

Fig. 4.5. Some plankton collection gear: **(A)** students deploying a plankton net from one side of their boat, beginning to tow the net in a circle to avoid sampling the propeller wash; **(B)** a jellyfish guard mesh attached to the bridle, when there are too many jellyfish to collect other zooplankton; **(C)** inspecting a fine phytoplankton collection; **(D)** if the sampling jar (i.e. cod jar) has no drainage mesh, then tip half the jar's water back out of the net's cod-end and splash up water (or hose-down) to rinse the mesh back into the jar; and **(E)** a square surface neuston net deployed from the side of a fishing trawler.

(Fig. 4.6A,B), to then determine the number or biomass of zooplankton per cubic metre. For plankton sampling, you should be concerned with speed through the water, rather than speed over the sea floor. You should tow for a constant period of time (between 3 and 10 minutes, depending on mesh size and the amount of debris in the water) for several practical reasons. A constant sampling interval reduces potential sources of error due to sleepiness or a variety of personnel. Sometime flow meters break during the tow, or jam or become tangled with debris and, rather than dumping an un-metered sample, the volume filtered can be estimated with reasonable precision from the tow duration.

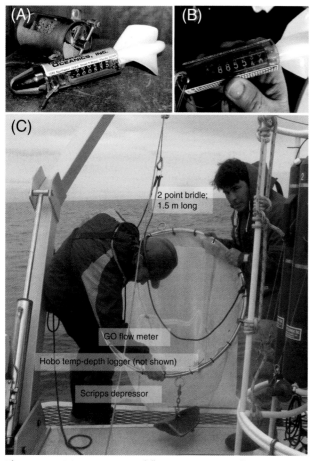

Fig. 4.6. **(A)** Two types of flow meter. **(B)** Where to read the flow meter. **(C)** A large 85 cm diameter net with 0.5 mm mesh to sample larval fish. Note the two-point bridle, the flow meter (obscured), and the depressor (photo C, Derrick Cruz).

4.7.2 Net design and construction, and some typical plans

The frame of a circular net can be made from stainless steel round rod or, more cheaply, from flat bar. Stainless flat bar is easier to bend into a circle than rod and is stronger (with respect to the incident flow) and cheaper per unit weight. A square frame can be welded from flat bar, but, again, stainless angle iron is stronger and cheaper per unit weight.

The bridle of the net is the harness, comprising ropes or strops that attach the mouth of the net to a tow rope. Three-point attachments on a ring net or four-point attachments on a square net are the most reliable, but may disturb plankton before they enter the net and may generate net avoidance. For a square neuston net, the lower strops may be longer than the upper ones. A circular net may have only

> ### Box 4.5 Manufacture of a simple ring net, 40 cm diameter, 0.2 mm mesh
>
> For a typical 40 cm diameter ring net of 0.2 mm mesh (with a generous filtration surface area in case of minor clogging), make the net 42 cm diameter to comfortably fit over the ring (i.e. radius or r = 0.21 m):
>
> - Mouth area = $\pi r^2 \approx 3.14 \times (0.21)^2 = 0.139$ m^2
> - Mesh porosity (i.e. free area) $\approx$50% >200 µm mesh; 30% (~100 µm)
> - Mesh area = 10 × mouth area ÷ porosity (0.4) = 3.5 m^2
> - Circumference of cylinder section = 2 πr = 1.32
> - Area of cylinder section 1.3 m long = 2.53 m^2
> - Area of cone section 1.5 m long = 0.5 × circumference × length = 0.99 m^2
>
> Adjust these dimensions to optimise the roll of mesh. The collar of the net should be 20 cm wide and made of a strong polyester canvas, with brass eyelets 1 cm diameter every 10 cm around the circumference to lash onto the ring. Seams should be reinforced with polyester tape inside and out. The polyester canvas cod-end should be ~10.5 cm diameter to accept a PVC pipe coupler (held in place with a stainless-steel hose clamp), that has a thread cut on the inside to match your plankton jars.

a two-point bridle, causing less net avoidance (Fig. 4.4, 4.6C).

The net may be sewn by a local sail maker, in the shape of a cylinder and a cone leading to the cod-end (Fig. 4.3F). The area of mesh must be nearly an order of magnitude greater than the mouth area of the net, so that there is a surplus of filtering surface area to cope with clogging and surface drag. A net is useless if there is a pressure wave in front of it resulting from insufficient surface area. This means that a typical 40 cm diameter net with 0.2 mm mesh is ~3 m long (Fig. 4.3F, Box 4.5). Nets and towing devices can be designed to investigate specific questions. For example, neuston nets can be used to collect plankton from the surface or epi-benthic sleds can be used to collect plankton just above the substrate.

4.7.3 Simple plankton net

The bridle attachment may be a three-point or, with a weight such as a depressor, you may use a two-point attachment (Fig. 4.4B). Attach the tow rope to a solid mid-point near the keel (a strong seat or thwart), and ensure the tow rope does not press onto the outboard motor (using a loop of twine to hold the tow rope away from the motor). Samples are collected by slowly towing the net behind the boat and turning in a slight circle so that the net is not in your propeller wash (Fig. 4.5A,E). The inside of the boat's circle is the side with the net, and you have little manoeuvrability. Without any depressor or weight, the net will remain just beneath the surface at slow speeds (~2 knots, or 1 m/s). The drag on the boat is substantial, especially with larger nets, and care should be taken by securely tying the tow rope to the boat's strongest points. The railings of a small boat, or a bollard on the side is not the best tow point, because being on the side away from the motor thrust makes steering even more difficult. In boats that are more than 5 m long this is less of a problem. The tow rope may simply be tied around the thwarts or seat of an open boat, or even at the front anchor attachment and passing it over the transom near the stern. The net is best retrieved by turning

off the engine and rapidly hauling it in hand over hand to prevent plankton from swimming out, or from dragging in the mud. If a winch is available, then it is best just to throttle back and haul the net up and into the flow. This is a simple, practical method, especially when working at night, but the boat's wash can still interfere with the net, potentially disturbing the zooplankton. Driving in a circle can be difficult in tidally flowing channels, and near fishing boats and pylons (Box 4.6).

The cod jar of a plankton net is a jar for draining and collecting the final sample (Fig. 4.4E, 4.5C,D). It needs to easily screw into a PVC fitting (or 'coupling') that is ring-clamped to the cod-end of the net. There are many individual designs for cod jars, but the simplest is to use one of your many sample containers – such as a 1 L PVC jar. A workshop can turn your standard jar's thread into the PVC coupling (Fig. 4.3H). The jar will be brim full of plankton, so before unscrewing and spilling it, tip the excess water back out through the mesh, and splash water back up onto the mesh as a quick rinse-down (Fig. 4.5D).

Box 4.6 Safety note when towing nets

When towing a large (>40 cm) diameter plankton net at 3 knots (1.5 m/s), the drag and pressure on the rope and boat are substantial – far more drag than simply dragging a similarly sized sheet perpendicularly through the water. Towing a large plankton net with a 5 kg depressor adds to the physical challenge. The small net from a small boat at only 1 m/s (Fig. 4.5A) is easier and, when the tow is complete, reduce the throttle and haul in as briskly as you can.

A small boat towing a plankton net is not a common sight and many tourists and trawlers will not expect you to be so slow and immobile. Sometimes they may come right at you out of curiosity, coming close to your tow rope. We generally avoid doing any plankton work during school holidays specifically for this reason. Towing at night in estuaries near trawlers can be dangerous, especially because you have limited mobility and some trawlers may turn off their navigation lights.

After a day's sampling, your gear just needs to be soaked in fresh water and dried. With gentle tows plankton is easily rinsed off, but detritus jammed in the mesh must be dislodged with a good blast, and even a little detergent.

4.8 Preparing and quantifying zooplankton

4.8.1 Observation of live plankton

Observing living plankton enables you to see how they use their swimming and feeding appendages and how they capture and consume food items. Phytoplankton and detritus are observed with a compound microscope (Fig. 4.7A) and using a wet mount (Fig. 4.7B). The colours and translucence of freshly caught zooplankton are amazing. You can capture live plankton around a bright light at night, or sample the contents of a gently towed plankton net. Live zooplankton cannot tolerate any trace of formalin or preservative or the heat of a lamp.

For large living zooplankton, use a wide-mouth pipette to place a small volume of the sample into a clean Petri dish. It is best to observe large copepods and cladocerans under the dissecting microscope at low magnifications (less than ×40, Fig. 4.8), because then they remain focused in the larger depth of field and they are less able to swim out of the microscope's field of view. A few drops of an anaesthetic ($MgCl_2$ solution, soda water or some ice) will slow the activity of larger zooplankton.

For small living zooplankton, use a pipette to place a small volume of the sample into a clean observation chamber, such as a counting chamber. A counting chamber can be made with two glass cover slips placed 3–10 mm apart, and a few drops of the sample placed between. Then gently place an intact cover slip over the sample, resting on the two beneath. The water will be held in the small chamber to prevent zooplankton specimens from being squashed between the slide and the cover slip. If an inverted microscope is available, you may observe living zooplankton held in a small volume of water on the glass slide without placing a cover slip.

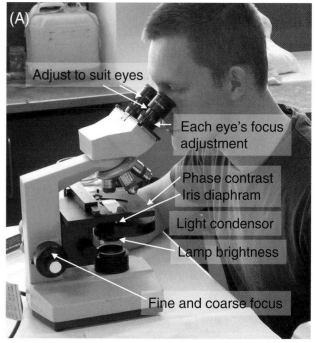

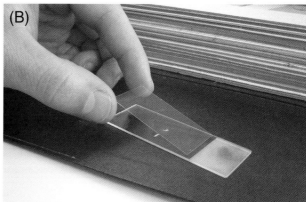

Fig. 4.7.   Compound microscope for phytoplankton. **(A)** Viewing a plankton sample should be relaxing, without squinting or using only one eye. By adjusting your chair height you should have a straight back and neck. Adjust the eye-pieces to suit the distance between your own eyes (see Fig. 4.8B; by closing each eye separately you should have an unobstructed view); after adjusting the coarse and fine focus knobs for one eye, you may also need to twist one of the eye-piece's individual focus adjustments. **(B)** Method for preparing a wet mount for a compound microscope.

4.8.2 Sorting a zooplankton sample

The laboratory analysis should also be guided by what the investigator requires, and by the budget. Sorting and identifying zooplankton to a reasonable degree of accuracy is arduous and may take

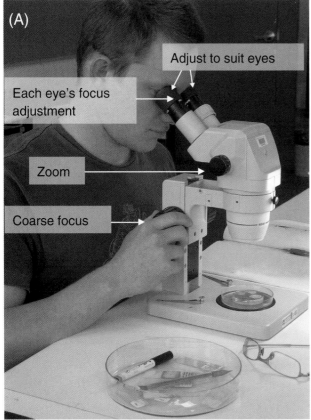

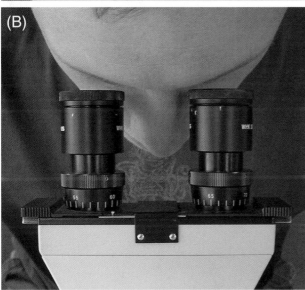

Fig. 4.8. (A) Dissecting microscope for zooplankton. (B) Ensure the eye pieces are adjusted to suit the distance between your own eyes.

1–4 hours per sample. Could your question be resolved by zooplankton biomass or by identifying to the level of phylum, family or genus? Perhaps only the copepods – the greatest phytoplankton consumers – need to be identified. Or is a size analysis sufficient? Will you sort two or three sub-samples, or do you plan to sort the entire sample for fish larvae only? You should prepare a sorting data sheet to complement the field data sheet (Figs 4.9, 4.10).

The sample should first be rinsed in a sieve (of the same or smaller mesh of the net) to remove grass and sticks, and to rinse off the fixative solution (if used). Rinsing with cold fresh water is perfectly adequate for preserved plankton (but be aware that your tap-water supply may contain small harmless microcrustaceans that may contaminate your sample; for most sampling this is not a problem). Gelatinous zooplankton should be counted and removed at this stage, and recorded on your field data sheet (Fig. 4.9). Then carefully rinse the plankton from the sieve into a beaker or a 100 mL volumetric cylinder (if necessary make up the volume to 100 mL, to take a quantitative sub-sample). With bulky samples, especially with detritus, a 200 or 500 mL cylinder may be necessary. Allow a uniform time period for the plankton to settle (~1 hour), and read off the approximate settled biovolume (i.e. the approximate volume in millilitres of zooplankton – normally with displacement volume zooplankton is added to the water). Detritus tends to sink slower than zooplankton, while any sand grains will sink faster, enabling you to estimate the actual zooplankton biomass.

After you have recorded the biovolume, thoroughly mix the zooplankton in the volumetric cylinder, and while still swirling remove an accurate 2 or 4 mL sub-sample with a pipette (with the fine tip cut off, Fig. 4.11). A Stempel pipette is a spring-loaded wide bore pipette specifically designed to take a quantitative sub-sample from the swirling, well-mixed plankton sample. A Folsom splitter (Fig. 4.11C) or a Motoda box are both simple devices designed to halve your sample, and then halve that sub-sample, and so on (i.e. sub-sample maybe 1/2, 1/4, 1/8, 1/16; any further splitting is not recommended). If you have removed 2 or 4% of the total sample, then you multiply your counts by 50 or 25

FIELD DATA SHEET

Crew: _____ **Sample ID code:**_____

Date:_____ **Time:**_____

Location/GPS:_____

Station:_____ **Depth:**_____

Weather:
Wind speed/direction:_____ Waves/tide/current:_____

Air Temp:_____ % cloud:_____

Moon phase: _____

Water @ start:
Temperature/Salinity:_____°C_____Secchi depth:_____

pH:_____ DO:_____
Comments:

Sampling gear:_____

a) Sample #:_____Time:_____Flowmeter:_____

b) Sample #:_____Time:_____Flowmeter:_____

c) Sample #:_____Time:_____Flowmeter:_____

d) Sample #:_____Time:_____Flowmeter:_____

Comments:

Water @ end:
Temperature/Salinity:_____°C_____Secchi depth:_____

pH:_____ DO:_____ Comments:

Fig. 4.9. A typical plankton field sampling data sheet.

LABORATORY DATA SHEET

LOCATION:_____ STATION #:_____

Sorter's name: _____ Date: _____

Sample #_____ Location: _____

Gear & mesh: _____ Tow duration/speed:_____

	Sample# _____	Comments: (sub-sample?)	Sample# _____	Comments: (sub-sample?)

copepods
    calanoid
 	
    cyclopoid
 	
    harpacticoid				

bivalved crustaceans
    ostracod
    cladoceran				

crab larvae				
amphipod isopod				
nauplii				

elongate crust
    krill
    mysids
    penaeids
    Jaxea				

polychaetes				
chaetognaths				
pelagic snails				
bivalve molluscs				
cnidaria *Obelia*				
larvaceans				
salps				
other gelatinous				
fish eggs				
fish larvae				
Large jellies, ctenos, algae?				

Fig. 4.10. A typical laboratory data sheet.

to get an estimate of total number. The volume of the sub-sample should be determined by the density of zooplankton (and the time it takes to sort). It is better to take two or three 1 mL sub-samples, rather than one 3 mL sub-sample, because the variance due to sub-sampling error can be incorporated into your analysis. (Remember to account for the fact that the second and third sub-samples are not the same proportion of the total as the first – although the error introduced by ignoring it is minor compared with other factors).

The sub-sample is best sorted and identified in a Bogarov tray or an S-tray (a perspex (plexiglas) square with a 1 cm deep trough milled into it, Fig. 4.11D), or in a plankton ring (a perspex ring that can be rotated under the microscope). Your laboratory data sheet should be beside you (Fig. 4.10). Some fine probes are useful for turning individuals to identify them (Box 4.7). Your counts could be dictated onto tape if you wish, and transferred onto the spreadsheet, where you can insert the necessary formulae to correct for sub-sampling and the total volume filtered (below). The remaining

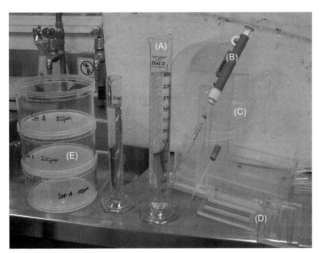

Fig. 4.11. Typical plankton sorter's equipment showing: **(A)** volumetric cylinders for determining settlement volume; **(B)** a blunt-ended pipette with a deliverer to take a quantitative sub-sample from the well-mixed sample thoroughly suspended in 100 mL or 250 mL of clean tap water (a non-quantitative Pasteur pipette is included); **(C)** a plankton splitter for dividing a plankton sample into half; **(D)** an S-tray for counting samples; and **(E)** a series of stacked home-made sieves to size-sort plankton with 300, 200 and 100 µm mesh.

Box 4.7 Fabrication of tungsten wire probes

Tungsten wire probes are very fine and firm needles for sorting tiny plankton. The wire may be sharpened by electrolysis. A mild electric current is passed between a 3 cm length of wire and an electrode immersed in a 1.0 M solution of sodium hydroxide (20 g of NaOH pellets in a litre of water). With an electric current, the tungsten tip is delicately dissolved only as it is dipped into the solution. You will need a dissecting microscope (Fig. 4.8) to observe and regulate the sharpening. The rate of electrolysis is proportional to the surface area of the wire, the amount of current and the concentration of NaOH. A microscope's AC light source can provide a variable current, with alligator clip leads. Once the wire is sharpened, the other end may be glued or fixed onto handles. Take care with all aspects of the process, including handling the moderately caustic solution, using the electric current and handling the sharp needles.

sample may be scanned for any large or interesting plankton, before storing it in 2% formaldehyde in fresh water.

4.8.3 Fixation and preservation of plankton

A fixative, such as formaldehyde, chemically treats the tissues, stopping biochemical activity and increases the mechanical strength of most plankton for handling (Table 4.1, Box 4.8). When the risks with formaldehyde solution are mitigated, it is an excellent, cheap preservative and requires less volume to be carried into the field. It is difficult to extract DNA sequences from formaldehyde fixed material.

A preservative, such as alcohol or salt, is a natural compound that reduces or stops decomposition without chemically fixing the tissue. Samples preserved in alcohol may shrink or become distorted more than in formaldehyde, but are safer and more pleasant to use and, providing the alcohol solution is replaced with fresh alcohol, the samples are suitable for DNA analysis. However, ethanol is flammable and a large volume is needed compared with the concentrated formaldehyde. The type and amount of preservative used should be determined by the sampling objective and the size of the samples being collected. If preservatives are not available, the samples should be kept cold – either stored in a refrigerator or stored in a portable icebox. Under these conditions, the samples are only viable for a period of 1–2 days.

Formaldehyde gas is usually made from the oxidation of methanol, using silver or copper as a catalyst, which is bubbled through water to produce a 40% saturated solution, referred to as concentrated formaldehyde solution. The concentrate has a trace of methanol to reduce polymerisation to paraformaldehyde (a white precipitate forms if stored in cool <16°C conditions – which may be cleared by warming or with a few pellets of sodium hydroxide). This concentrated solution is volatile, pungent and carcinogenic (see safety tips in Box 4.8).

Box 4.8 The safe use of fixatives, preservatives, and especially 4% formaldehyde

Most fixatives and preservatives (Table 4.1) are poisonous and carcinogenic. Adequate care should be taken at all times following approved, standard work procedures. Material safety data sheets (MSDS) and fume hoods are a modern necessity in the laboratory. Examination of live, non-preserved samples is best for many teaching situations. Otherwise, all samples should be preserved immediately, or should be placed in dark cool containers to ensure that no further primary production or grazing can take place. Ask the personnel at any identification laboratories regarding the method they require and remember that some researchers also like to get a separate live sample that can aid them with the identification of small flagellates and ciliates.

Concentrated formaldehyde is volatile, pungent, flammable and a rated carcinogen, so why would you use it at all? Because it is an excellent, cheap, convenient fixative in the field, and the fixed plankton hold their shape. In the field, away from laboratory coats and fume hoods, the risks can be greatly reduced by following the MSDS, which includes using safety gear such as safety glasses, blue nitrile gloves and in a well-ventilated space.

It is unfortunate that formaldehyde was previously used in confined spaces without any ventilation or safety gear. In the past, we'd take a squirt-bottle in the boat with concentrated formaldehyde, but the sunshine soon warmed the air space and forced a dribble of pungent formaldehyde into the boat! Do not pre-load your sample jars with formaldehyde either. You may wish to incorporate the following suggestions for field work into the safe working procedures of your organisation:

- Use 2 L plastic stock bottles with a handle and dripless spout to take a 50% solution of formaldehyde into the field (i.e. half of concentrated formaldehyde and half water), which reduces the volatility and flammability concerns. You may add your buffer to this stock solution.
- Therefore, with a ~100 mL sample of zooplankton and water, add ~10 mL of 50% stock solution to yield a 4% to 5% solution (or add ~100 mL of a 10% stock solution – which is even safer, but depends on how many samples and how much stock solution you need to transport). Consider where the breeze is blowing so your colleagues are upwind.
- Sometimes it is hard to tell (during arduous or sleepless field conditions) if formaldehyde has been added to a precious sample. Never sniff for formaldehyde. Add a few drops of an indicator stain such as eosin or rose bengal to the stock solution, which also stains the chitinous exoskeletons of zooplankton.
- If you are sorting samples at a field station, take a fine mesh sieve and waste container to pour off the formaldehyde solution into a waste cube, and then gently rinse under a tap (beware of plankton in the tap water). Waste formaldehyde solution should be properly disposed, and the dilute rinse solution does break down under sunlight and by bacteria.

In contrast to the opening sentiments of this section, it is important to acknowledge that formaldehyde is a naturally occurring compound. It is an oxidation state between methanol and formic acid and we metabolically produce formaldehyde in trace quantities. The difference is the concentration we use. So long as we follow the guidelines, we should not be scared to use 4% formaldehyde in plankton studies.

You need only a dilute solution to preserve plankton, and such dilutions are sometimes termed 'formal', 'formol' or 'formalin', but these are imprecise (formalin is an old trade name). A 4% solution of formaldehyde suitable for macrozooplankton is made up from 10 mL of the 37% commercial or concentrated grade and 90 mL of sea water or fresh water. For preserving smaller zooplankton, a 2% formaldehyde solution is made from ~50 mL of concentrated formaldehyde and made up to ~1 L. This may also be buffered with a few marble chips. In a tightly sealed jar, this solution is stable for decades if stored in a cool and dark location. Do not squeeze too much plankton into a sample jar: the volume of plankton is ~10–20% of the final solution.

Before sorting the sample, it is best to gently rinse it thoroughly in fresh water, and transfer to 70% alcohol as a preservative. Alcohol is a good long-term preservative, but it does not fix animal protein histologically. Formaldehyde solution may be buffered with sodium carbonate ($NaCO_3$, or 'soda ash') because a 5% formaldehyde solution becomes slightly acidic, which dissolves calcium carbonate, including snail shells, forams, and larval fish otoliths (used to determine age and daily growth). After a few weeks, buffered larvae suffer bleaching of their black spots (melanophores). Therefore, it is best to transfer fish larvae to 95%

Table 4.1.   List of possible plankton fixatives

Phytoplankton fixative	30% methylated spirits 5% glutaraldehyde Lugol's solution* Tincture of iodine* Acid Lugol's 2% formaldehyde
Microzooplankton fixative	2% formaldehyde
Macrozooplankton fixative	5% buffered formaldehyde (37% formaldehyde with sodium tetraborate or hexamine). Rinse and transfer to 70% alcohol for long-term preservation.

*NB: When adding tincture of iodine or Lugol's to the sample, do so 'drop by drop' until the sample turns a dark tea colour.

alcohol within weeks of capture (70% alcohol is also slightly acidic).

4.9 Automated methods for zooplankton sampling: size structure

Recent image analysis and video analysis instruments (such as Zooscan, Flowcam or Video Plankton Recorder) can make automated identification of plankton (see the Epilogue). With Zooscan, the plankton sample is poured onto a tray over a high-resolution flatbed scanner, and the resulting images are classified according to a training library of images relevant to your local area. Depending on the size of the library, the images can be classified to copepod, chaetognath, and other distinctive shapes. So long as particles are not touching, and detritus or fine bubbles are removed, the result is a coarse taxonomic composition and a size-frequency distribution.

Identifying plankton to genus and species is a specialist's job. Therefore some plankton ecologists resort to classifying plankton quickly and cheaply by size. Small particles are very abundant, while large particles are exceedingly rare – a general phenomenon known as the biomass size spectrum. Size is correlated with many ecological rates (Section 3.10) and the size frequency distribution can, for example, indicate the overall productivity in response to nutrients. A size-based analysis is based on the assumption that biomass is transferred from smaller to larger sizes by predation.

One limitation of size analysis is that debris, which may be abundant in estuarine and coastal waters, may be counted along with the zooplankton. Also, knowledge of certain key species or indicator species will not be known unless some calibration samples are visually inspected. At high zooplankton concentrations, the instrument will suffer from two simultaneous particles being counted as one – which is termed 'co-incidence'.

There are several size-based plankton counters, particularly for small particles such as bacteria and phytoplankton (such as the Coulter counter, flow cytometry and HIAC particle counters), but these are specialised instruments operating from a laboratory.

One of the major field instruments for counting and sizing zooplankton in the 0.3–3 mm size range is the Laser Optical Plankton Counter. The instrument counts and sizes plankton as it flows through a small sampling tunnel and interrupts a thin red light. The decrease in light intensity received by the sensor is recorded as a particle and converted to an area and thus an equivalent spherical diameter. Size is converted to biomass using the volume of a sphere and assuming a density of water. The sensor must receive a constant illumination, such that in turbid water, the light output must be increased, which is recorded as light attenuance (so one records counts, sizes and turbidity). The size categories can then be cross referenced with some typical taxa.

A cheaper semi-automated method is to count and measure the individual areas of your preserved plankton sample with image analysis – using a digital camera mounted onto a dissecting microscope. There are several public domain image analysis packages. The plankton sample is stained with any histological dye (such as rose bengal, lactophenol blue or chlorazol black), and a sub-sample placed into a Petri dish. Several images of different areas of the sample are recorded, which are then contrasted and the resultant blobs on the screen are counted and sized. A critical stage in this analysis is to optimise the appropriate sub-sample and dilutions, to reduce co-incidence and yet to have reasonable number of counts per grab. The actual particle concentration of each size category is determined by multiplying up from the sub-sample volume and the actual volume filtered.

The intercept and negative slope of the log-based biomass size distribution is a useful parameter of the plankton population dynamics (see Figs 3.3, 3.4). For analysis, any particular size intervals may be used, beginning at around three times larger than the net's mesh size. This is because many zooplankton species are shaped like an oblate spheroid, so that the smallest equivalent spherical diameter fully sampled by a certain mesh size is around three-fold larger. Any size classes – linear or logarithmic – may be used, providing one converts the biomass to 'normalised biomass size spectrum' (NBSS, dividing the biomass concentration (mg/m^3) of each size class by the biomass size interval). For smaller estuarine zooplankton with 100 μm mesh, we use 24 size limits set at 18^2 to 40^2 (i.e. 324 to 1600 μm equivalent spherical diameter).

Box 4.9 Calculating copepods per cubic metre

1. Calculate the distance through the water by the flow meter
2. Sampled volume (V) for a 40 cm diameter net towed at 1 m/s for 5 minutes or 300 seconds equals the volume of a cylinder, which is the mouth area times the distance towed:

$$V = \pi \times (0.4/2)^2 \times 1 \text{ m/s} \times 300 \text{ s} = 37.7 \text{ m}^3$$

3. If the average number of copepods in your 2% sub-samples = 45, then the concentration of copepods per cubic meter of lake water (C) is:

$$C = (45 \times 100/2)/37.7 = 59.7 \text{ copepods/m}^3$$

Numbers per m^3 or m^2? Survey results of phytoplankton or larval fish may be reported in numbers per m^3 or numbers per m^2. The former value is a concentration, while the latter is an overall abundance across the water column for that station. The areal abundance is calculated by multiplying the concentration by the bathymetric depth of the station (provided that you made an oblique or vertical haul, sampling the whole water column). In estuaries where you usually sample a fixed depth (surface or at 3 m), and where bathymetric depths can vary substantially, it is best to use a concentration.

4.10 Methods: analysis, quality control and presentation

Your data should be standardised as numbers per unit volume filtered – as indicated by the flow meter (litres, m^3, $10\ m^3$, $100\ m^3$, $1000\ m^3$, and so on). Generally the standard unit of volume should be similar to the actual volume of water filtered. For example, many of our neuston tows filter 200–300 m^3, so we would report our results as numbers per $100\ m^3$. Some surveys quote numbers per unit area, by multiplying the concentration by the maximum tow depth, thus estimating the numbers of larvae per unit area of ocean (see Box 4.9).

A flow meter to estimate the number of zooplankton per cubic metre of water filtered (m^3) is necessary for nearly all plankton work. The mouth area of the net (πr^2, where r is the radius of the mouth), times the distance will provide the maximum volume filtered (spillage around the mouth of the net is inevitable, depending on tow speed and clogging). This volume-filtered may be visualised as a column of water: the diameter of the net and the length of the tow.

There are two basic types – the General Oceanics (GO) or the barrel type (Tsurumi-Seiki Co., (TSK) or Rigosha & Co. (Fig. 4.6A,B). The GO flow meters have a six-digit number that increments by 10 for every revolution, and the number must be recorded at the beginning and end of each tow. The difference is used to calculate the volume filtered.

The formulae for calculation of volume (from their manual) are as follows:

1. Distance (m) = (difference × rotor constant)/999 999
2. Speed (cm/s) = (distance (m) × 100)/duration of tow(s)
3. Volume (m^3) = (3.14 × r^2) × distance (m)

Putting formula (3) into a spreadsheet is simple, and does not require you to time the duration of the tow (but a standard 5 or 10 minute tow is a good safety standard, if the flow meter jams). The rotor constant for a new standard rotor is 26 873 and this should be checked by attaching it to a rod and walking it briskly down a 100 m swimming pool. The axle of the propeller is delicate and prone to being bent, while corrosion can affect the internal mechanism if the meter is not flushed and dried after use. Seaweed may jam the rotor during a particular tow (and hence the standard 5 or 10 minute tow).

Sampling plankton entails the use of many vials and jars, which when sampled in various impact and control sites requires a good system to be in place to ensure that data are not mixed up. Label your jars with a unique number, which should travel through to the field data sheet (Fig. 4.9), spreadsheet and analysis (Fig. 4.10). To ensure compatibility and accuracy, also record:

- water collecting device and dimensions
- depth of water samples (m) and their volume (mL)
- number of stations sampled and number of samples collected
- analyses performed and laboratory methods used
- water temperature, nutrients, light and salinity
- identity of species using field guide
- preservative used and volumes of sub-samples
- station/transect/grid location
- date, time of day sampling conducted.

The individuals/teams collecting data should undergo training and should be provided with a comprehensive list of actions and requirements while sampling (Box 4.10). This ensures consistency among, and between, teams. Field notes and data sheets are essential and a chain of custody should be in place through which the sample can be tracked back to the collection stage. Information about detection limits, methods and standards used should be provided and should be consistent with the objectives and hypotheses of the management plan/monitoring program. With certain types of variables it is often useful to conduct inter-laboratory comparisons.

Box 4.10 Safety and care

Legislation

In many places, you may be required to obtain a permit from a government authority to collect samples. Make sure you have considered this before going into the field. It is useful to let local authorities know about your activities as community members may be alarmed if they see you sampling, particularly if you are using a fine mesh net.

Safety procedures while plankton sampling

- Conform with your organisation's safety procedures, such as a boat driver's licence and use common sense. Always notify someone of your proposed boating activities before leaving and notify them again when you return. Provide them with an estimated time of return and let them know the approximate areas you will be sampling. Look at a map of the site and select appropriate boat ramps.
- Ensure your boat and engine are properly maintained and ensure that you have sufficient fuel.
- Use plenty of sunscreen (water reflects back additional radiation).
- Take maps, mobile phones and/or radios.
- Entrance bars can cause extremely dangerous conditions – never leave the mouth of an estuary to enter the ocean unless you are with an experienced boat handler and in a suitable boat.
- Depending on the boat you are using, keep checking the weather as swells can develop rapidly in some systems and can cause problems with small boats.
- Carry appropriate equipment, such as a personal locator beacon (PLB), or an emergency position-indicating radio beacon (EPIRB), life jackets, oars, rope, anchor, torch, bucket and water, and don't overload your boat.

4.11 References

Bowling LC, Shaikh M, Brayan J, Malthus T (2017) An evaluation of a handheld spectroradiometer for the near real-time measurement of cyanobacteria for bloom management purposes. *Environmental Monitoring and Assessment* **189**, 495. doi:10.1007/s10661-017-6205-y

Crawford A, Holliday J, Merrick C, Brayan J, van Asten M, Bowling L (2017) Use of three monitoring approaches to manage a major *Chrysosporum ovalisporum* bloom in the Murray River, Australia, 2016. *Environmental Monitoring and Assessment* **189**, 202. doi:10.1007/s10661-017-5916-4

Harris R, Wiebe P, Lenz J, Skjoldal HR, Huntley M (2000) *ICES Zooplankton Methodology Manual.* Academic Press, London, UK.

Hillebrand H, Dürselen CD, Kirschtel D, Pollingher U, Zohary T (1999) Biovolume calculation for pelagic and benthic microalgae. *Journal of Phycology* **35**, 403–424. doi:10.1046/j.1529-8817.1999.3520403.x

Hansson M, Håkansson B (2007) The Baltic Algal Watch System – a remote sensing application for monitoring cyanobacterial blooms in the Baltic Sea. *Journal of Applied Remote Sensing* **1**, 011507. doi:10.1117/1.2834769

Kaushik R, Balasubramanian R (2013) Methods and approaches used for detection of cyanotoxins in environmental samples: a review. *Critical Reviews in Environmental Science and Technology* **43**, 1349–1383. doi:10.1080/10643389.2011.644224

Kingsford MJ, Battershill CN (1998) *Studying Temperate Marine Environments.* University of Canterbury Press, Christchurch, New Zealand.

Moore SK, Baird ME, Suthers IM (2006) Relative effects of physical and biological processes on nutrient and phytoplankton dynamics in a shallow estuary after a storm event. *Estuaries and Coasts* **29**, 81–95. doi:10.1007/BF02784701

Pacheco ABF, Guedes IA, Azevedo SMFO (2016) Is qPCR a reliable indicator of cyanotoxin risk in freshwater? *Toxins* **8**, 172. doi:10.3390/toxins8060172

Ritchie RJ (2008) Universal chlorophyll equations for estimating chlorophylls a, b, c, and d and total chlorophylls in natural assemblages of photosynthetic organisms using acetone, methanol, or ethanol solvents. *Photosynthetica* **46**, 115–126. doi:10.1007/s11099-008-0019-7

Strickland JDH, Parsons TR (1972) *A Practical Handbook of Seawater Analysis*. Bulletin 167, Fisheries Research Board of Canada, Ottawa, Canada.

Whitfield AK, Elliot M, Basset A, Blaber SJM, West RJ (2012) Paradigms in estuarine ecology – a review of the Remane diagram with a suggested revised model for estuaries. *Estuarine, Coastal and Shelf Science* **97**, 78–90. doi:10.1016/j.ecss.2011.11.026

Wynne TT, Stumpf RP (2015) Spatial and temporal patterns in the seasonal distribution of toxic cyanobacteria in western Lake Erie from 2002–2014. *Toxins* **7**, 1649–1663. doi:10.3390/toxins7051649

Wynne TT, Stumpf RP, Tomlinson MC, Fahnenstiel GL, Dybly J, Schwab DJ, *et al.* (2013) Evolution of a cyanobacterial bloom forecast system in western Lake Erie: development and initial evaluation. *Journal of Great Lakes Research* **39**, 90–99. doi:10.1016/j.jglr.2012.10.003

Zamyadi A, Choo F, Newcombe G, Stuetz R, Henderson RK (2016) A review of monitoring technologies for real-time management of cyanobacteria: recent advances and future directions. *Trends in Analytical Chemistry* **85**, 83–96. doi:10.1016/j.trac.2016.06.023

4.12 Further reading

Omori M (1991) *Methods in Marine Zooplankton Ecology*. Krieger Publishing Company, Malabar FL, USA.

Parsons TR, Takashashi M, Hargrave B (1984) *Biological Oceanographic Processes*. 3rd edn. Pergamon Press, Oxford, UK.

Tranter DJ (Ed.) (1968) *Zooplankton Sampling. UNESCO Monographs in Oceanic Methodology Vol. 2*. UNESCO Press, Paris, France.

5

Freshwater phytoplankton: diversity and biology

Lee Bowling

5.1 Identifying freshwater phytoplankton

The group commonly referred to as 'algae' constitutes a large and very diverse assemblage of organisms. Up to 15 different groups or 'divisions' are recognised, depending on the system of classification used. Although there may be some superficial similarities between these divisions, they can differ greatly from each other, especially in regards to their pigment arrays and their cellular ultrastructure.

Several these algal divisions occur predominantly in fresh water and have only a few marine representatives, while others are well represented in both the marine and freshwater environments, albeit by different genera. Additionally, even though some divisions may be present in fresh water, they do not form part of the phytoplankton community, but instead grow attached to various substrates: stonewarts (Charophyta), and freshwater species of red algae (Rhodophyta), provide examples of these.

Some phytoplankton are extremely small, with cells of less than 1 μm in diameter. Even the larger freshwater phytoplankton cells may be only up to 500 μm in their maximum dimension. The majority, however, fall within the nanoplankton (2–20 μm) and microplankton (20–2000 μm) size ranges, although the abundance, role and importance of freshwater picoplankton algae may be

often overlooked because of their small size. Some colonial and filamentous phytoplankton species may form aggregations more than 2 mm in diameter and be visible to the naked eye.

Today there is an increasing reliance on DNA-based molecular techniques for identifying phytoplankton species, especially for toxigenic species where reliable identification is necessary for the protection of public health. However, a range of morphological features have traditionally been used in the microscopic identification of freshwater phytoplankton including:

- the size, shape and colour of cells
- the arrangement of cells (single, filamentous, colonial)
- the type of cell wall
- the presence, absence and positioning of flagella and other distinguishing organelles and specialised cells.

Many of these features are distinctive to each division of algae. This chapter presents summary descriptions of the main divisions of phytoplankton that occur in fresh waters to illustrate the diversity found within these organisms in this environment (see Chapter 6 for the marine phytoplankton). Far more detailed descriptions and references to original research can be found in specialist textbooks on algae (e.g. Bold and Wynne

1986; South and Whittick 1987; Van Den Hoek *et al.* 1995; Lee 1999). Details of the ecology and reproductive strategies of many of the different divisions of freshwater phytoplankton may be found in Sandgren (1988a).

5.2 Cyanobacteria (blue-green algae)

The most striking example of the great variation and differences between phytoplankton comes when the cyanobacteria – or 'blue-green algae' – are compared with all the other algae (Boxes 5.1, 5.2). Cyanobacteria belong to the Kingdom Eubacteria, which, together with the Archaebacteria, makes up the Prokaryota. Prokaryotes are organisms whose cells possess little internal organisation and lack organelles (such as a nucleus or mitochondria) that characterise the eukaryotes.

All other types of algae (and indeed all other cells) are eukaryotic organisms, in which there is separation of different cellular functions into distinct membrane-bound organelles within the cell. These types of algae have a closer affinity to the higher plants than to the bacteria. Cyanobacteria also have other features that they share with bacteria. Under certain conditions – especially when there are low concentrations of nitrogenous nutrients present in the water column – many of them can fix atmospheric nitrogen into organic nitrogen (Box 5.3). This is a feature that they share with some other bacteria, such as those that live in the roots of leguminous plants (such as lupins and clover) – and the same biochemical pathways to fix atmospheric nitrogen are used by both. They also have a cell wall structure similar to that of the Gram-negative bacteria, including the presence of substances known as lipopolysaccharides. These can be potent

Box 5.1 Cyanobacteria and other photosynthetic bacteria

As well as cyanobacteria, red and purple photosynthetic bacteria also occur in some lakes and ponds. However, there are marked differences between the two. Cyanobacteria have in common with eukaryotic algae the presence of the pigment chlorophyll-*a*, which is used to trap light energy for photosynthesis. The biochemical pathway for photosynthesis in cyanobacteria is exactly the same as that in other algae and the higher plants – where carbon dioxide and water are used as the basic ingredients to manufacture carbohydrates, and oxygen is liberated in the process. In addition to chlorophyll-*a*, cyanobacteria also possess the accessory light-trapping pigments phycocyanin and phycoerythrin, which are blue and red coloured, respectively, and give the cyanobacteria their distinctive blue-green colouration. In contrast, the photosynthetic bacteria possess pigments other than chlorophyll-*a* (they have instead bacteriochlorophylls), are obligate anaerobes (must live in environments devoid of oxygen), and they do not release oxygen as a result of their photosynthetic processes (unlike cyanobacteria).

Box 5.2 Buoyancy regulation in cyanobacteria

Although the cells of cyanobacteria do not possess any internal structure or flagella, many planktonic species, but not all, contain gas vesicles, which can form larger aggregations known as gas vacuoles, and which may be observable under light microscopy as black speckles within the cell. Gas vesicle production provides the cells with positive buoyancy, enabling them to float up through the water column towards the surface to obtain additional light for photosynthesis. Photosynthesis leads to the accumulation of denser carbohydrate metabolites that increase ballast, and also increases turgor pressure within the cells that will collapse the gas vesicles. These mechanisms lead to the cells sinking again (Oliver 1994). Using their buoyancy regulation mechanisms, cyanobacteria can actively migrate up and down the water column – usually rising towards the surface in the early morning and sinking during the afternoon. It has been proposed that sinking into deeper waters may allow the cells to obtain additional soluble nutrients that can accumulate at depth. However, Bormans *et al.* (1999) consider that vertical migrations only occur within the surface mixed layer, and do not extend down into these deeper nutrient-enhanced waters.

toxins in some Gram-negative bacteria (such as *Salmonella*), but in cyanobacteria they are more benign, but still present a potential public health hazard as they may act as contact irritants (see Section 3.6).

Cyanobacteria commonly comprise a portion of the phytoplankton community of most freshwater bodies, including even the most pristine, although in these cases they may be only minor components. They also occur in marine (Section 6.4) and terrestrial environments.

Species from three taxonomic orders of cyanobacteria are commonly found within the freshwater phytoplankton of Australia, although species from other orders may also occur occasionally. These three orders are the Chroococcales, the Nostocales and the Oscillatoriales. The distinguishing features of each order are summarised in Table 5.1.

A commonly occurring member of the Chroococcales worldwide is *Microcystis aeruginosa* (Fig. 5.1A). This species is of particular concern because some strains produce a potent hepatotoxin – a toxic compound that typically attacks the liver (Falconer 2001). Major blooms of this species have been occurring in recent years in the western end of Lake Erie in North America, covering an area of approximately 3350 km^2 (1300 square miles) of the lake in 2015. That same year, the 400 000 residents of the city of Toledo, Ohio were required to use bottled water for 3 days because the town water supply was impacted with high concentrations of the hepatotoxin microcystin. *Microcystis flos-aquae* (Fig. 5.1B) is a similar species that is also potentially toxic. There are also many tiny picoplanktonic (<2 μm in diameter) species within the Chroococcales, including species from the genera *Chroococcus, Merismopedia, Aphanocapsa, Anathece* and *Coelosphaerium*, all of which are commonly encountered in slow flowing rivers, lakes and reservoirs.

In many parts of southern Australia, the most common problem-causing freshwater species is *Dolichospermum circinale* (former name *Anabaena circinalis*) (Fig. 5.2A). This cyanobacterium belongs to the Order Nostocales and may produce neurotoxins (toxins that affect the nervous system) (Baker and Humpage 1994). It was the main cyanobacterium that caused the bloom that occurred over 1000 km of the Barwon–Darling River in New South Wales in 1991 (Bowling and Baker 1996). Several other species of *Dolichospermum* also occur in Australian freshwaters, including the tightly spiralled *Dolichospermum spiroides* (Fig. 5.2B). A number of neurotoxin producing species of *Dolichospermum* occur in many other parts of the world. Other problem cyanobacteria from the Order Nostocales include *Cylindrospermopsis raciborskii* (Fig. 5.2C) – a tropical species that produces a very potent hepatotoxin (Hawkins *et al.* 1985) and is especially common in Queensland. It is also considered an invasive species, spreading into warm temperate water bodies with global warming (Sukenik *et al.* 2012), including lakes in central Europe. Another, *Nodularia spumigena*, produces yet another kind of hepatotoxin, and has been responsible for stock deaths in South Australia (Francis 1878; Codd *et al.* 1994). It is common in the

Box 5.3 Heterocytes and akinetes

Cyanobacteria within the Order Nostocales can produce two types of specialised cells that are not found in the other two orders discussed here. The first are the heterocytes, where nitrogen fixation takes place. Heterocytes usually have thickened walls to exclude oxygen, the presence of which prevents nitrogen fixation. However, heterocytes may not be present if there is plenty of bioavailable nitrogen present within the water column, because fixation is therefore not necessary. The other type of specialised cells – called akinetes – are resting cells or spores produced from vegetative cells. These also develop thick walls, have concentrated food reserves and sink and remain in the bottom sediments until environmental conditions suited to a renewed bloom. The akinetes then germinate and commence a new bloom. Akinetes also may not always be present, but frequently develop when environmental conditions become unfavourable for the continuation of an existing bloom. The location of the heterocytes and akinetes within the filament are some of the morphological features used to distinguish different genera and species within the Nostocales.

Table 5.1.   Summary of distinguishing features of cyanobacteria

Classification follows that of Baker (1991, 1992).

Order	Distinguishing features	Cell shape	Typical freshwater genera
Chroococcales	Unicellular and colonial species with no physiological connection between the cells (Komárek and Anagnostidis 1999). In colonial species, the cells are embedded within a clear mucilaginous envelope, or are located at the ends of fine, thread-like gelatinous strands that radiate from the centre of the colony. Cell numbers in colonies range from a few to many thousands.	Spherical, oval to rod shaped, depending on species, but many are coccoid.	*Microcystis* *Chroococcus* *Merismopedia* *Aphanocapsa* *Anathece* *Coelosphaerium*
Nostocales	Multicellular filamentous species that contain some specialised cells (heterocytes, akinetes) within the filament or trichome. The filaments do not branch (false branching may occur in some genera).	The shape of the vegetative cells ranges from spherical, ovate, cylindrical to barrel shaped.	*Dolichospermum* *Cylindrospermopsis* *Nodularia* *Aphanizomenon* *Anabaenopsis* *Chrysosporum*
Oscillatoriales	Filamentous and multicellular, but without specialised cells such as heterocytes and akinetes. The filaments are without true branching. In some genera, the filaments are enclosed within a fibrillar sheath.	The vegetative cells of some genera are often discoid – being wider than they are long – so that a filament viewed lengthwise may resemble a stack of coins. Other genera have squarish to rectangular cells. Terminal cells may differ slightly (e.g. more rounded) from those within the filament.	*Planktothrix* *Planktolyngbya* *Pseudanabaena* *Spirulina* *Limnothrix* *Phormidium* (mostly benthic) *Lyngbya*

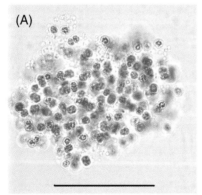 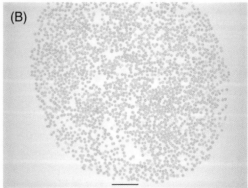

Fig. 5.1.   **(A)** Colony of *Microcystis aeruginosa*. Note almost spherical cells – often in doublets – within a gelatinous matrix. Scale bar 50 µm. **(B)** Colony of *Microcystis flos-aquae*. Similar to *M. aeruginosa*, but cells are generally more dispersed within the gelatinous matrix, which has a more compact shape. Scale bar 50 µm.

freshwater sections of the lower Murray River, and also occurs in brackish through to hypersaline coastal lakes. It also forms extensive blooms in the Baltic Sea. A major bloom of *Chrysosporum ovalisporum*, also from this order, contaminated hundreds of kilometres of the River Murray in 2016. It had been uncommon in the river up to this time. Blooms of this species also occur regularly in Lake Kinneret, Israel. Other genera of Nostocales commonly encountered in freshwater environments

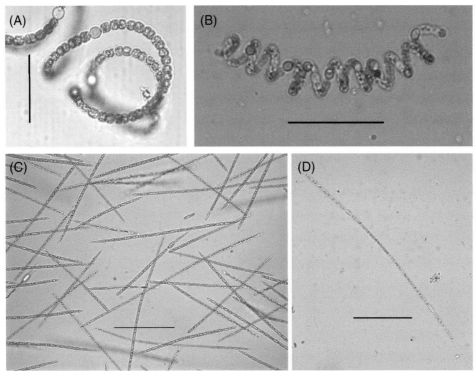

Fig. 5.2. **(A)** Filament of *Dolichospermum circinale*. Note the specialised cells – known as heterocytes – within the filament. These are sites of nitrogen fixation. Scale bar 50 μm. **(B)** Filament of *Dolichospermum spiroides*, also with heterocytes. Compare the tight spirals with the open spirals of *D. circinale*. Scale bar 50 μm. **(C)** Filaments of *Cylindrospermopsis raciborskii*. The specialised cells within the filaments are akinetes (resting spores). Tiny conical heterocytes occur at the ends of some filaments. Scale bar 50 μm. **(D)** A filament of *Cuspidothrix issatschenkoi* containing heterocytes. The terminal cells are long, tapering and colourless. Scale bar 50 μm.

include *Cuspidothrix* (Fig. 5.2D), *Aphanizomenon* and *Anabaenopsis*.

No hepatotoxin- or neurotoxin-producing planktonic species of cyanobacteria from the Order Oscillatoriales have so far been reported from Australian fresh waters, although a toxic benthic species of *Phormidium* has been reported from South Australia (Baker *et al.* 2001), and toxic *Lyngbya wollei* have been reported from Queensland (Seifert *et al.*

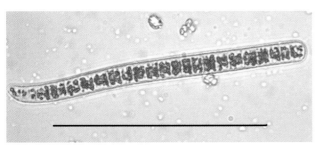

Fig. 5.3. A filament of *Planktothrix isothrix*, with rounded terminal cells. Scale bar 50 μm.

2007). Toxin-producing species from the Order Oscillatoriales are, however, common elsewhere in the world, both within the phytoplankton community and growing as benthic mats on the bottom of shallow water bodies (Sivonen and Jones 1999). In particular, *Planktothrix agardhi* and *P. rubescens* form massive blooms in freshwater lakes and reservoirs in the Northern Hemisphere and can produce the hepatotoxin microcystin. Common freshwater planktonic genera in Australia include *Planktothrix* (Fig. 5.3), *Planktolyngbya*, *Pseudanabaena*, and occasionally, *Geitlerinema* and *Planktotrichoides*.

5.3 Chlorophyceae (green algae)

Green algae, or Chlorophyceae, are among the most numerous and diverse of all freshwater algae. At least 11 orders of green algae are recognised – and sometimes up to 19 – depending on the author.

They often comprise the majority of the planktonic species of algae present in healthy freshwater ecosystems. Although some species can form blooms at times in nutrient-enriched waters, none are toxic. The Chlorophyceae are primarily a freshwater group, with ~90% of representatives occurring in freshwater environments. Attached and benthic species are common in many shallow streams and rivers, while planktonic species occur in lakes, reservoirs, ponds and other open water environments, as well as in rivers and streams (Box 5.4).

Some commonly occurring flagellated freshwater green algae belonging to the Order Volvocales include the single-celled *Chlamydomonas* and the colonial *Gonium* (Fig. 5.4A), *Pandorina* (Fig. 5.4B) and *Eudorina*, which contain small flat or spherical colonies of up to 32 or 64 cells (occasionally more), depending on species. The genus *Volvox* has hollow spherical colonies up to 2 mm in diameter that consist of several thousand small biflagellated cells. Common non-flagellated colonial green algae include *Pediastrum* (Fig. 5.5A) – which consists of a flat circular plate of cells that often have horn-like extensions – and *Scenedesmus* (Fig. 5.5B), which has cylindrical cells that are joined laterally in groups of four or eight. *Desmodesmus* is a similar genus

Box 5.4 Distinctive features of Chlorophyceae

Chlorophycean algae are eukaryotic organisms. The planktonic species can be present as single-celled species, as colonial species and as filamentous species. Many of the colonial species have a set number of cells per colony, with 4, 8, 16, 32 or 64 cells being present. Chlorophycean cells typically have a single nucleus and a large chloroplast in relation to the cell size. The chloroplasts can display a great variety of shapes among different genera and may also contain pyrenoids, which are associated with starch storage. Green algae contain both chlorophylls *a* and *b*, as well as carotene and xanthophyll accessory pigments. The protoplast usually fills the entire cell, but some species possess large, central aqueous vacuoles. The cell walls are generally (but not always) composed of cellulose, which is surrounded by a layer of mucilage. One group of green algae – the Order Volvocales – is normally actively motile, and swim with the aid of one, two, or occasionally four or eight flagella. All other orders have non-motile vegetative cells, but many still have a flagellated motile stage during their life cycle – either as gametes or as zoospores. Many of the non-flagellated planktonic forms have flattened colonial forms – or flattened cells with spines and other protuberances – that optimise the cell or colony's surface-area-to-volume ratio, increasing their friction against the surrounding water medium, and thus reducing their sinking rates. By this means, they remain within the circulating surface waters where they can obtain light for photosynthesis.

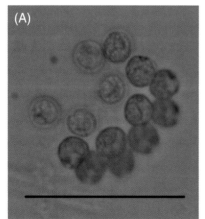 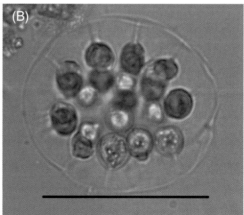

Fig. 5.4. **(A)** Part of a colony of *Gonium* sp. showing the almost spherical biflagellated cells in a flat plate arrangement. Scale bar 50 µm. **(B)** Colony of *Pandorina* sp. The colony has a spherical structure with the flagella of each cell radiating outwards. Scale bar 50 µm.

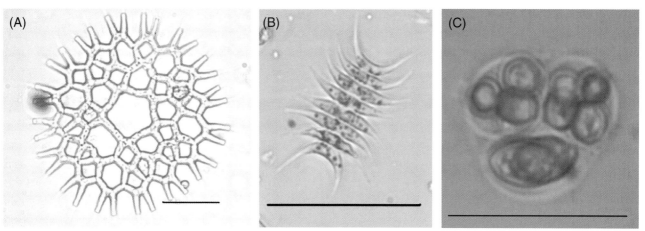

Fig. 5.5. **(A)** A colony of *Pediastrum duplex*, composed of approximately X- or H-shaped cells joined at the tips. Scale bar 50 μm. **(B)** *Scenedesmus dimorphis* – a colonial green alga composed of eight crescent-shaped cells. Scale bar 50 μm. **(C)** Colonies of ovoid-shaped *Oocystis* sp. cells. Three new colonies are contained within the original parent cell wall. Scale bar 50 μm.

where the terminal cells have spines. *Chlorella* and *Oocystis* (Fig. 5.5C) are also commonly found among the freshwater phytoplankton of lakes and reservoirs. These may be present as single cells, or as colonies of four to eight cells formed by the cellular division of a single parent cell and contained within the stretched original cell wall of that parent.

The desmids are a very distinctive group of freshwater green algae, which occur either as single cells or as filaments of cells within the water column. The cells of desmids are composed of two mirror-image halves – each with a chloroplast and pyrenoids – which are joined at the centre of the cell. In many species the junction between the half cells is deeply incised to form an isthmus and is the

location of a large nucleus. Asexual reproduction is by cell division at the isthmus, with each half-cell separating and growing a new half-cell. Thus, one half of the desmid cell is always older than the other half. Desmids also reproduce sexually via the conjugation of two vegetative cells to form a zygote. There is a great variation in cell morphology between the common genera of desmids. *Closterium* (Fig. 5.6A) are frequently elongate and crescent shaped, *Cosmarium* (Fig. 5.6B) has an incised isthmus and hemispherical or lobed half cells, while *Micrasterias* have laterally flattened half-cells with deep incisions, so that the complete cell resembles a little star. The genus *Staurastrum* contains very many different species. This genus is typified by the usual bilateral symmetry of desmids in lateral

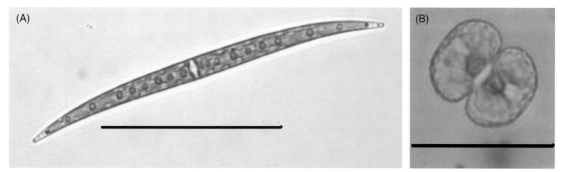

Fig. 5.6. **(A)** *Closterium* sp. – a crescent-shaped desmid. Note the two half cells with large chloroplasts containing pyrenoids. Scale bar 50 μm. **(B)** *Cosmerium* sp. – a desmid with two distinct half cells joined at a central isthmus. Scale bar 50 μm.

view, while in polar view the cells have tri-radial or hexa-radial symmetry. The half-cells are ornamented with spines and other appendages.

5.4 Bacillariophyceae (diatoms)

Diatoms are widely distributed in both freshwater and marine habitats (Section 6.2). There are many planktonic species, but also many benthic and epiphytic (growing on plants) species as well (Box 5.5).

Many planktonic species of diatoms occur as single cells or as colonies, although some are filamentous. The most marked distinguishing feature of diatoms is their cell wall, which is composed of silica. These siliceous cell walls are composed of two overlapping halves, known as valves. One valve, the hypovalve, is smaller than the other (the epivalve), so that it fits inside the larger valve. The two valves are joined together by a girdle band that runs around the centre of the cell. When viewed

under a microscope, cells from the same species may look entirely different, depending on the orientation of the cell, and whether it is seen in valve view, or girdle view. There are two main forms of diatoms: centric diatoms and pennate diatoms. When viewed in valve view, centric diatoms appear circular, with radial symmetry. In comparison, pennate diatoms have long narrow cells and have bilateral symmetry. Some diatoms also have a longitudinal opening in one or both valves, known as a raphe. In addition, the siliceous cell walls are often decorated with small holes, or pores, that may form lines or patterns on the cell wall. The cell walls may also have areas of heavy silica deposition that form strengthening ribs known as costa. The taxonomy of diatoms is based to a great degree on the pattern and structure of the cell wall. Their reproductive strategies are discussed in Box 5.6.

Some of the more common centric diatoms that occur in freshwater ecosystems include *Cyclotella* and *Coscinodiscus*, which have flattened disc shaped cells, and generally occur as single cells entrained in the water. *Aulacoseira* (Fig. 5.7A) is a filamentous

Box 5.5 Other distinctive features of diatoms

The living cells of diatoms contain a single nucleus, and from one to many chloroplasts, the shape of which varies greatly from genus to genus. Most chloroplasts have a central pyrenoid. Diatoms contain chlorophylls *a*, c_1 and c_2 as their main photosynthetic pigments, plus the accessory pigment fucoxanthin, which give the diatoms their typical golden-brown colouration. Diatom cells do not possess flagella, and thus planktonic species are reliant on turbulence within the water column to keep them from sinking. The silica cell wall is a disadvantage with regard to remaining suspended in the water column, and many planktonic species have adopted flattened or needle-like cell morphologies, spines, or colonial or filamentous growth habits, to increase their surface to volume ratio. By doing so, the cells present more resistance to the water, and sinking rates are reduced. Some non-planktonic species are, however, motile and move with a gliding motion over the substrate to which they are attached. This is done by extruding substances from their raphes.

Box 5.6 Vegetative reproduction in diatoms

Vegetative reproduction involves the separation of the two valves of the parent cell, along with nuclear and protoplast division. A new valve then forms within the existing original valve (i.e. the new valve is always the smaller of the two). This results in the daughter cell that originated from the parental hypovalve always being slightly smaller than the parent. With continued cell division, a progressive reduction in cell size within the population occurs. Once a minimum size is reached, sexual reproduction will take place to produce an auxospore, which characteristically increases its size immediately to retain maximum size. Diatoms can also produce resting spores, which sink to the bottom and remain there until conditions for germination are suitable. Upon germination, the size increases and new vegetative cells are formed that are much larger than the original parent resting spore.

centric diatom where the cells within the filaments appear in girdle view like miniature oil drums stacked end to end. Examples of unicellular pennate freshwater diatoms include the long skinny *Synedra* (Fig. 5.7B), and the spined *Urosolenia*. *Navicula* (Fig. 5.7C) is a genus with very many different species, both planktonic and benthic, and which typically has an elongated oval shape in valve view and has a raphe in both valves. Colonial pennate diatoms include *Asterionella*, where one end of each of the cells are joined at a common centre to form a spoke or star-like arrangement, and *Fragilaria* (Fig. 5.7D), where the long narrow cells lie side by side to form rafts of cells. *Tabellaria* is another colonial freshwater pennate diatom, where the cells are joined at the corners to form zigzag chains. Some benthic species also commonly occur within the plankton community at

times, especially after stormwater inflows where they have been washed off the substrate that they were growing on. These include not only small species such as the oval shaped *Cocconeis* (Fig. 5.7E), and also some of the large thick-walled heavy species of pennate diatoms such as *Surirella* and *Pinularia* (Fig. 5.7F).

5.5 Pyrrhophyceae (or Dinophyceae) (dinoflagellates)

The 'dinos' are also common members of freshwater phytoplankton communities, although there are fewer freshwater forms than marine species (Section 6.3). Although some marine species are known to produce a range of different toxins, freshwater species are presently considered harmless (Box 5.7). Nevertheless, blooms can cause problems to

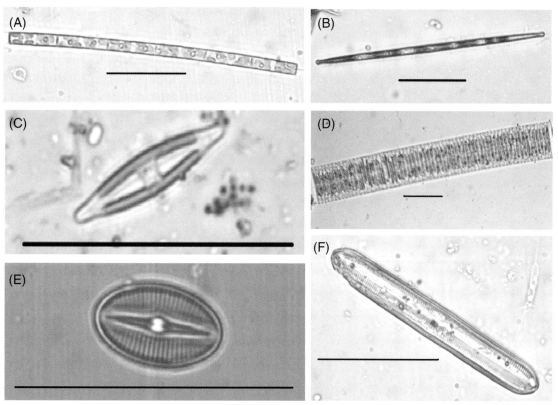

Fig. 5.7. **(A)** A filament of the diatom *Aulacoseira* sp. Note the number of chloroplasts within each cell. Scale bar 50 μm. **(B)** *Synedra* sp. – a long, needle-shaped, pennate diatom. Scale bar 50 μm. **(C)** A small cell of *Navicula* sp. There are several hundred of species within this genus. Scale bar 50 μm. **(D)** A colony of the diatom *Fragilaria* sp. The pennate shaped cells join together lengthwise to form a raft of cells. Scale bar 50 μm. **(E)** A small ovoid shaped cell of *Cocconeis* sp. – in valve view – illustrating the patterned silica cell wall. Scale bar 50 μm. **(F)** A cell wall from *Pinularia* sp. These are large diatoms that have a heavy silica cell wall and are usually found in benthic habitats. Scale bar 50 μm.

Box 5.7 Distinctive features of dinoflagellates

Dinoflagellates have a wide range of nutritional strategies, ranging from phototrophic, heterotrophic (consuming other cells) and saprophytic (consume dissolved organic substances). The cells of phototrophic dinoflagellates can contain several, to many, chloroplasts, which often radiate outwards from the centre of the cell. The main pigments for photosynthesis are chlorophylls a and c_2, but there are also several unique carotenoids present – the main one of which is peridinin. Pyrenoids are sometimes present and starch is stored as a food reserve. The dinophycean nucleus is distinct from that of all other eukaryotic organisms in having chromosomes that are permanently condensed – and a particular form of division during cell division. Reproduction is by simple cell division. Sexual reproduction also occurs, when the zygote can form into a resting cyst. However, resting cysts can also form from vegetative cells, and are considered to be part of the natural life cycle of these organisms. Dinoflagellates also have a specialised organelle that fire projectiles if the cell is irritated. Other distinctive features of dinoflagellates are their bioluminescence and circadian rhythms.

motile – swimming with the aid of two flagella – although other variants also occur. The typical planktonic form consists of a cell with an upper hemisphere (epicone) and a lower hemisphere (hypocone) that are separated by a groove that encircles the cell in its equatorial region, known as the cingulum. A second groove – the sulcus – runs transversally down the lower hemisphere from the cingulum to the pole. One flagellum encircles the cell within the cingulum; the second projects backwards from the sulcus. Many species – known as armoured dinoflagellates – have thecal plates made of cellulose that cover the entire cell. Both the number and arrangement of these plates are used to distinguish between genera and species by taxonomists. Not all dinoflagellates are armoured, however. Some – known as naked dinoflagellates – lack (or have only very thin) transparent thecal plates, but they still display the typical cellular organisation and morphology of this division of algae.

Common freshwater genera of armoured dinoflagellates include *Peridinium* (Fig. 5.8A) and *Ceratium* (Fig. 5.8B). Naked freshwater dinoflagellates, such as *Gymnodinium* (Fig. 5.8C), are less common.

5.6 Other algae

Several other groups of flagellated, motile algae – including the euglenids (or euglenoids, Euglenophyceae), cryptomonads (Cryptophyceae) and golden-brown algae or chrysophytes (Chrysophyceae) – are components of the freshwater phytoplankton. Euglenids are common in fresh waters,

water managers, especially for town supply, due to the fishy tastes and odours that they produce, and by blocking water filtration equipment.

Most freshwater dinoflagellates occur as single-celled species, although some filamentous species do exist. As the name suggests, they are typically

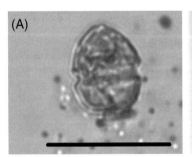 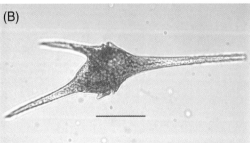 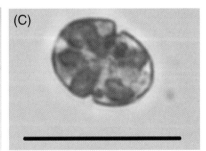

Fig. 5.8. **(A)** A small cell of *Peridinium* sp., illustrating the epicone, hypocone and cingulum. Scale bar 50 μm. **(B)** *Ceratium hirundinella* – a large dinoflagellate often found in nutrient enriched waters, which can cause fishy tastes and odours and block filtration equipment in town water supplies. Scale bar 50 μm. **(C)** *Gymnodinium* sp. – a naked dinoflagellate. Note the cingulum and the multiple chloroplasts within the cell. Scale bar 50 μm.

Box 5.8 Distinctive features of euglenids

Euglenids are single-celled, motile algae. They usually have at least two flagella, but in many cases – especially in the freshwater species – only one is emergent, from a canal at the anterior end of the cell. Euglenids often appear bright green under a microscope, due to the presence of both chlorophyll-*a* and *b*. Chlorophyll-*b* is something that euglenids share in common with the Chlorophyceae, but not with any other division of algae. Other pigments include β carotenes and xanthophylls, which can at times give blooms of euglenids a brick-red appearance. Many other euglenids are colourless – lacking any photosynthetic pigmentation – and they survive by purely heterotrophic means. Even pigmented euglenids can exhibit both photosynthetic and heterotrophic nutrition and, if placed in the dark, can lose their photosynthetic pigmentation, or become 'bleached'.

Many euglenids are naked – lacking a cell wall as such. They do, however, contain a structure known as a pellicle just inside the exterior cellular membrane, which is composed of overlapping proteinaceous strips that wind helically around the cell, and provide considerable flexibility to change shape. There is also a group of euglenids where the naked cells are enclosed in a non-living outer layer surrounding the cell, known as a lorica. These are often ornamented with spines, and have a short neck or pore, through which the flagella emerge.

There are often numerous disc-shaped chloroplasts scattered throughout the cells of photosynthetic species, which may have paramylon – a carbohydrate storage product – associated with them. Eyespots are present in the anterior part of the cell, near the base of the flagella. The anterior of the cell also contains a contractile vacuole that assists with osmotic regulation within the cell. The nucleus is also sometimes visible under light microscopy in the centre of the cell. Reproduction is asexual – occurring by cell division. Sexual reproduction has yet to be demonstrated. Some euglenids can form cysts to withstand periods of unfavourable environmental conditions. Some species also have phototaxic circadian rhythms: moving up and down the water column in response to light and at times, forming scums on the surface of the water. Common genera include *Euglena*, *Phacus*, *Lepocinclis*, *Trachelomonas* and *Strombomonas*. *Euglena* is relatively easy to culture and is a favourite in senior biology classes, including the use of a compound microscope.

especially in small ponds and farm dams where there is considerable organic pollution from animals, although members of this group also occur in brackish and marine waters. Cryptomonads also occur across a range of freshwater, brackish and marine environments, and are common components of most phytoplankton communities in non-flowing waters, although they are seldom present at high cell densities. In comparison, chrysophytes are a predominantly freshwater group of phytoplankton. Many species have a preference for cool, unpolluted soft waters that may be slightly acidic. They may be common in such locations, and form blooms sufficient to turn the water brown. They also tend to occur more in waters with low nutrient concentrations, rather than in phosphorus-enriched waters. In Australia, such situations include the dilute humic-acid stained coastal dune lakes of western Tasmania, and in wetlands in the coastal and tableland regions of New South Wales. They are less common in the warmer, harder waters of

the Murray–Darling Basin, although they still occur as minor components of the phytoplankton communities of these ecosystems. Elsewhere, chrysophytes are found in the higher latitude forest regions of the Northern Hemisphere, such as in small humus-stained lakes in Finland. However, one genus, *Dinobryon*, is common in tropical and subtropical reservoirs. Populations may also have seasonally restricted growing seasons (Sandgren 1988b), so cells may not always be present within the phytoplankton community.

The distinctive features of euglenids, cryptomonads and chrysophytes are provided in Boxes 5.8, 5.9 and 5.10, respectively.

Free-swimming naked euglenids typically have long cigar-shaped to oval-shaped or pear-shaped cells (such as *Euglena* Fig. 5.9A), or a flattened leaf-shaped cell (such as *Phacus* Fig. 5.9B) and move with a spiralling motion through the water. Their flexible cells allow them to change shape, especially under high light intensity under a

Box 5.9 Distinctive features of cryptomonads

The cells of cryptomonads are flattened dorsiventrally, giving them a bean- or heart-shaped appearance when viewed from the side. They are mainly single-celled, free-living and highly motile flagellates – having two flagella, one of which may be slightly shorter than the other. These typically emerge from a ventrally located depression or gullet, which, if present, opens towards the anterior end of the cell. The gullet is often lined with small organelles known as ejectosomes, which are discharged when the cell experiences some disturbance, unreeling long threads. These ejectosomes also occur on other parts of the cell.

The cells of cryptomonads are naked – lacking a cell wall. The cell itself most usually contains either one or two chloroplasts. In most cells, a single chloroplast is present, which contains two lobes joined in the middle by a pyrenoid. Cryptomonads possess both chlorophyll-a and c_2, plus several other distinctive accessory pigments including carotenes, xanthophylls, phycocyanin and phycoerythrin. Cryptomonads can therefore display a variation in colouration, including red, blue, yellow, brown and green. Some are colourless (because they lack a chloroplast) and are heterotrophic. Starch is the main storage product. Asexual reproduction occurs with the cell dividing longitudinally, but no sexual reproduction has been recorded.

microscope when they may withdraw their flagella and form into a spherical shape. When not swimming, the flexible pellicle also allows the cells to move across a surface by expanding parts of the cell while other parts contract. Armoured euglenids – which have cells enclosed in a lorica – are typified by *Trachelomonas* (Fig. 5.9C).

Commonly occurring freshwater cryptomonads include *Cryptomonas* and *Rhodomonas*.

Common genera of chrysophytes that illustrate the diversity in morphology within this algal division include the unicellular *Mallomonas* and *Synura*, which forms spherical to ovate colonies. Both genera have small siliceous scales and some species have spines or bristles. Another genus, *Dinobryon*, has cells enclosed in loricas that form linear or branching colonies.

5.7 Conclusions

There is considerable diversity found among freshwater phytoplankton. At least seven algal divisions are commonly represented within freshwater phytoplankton communities – each differing from the

Box 5.10 Distinctive features of chrysophytes

Planktonic chrysophytes are motile, and swim with the aid of two flagella – although in many species the second of these may be reduced to only a short stub. An eyespot may be present in the cell near the base of the flagella. Some chrysophytes may also undergo diurnal migrations up and down the water column of water bodies, indicating that they may be responsive to light availability within the water body.

In general, planktonic chrysophyte cells are ovate to tear drop in shape. The outside of the cell varies considerably, with some genera being naked – with nothing covering the cell membrane – while other genera have coverings of ornate siliceous scales and spines. In yet others, the cells are contained within a funnel- or urn-shaped lorica secreted by the cell itself. There may be one or a few chloroplasts present within the cell. Chrysophyte pigmentation includes chlorophyll-a and both c_1 and c_2, and also fucoxanthin, which gives the typical golden-brown colour. Pyrenoids occur within the chloroplasts, and the cells contain a storage product know as chrysolaminarin. In addition to being photosynthetic, many chrysophytes have been shown to also be heterotrophic – actively ingesting bacteria, and even other algae. The chrysophyte nucleus is located in the anterior section of the cell. Asexual reproduction takes place through the binary fission of cells. Sexual reproduction has been reported for only a few species, with two vegetative cells fusing to form a zygote. Chrysophyte vegetative cells can also form resting cysts, which have ornamented siliceous external walls.

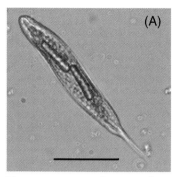
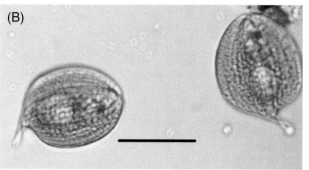
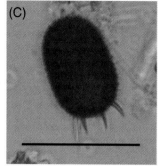

Fig. 5.9. **(A)** *Euglena* sp., showing numerous small disc-shaped chloroplasts and other internal structures. Scale bar 50 μm. **(B)** *Phacus* sp. – a flattened leaf shaped euglenid. Scale bar 50 μm. **(C)** A cell of *Trachelomonas* sp. – an armoured euglenid. Scale bar 50 μm.

other in their cellular structure, pigment arrays and the presence or absence of motile structures such as flagella. Within each division there is further variability. Examples of this include:

- the three commonly found orders of cyanobacteria
- the great diversity within the green algae, including both flagellated and non-flagellated forms
- the centric and the pennate forms of diatoms
- the armoured and naked forms of dinoflagellates and euglenids.

Superimposed on this is the variation in growth form throughout the cell cycle, with single-celled, filamentous and colonial species within many of the divisions.

Freshwater phytoplankton are an integral part of all freshwater ecosystems, with representatives found from pristine to polluted water bodies. They contribute to the food webs of these systems, along with benthic algae, other aquatic macrophytes and inputs from terrestrial sources. In most systems, freshwater phytoplankton do not cause environmental problems. It is only when conditions are suitable for explosive growth, such as an excess in nutrients, that algal blooms cause water-quality problems that may affect both the ecosystem in which this occurs and anthropogenic uses of the water. Of all the types of freshwater phytoplankton that may bloom, the cyanobacteria are of most concern because of the potential hazard these create

through the ability of some species to produce potent toxins. Because of this, considerable effort must be put into sampling freshwater phytoplankton communities – especially for public health surveillance – and adequate sampling methods must be employed to obtain a representative measure of phytoplankton presence within particular water bodies.

5.8 References

Baker PD, Humpage AR (1994) Toxicity associated with commonly occurring cyanobacteria in surface waters of the Murray-Darling Basin, Australia. *Australian Journal of Marine and Freshwater Research* **45**, 773–786. doi:10.1071/MF9940773

Baker PD, Steffensen DA, Humpage AR, Nicholson BC, Falconer IR, Lanthois B, *et al.* (2001) Preliminary evidence of toxicity associated with the benthic cyanobacterium *Phormidium* in South Australia. *Environmental Toxicology* **16**, 506–511. doi:10.1002/tox.10009

Bowling LC, Baker PD (1996) Major cyanobacterial bloom in the Barwon-Darling River, Australia, in 1991, and underlying limnological conditions. *Marine and Freshwater Research* **47**, 643–657. doi:10.1071/MF9960643

Bold HC, Wynne MJ (1986) *Introduction to the Algae. Structure and Reproduction.* 2nd edn. Prentice-Hall, Edgewood Cliffs NJ, USA.

Bormans M, Sherman BS, Webster IT (1999) Is buoyancy regulation in cyanobacteria an adaptation to exploit separation of light and nutrients? *Marine and Freshwater Research* **50**, 897–906. doi:10.1071/MF99105

Codd GA, Steffensen DA, Burch MD, Baker PD (1994) Toxic blooms of cyanobacteria in Lake Alexandrina, South Australia – learning from history. *Australian Journal of Marine and Freshwater Research* **45**, 731–736. doi:10.1071/MF9940731

Falconer IR (2001) Toxic cyanobacterial bloom problems in Australian waters: risks and impacts on human health. *Phycologia* **40**, 228–233. doi:10.2216/i0031-8884-40-3-228.1

Francis G (1878) Poisonous Australian lake. *Nature* **18**, 11–12. doi:10.1038/018011d0

Hawkins PR, Runnegar MTC, Jackson ARB, Falconer IR (1985) Severe hepatotoxicity caused by the tropical cyanobacterium (blue-green alga) *Cylindrospermopsis raciborskii* (Woloszynska) Seenaya and Subba Raju isolated from a domestic water supply reservoir. *Applied and Environmental Microbiology* **50**, 1292–1295.

Komárek J, Anagnostidis K (1999) *Cyanoprokaroyota 1. Teil Chroococcales. Süßwasserflora von Mitteleuropa Band 19/1.* Gustav Fischer, Stuttgart, Germany.

Lee RE (1999) *Phycology.* 3rd edn. Cambridge University Press, Cambridge, UK.

Oliver RL (1994) Floating and sinking in gas-vacuolate cyanobacteria. *Journal of Phycology* **30**, 161–173. doi:10.1111/j.0022-3646.1994.00161.x

Sandgren CD (Ed.) (1988a) *Growth and Reproductive Strategies of Freshwater Phytoplankton.* Cambridge University Press, Cambridge, UK.

Sandgren CD (1988b) The ecology of chrysophyte flagellates: their growth and perennation strategies as freshwater phytoplankton. In *Growth and Reproductive Strategies of Freshwater Phytoplankton.* (Ed. CD Sandgren) pp. 9–104. Cambridge University Press, Cambridge, UK.

Seifert M, McGregor G, Eaglesham G, Wickramasinghe W, Shaw G (2007) First evidence for the production of cylindrospermopsin and deoxy-cylindrospermopsin by the freshwater benthic cyanobacterium, *Lyngbya wollei* (Farlow ex Gomont) Speziale and Dyck. *Harmful Algae* **6**, 73–80. doi:10.1016/j.hal.2006.07.001

Sivonen K, Jones G (1999) Cyanobacterial toxins. In *Toxic Cyanobacteria in Water. A Guide to their Public Health Consequences, Monitoring and Management.* (Eds I Chorus and J Bartram) pp. 41–111. E & FN Spon, London, UK.

South G, Whittick A (1987) *Introduction to Phycology.* Blackwell Scientific, Oxford, UK.

Sukenik A, Hadas O, Kaplan A, Quesada A (2012) Invasion of Nostocales (cyanobacteria) to subtropical and temperate freshwater lakes – physiological, regional, and global driving forces. *Frontiers in Microbiology* **3**, 1–9. doi:10.3389/fmicb.2012.00086

Van Den Hoek C, Mann DG, Jahns HM (1995) *Algae: An Introduction to Phycology.* Cambridge University Press, Cambridge, UK.

5.9 Further reading

Chorus I, Bartram J (Eds) (1999) *Toxic Cyanobacteria in Water. A Guide to their Public Health Consequences, Monitoring and Management.* E & FN Spon, London, UK.

Hötzel G, Croome R (1999) 'A phytoplankton methods manual for Australian freshwaters'. LWRRDC Occasional paper 22/99. Land and Water Resources Research and Development Corporation, Canberra, Australia.

Pilotto LS, Douglas RM, Burch MD, Cameron S, Beers M, Rouch GR, *et al.* (1997) Health effects of recreational exposure to cyanobacteria (blue-green algae) during recreational water-related activities. *Australian and New Zealand Journal of Public Health* **21**, 562–566. doi:10.1111/j.1467-842X.1997.tb01755.x

Pilotto L, Hobson P, Burch MD, Ranmuthugala G, Attewell R, Weightman W (2004) Acute skin irritant effects of cyanobacteria (blue-green algae) in healthy volunteers. *Australian and New Zealand Journal of Public Health* **28**, 220–224. doi:10.1111/j.1467-842X.2004.tb00699.x

Sukenik A, Hadas O, Kaplan A, Quesada A (2012) Invasion of Nostocales (cyanobacteria) to subtropical and temperate freshwater lakes – physiological, regional, and global driving forces.

Frontiers in Microbiology **3**, 1–9. doi:10.3389/fmicb.2012.00086

Tyler PA (1996) Endemism in freshwater algae, with special reference to the Australian region. *Hydrobiologia* **336**, 127–135. doi:10.1007/BF00010826

Whiterod N, Bice C, Zukowski S, Meredith S (2004) 'Cyanobacteria mitigation in the Mildura Weir Pool'. Murray-Darling Freshwater Research Centre Lower Basin Laboratory (MDFRCLBL), Report No. 8/2004. MDFRCLBL, Mildura, Australia.

5.9.1 Taxonomic guides and texts for the laboratory identification of Australian freshwater phytoplankton

Baker PD (1991) 'Identification of common noxious cyanobacteria. Part I – Nostocales'. Urban Water Research Association of Australia, Research Report No. 29. UWRAA, Melbourne, Australia.

Baker PD (1992) 'Identification of common noxious cyanobacteria. Part II – Chroococcales, Oscillatoriales'. Urban Water Research Association of Australia, Research Report No. 46. UWRAA, Melbourne, Australia.

Baker P, Fabbro L (2002) *A Guide to the Identification of Common Blue-Green Algae (Cyanoprokaryotes) in Australian Freshwaters.* Identification Guide No. 25, 2nd edn. Murray Darling Freshwater Research Centre, Albury, Australia.

Entwisle TJ, Sonnerman JA, Lewis SH (1997) *Freshwater Algae in Australia.* Sainty and Associates, Potts Point, Australia.

Foged N (1978) Diatoms in eastern Australia. *Bibliotheca Phycologica* **47**, 1–225.

Gell P, Sonneman J, Reid M, Illman M, Sincock A (1999) *An Illustrated Key to Common Diatom Genera from Southern Australia.* Identification Guide No. 26. Murray Darling Freshwater Research Centre, Albury, Australia.

Ling HU, Tyler PA (1986) *A Limnological Survey of the Alligator Rivers Region. Part 2: Freshwater Algae, Exclusive of Diatoms.* Australian Government Publishing Service, Canberra, Australia.

Ling HU, Tyler PA (2000) *Australian Freshwater Algae (exclusive of diatoms).* J. Cramer, Berlin, Germany.

Ling HU, Croome RL, Tyler PA (1989) Freshwater dinoflagellates of Tasmania, a survey of taxonomy and distribution. *British Phycological Journal* **24**, 111–129. doi:10.1080/00071618900650111

McGregor GB (2007) *Freshwater Cyanoprokaryota of North-eastern Australia 1: Oscillatoriales.* Flora of Australia Supplementary Series Number 24. Australian Biological Resources Study, Canberra, Australia.

McGregor GB, Fabbro LD (2001) *A Guide to the Identification of Australian Freshwater Planktonic Chroococcales (Cyanoprokaryota/Cyanobacteria).* Identification Guide No. 39. Murray Darling Freshwater Research Centre, Albury, Australia.

McLeod JA (1975) The freshwater algae of South-Eastern Queensland. PhD Thesis, University of Queensland, Brisbane, Australia.

Prescott GW (1978) *How to Know the Freshwater Algae.* Wm. C. Brown Co., Dubuque IA, USA.

Sonneman JA, Sincock A, Fluin J, Reid M, Newall P, Tibby J, Gell P (2000) *An Illustrated Guide to Common Stream Diatom Species from Temperate Australia.* Identification Guide No. 33. Murray Daring Freshwater Research Centre, Albury, Australia.

Thomas DP (1983) *A Limnological Survey of the Alligator Rivers Region, Northern Territory. Part 1. Diatoms (Bacillariophyceae) of the Region.* Australian Government Publishing Service, Canberra, Australia.

6

Coastal and marine phytoplankton: diversity and ecology

Penelope Ajani, Ruth Eriksen and David Rissik

6.1 Marine phytoplankton: diversity and ecology

Phytoplankton (*phyto* = *plant*, *planktos* = made to wander) consist of microscopic algae, cyanobacteria (blue-green algae) and other protists that live suspended in the water. With tens of thousands of species identified in coastal and oceanic waters, microalgae are highly diverse, yet largely underexplored, with up to nine major functional groups (having similar ecological or biogeochemical roles). Most phytoplankton are unicellular and range in size from 0.2 to 200 μm, with a few taxa attaining up to 2 mm in length. Many phytoplankton species are autotrophic and produce energy for growth and reproduction through photosynthesis, by fixing carbon and converting it to chemical energy. Carbon fixed by photosynthesis is termed 'primary production', which uses sunlight, atmospheric carbon dioxide and dissolved nutrients such as nitrate and phosphate to provide energy. Calcium or silicate is used for development of the structure of the cell in some species. A by-product of this process is the generation of oxygen, and approximately half of the oxygen in the atmosphere is produced by phytoplankton. Although the majority of the phytoplankton are autotrophic (i.e. 'self-feeding'), some are heterotrophic where cells depend on the uptake of organic material from their environment, by ingesting other cells and particles, or

by taking in soluble material (Horner 2002). Other species are mixotrophic (combining autotrophic and heterotrophic modes), and symbiotic relationships of photosynthetic cells with many tiny bacteria and cyanobacteria abound. Either way, phytoplankton form the foundation of the marine food web, providing nutrition either directly or indirectly to higher trophic levels, including important fisheries and ourselves.

Pigments are chemical compounds contained in the chloroplasts of microalgae that assist in capturing solar energy for photosynthesis. In order to acquire more of the sun's energy, different phytoplankton produce several different kinds of pigments to absorb a broader range of wavelengths. As well as distinguishing between the major functional groups, pigments reflect evolutionary relationships between these groups and provide us with a method of measuring phytoplankton biomass and determining production rates (growth) of different phytoplankton communities.

In the following sections we will discuss the major functional groups of phytoplankton found in temperate coastal waters and give a brief description of each group:

- Bacillariophyta (diatoms) (Section 6.2)
- Dinoflagellata (dinoflagellates) (Section 6.3)
- Cyanobacteria (blue-green algae) (Section 6.4)
- Other marine phytoplankton (Section 6.5)

> ➤ Raphidophyceae (raphidophytes)
> ➤ Dictyochophyceae (silicoflagellates)
> ➤ Prymnesiophyceae = Haptophyta (coccolithophorids, prymnesiophytes, golden brown flagellates)
> ➤ Cryptophyceae (cryptomonads)
> ➤ Euglenophyceae (euglenids)
> ➤ Chlorophyceae (green algae – prasinophytes, chlorophytes)

Phytoplankton are classified into these functional or taxonomic groups based on a combination of their photosynthetic pigments, as well as other characteristics such as the way in which they store energy (lipid or carbohydrate) and the structure of their cell walls. Other distinguishing features include the presence or absence of flagella, the structure of the flagella or flagella roots, the pattern and course of mitosis (cellular division) and cytokensis (cell division) or other morphological attributes such as symmetry and size.

Many of these groups are represented in the microplankton (20–200 µm), the nanoplankton (2–20 µm) and the picoplankton (0.2–2 µm) – with some occurring in all three-size classes. In temperate coastal waters, the nanoplankton can account for 80% of the total phytoplankton biomass, while in tropical waters the picoplankton can account for 80% of the total phytoplankton biomass. Green flagellates, small non-thecate dinoflagellates, cryptomonads, prymnesiophytes, coccolithophorids and other colourless flagellates are all common representatives of the nanoplankton in coastal waters. Picoplankton are represented by the cyanobacteria and chrysophytes.

6.2 Bacillariophyta (diatoms)

Diatoms are abundant single-celled phytoplankton containing membrane-bound cell organelles, with around 15 000 species described in fresh waters (Section 5.4) and marine waters. They are distinguished from other groups by two highly ornamented siliceous plates or valves (sometimes termed theca or frustules) that enclose the protoplast, with the hypotheca slightly smaller and fitting inside the epitheca. They may occur as solitary cells or in stunning beautiful colonial or chain forms, linked by siliceous structures such as setae or spines, or by mucilage threads of a sugary polysaccharide secreted through specialised structures. The structure, patterns and processes of the cell wall, but more importantly, valve symmetry, form the basis for the two major groups within the diatoms: the pennate and the centric diatoms. Pennate diatoms are elongate and usually bilaterally symmetrical, with up to 800 marine species identified. Centric diatoms are usually round or 'radially symmetrical' (with the frustule often compared to a Petri dish or pillbox) and there are up to ~1000 species in marine waters (Fig. 6.1).

Diatoms, unlike nearly all other phytoplankton, have no flagella and are in most cases non-motile. Pennate forms can achieve a gliding motion via mucilage secretion through their raphe system (a longitudinal slit in the valve), while centric diatoms can exude mucilage through their labiate processes (tubes or openings through the valve wall), allowing limited movement. Diatoms tend to sink because of their density, relying on the turbulence of marine systems to keep them suspended in the sunlit upper layers of the ocean where they can photosynthesise, but recent studies have shown

Fig. 6.1. Transmission electron microscopy of common diatom species found in temperate coastal waters of New South Wales, Australia. Both centric (radially symmetrical) and pennate (bilaterally symmetrical) forms can be seen (photo: P. Ajani).

that they are able to control the rate of sinking to optimise nutrient exchange.

Diatoms may also be benthic: living in or growing on sediments, rocks and larger plants (Box 6.1). In coastal waters, limiting factors – such as silicate (and other nutrient) availability, water stability, light climate, parasitism and grazing – affect which species are present in the water column at particular times. Diatom blooms often occur in coastal waters when episodic upwelling brings nutrient-rich water to the surface, where there is better access to light and subsequent increased production (Box 6.2).

Some diatom blooms can become so dense that they can kill fish and invertebrates due to either

---

Box 6.1 Benthic phytoplankton

Benthic phytoplankton or microphytobenthos (MPB) are important communities in terms of estuarine and coastal ecology. MPB assemblages play a central role in the production and cycling of organic matter in these environments, as well as stabilising sediments by excreting mucilaginous substances into the sediment and thus preventing erosion. Microphytobenthos assemblages usually include bacteria, flagellates, ciliates, diatoms, dinoflagellates and other algae, as well as foraminifers and nematodes. Further groupings can be found within the diatoms – some live freely on (epipelic) or in the sediments (endopelic). Those living attached to the substratum are classified according to their substrata preference: sand grains (epipsammic), rock or stones (epilithic), plants (epiphytic) and epizoic (animals).

Although phytoplankton communities in coastal waters have received much attention, very few studies have been carried out on the MPB communities. This is probably because of the difficulties in extracting and enumerating the cells from the sediments. The few studies that have been carried out in our coastal waters list the abundant MPB genera as the diatoms *Amphora*, *Cocconeis*, *Bacillaria*, *Navicula*, *Nitzschia*, *Gyrosigma*, *Mastogloia*, and *Pleurosigma* (Saunders *et al.* 2010) as well as the dinoflagellates *Amphidinium*, *Prorocentrum Amphidiniopsis*, *Ostreopsis*, *Gambierdiscus* and *Gyrodinium* (Hallegraeff *et al.* 2010). Green euglenids, such as *Eutreptia*, are also common.

Box 6.2 Diatom blooms

Seasonal signals in phytoplankton biomass along the east coast of Australia coincide with intermittent slope water intrusions bringing cold, nutrient-rich water into the euphotic zone (Ajani *et al.* 2001b). In addition to these sporadic upwelling/downwelling events, a distinct seasonal cycle in phytoplankton community composition has been observed with species richness peaking in the winter. The spring and summer blooms are characterised by a clear successional pattern, beginning with small diatoms, such as *Asterionellopsis*, *Thalassiosira*, *Skeletonema*, *Pseudo-nitzschia* and *Chaetoceros*, followed by larger diatoms, such as *Ditylum*, *Leptocylindrus*, *Eucampia*, *Rhizosolenia*, *Melosira* and *Thalassiothrix*, and concluding with dinoflagellates, most notably species of the genera *Tripos* and *Protoperidinium*

---

oxygen depletion or by abrasion damage to their gills (such as *Thalassiosira* spp. and *Chaetoceros* spp.). Certain diatom species belonging to the genera *Pseudo-nitzschia* and *Nitzschia* produce the potent neurotoxin domoic acid and are implicated as causing amnesic shellfish poisoning (ASP) (Box 6.3).

Diatom frustules have a slow rate of decay, which has resulted in massive geological deposits known as diatomaceous earth. This material is used in filtration, cosmetics, toothpaste and even forensic science. Diatoms also present an important record of past environmental conditions, through accumulation of their silica-based frustules in ocean sediments.

6.3 Dinoflagellata (dinoflagellates)

Dinoflagellates are a group of unicellular algae with membrane bound organelles (and therefore eukaryotes) and two flagella. There are ~2000 living species known (130 genera, Fig. 6.2; see Section 5.5 for freshwater species). About half the dinoflagellates feed on organic matter only (i.e. they are heterotrophs, including some carnivores) and the other half either photosynthesise or are partly autotrophic and partly heterotrophic.

Box 6.3 Species belonging to the *Pseudo-nitzschia* genus

Pseudo-nitzschia is a pennate diatom genus with a global marine distribution. Of the 49 species described to date worldwide, 26 have been found to produce domoic acid (DA): a potent neurotoxin that can accumulate in the marine food web and cause both ecosystem and human health effects such as amnesic shellfish poisoning (ASP) (https://en.wikipedia.org/wiki/Pseudo-nitzschia). For this reason, the routine monitoring of *Pseudo-nitzschia* cell densities and the concentration of the toxic compound DA is the focus of many seafood safety programs globally.

Species belonging to the genus *Pseudo-nitzschia* are a significant component of the phytoplankton community in Australian waters (Hallegraeff 1994, Ajani *et al.* 2013a,b; 2016a). To date, only three species of *Pseudo-nitzschia* have been found to produce DA: *P. australis, P. cuspidata* and *P. multistriata. Pseudo-nitzchia cuspidata* was responsible for a significant toxic bloom in southeastern Australia in 2010, where maximum cell densities of >6 × 10⁶ cells/L and DA in oyster tissue of 34 mg DA/kg were reported (Ajani *et al.* 2013b).

Toxic *Pseudo-nitzschia* blooms are also a common reoccurrence along the west coast of North America. A major bloom occurring in 2015–2016 was caused by anomalous ocean conditions and resulted in high levels of domoic acid in the food web (McCabe *et al.* 2016). While a sophisticated harmful algal bloom monitoring program and response network meant that no human illnesses were reported, there were significant impacts to commercial and recreational fisheries.

Dinoflagellates are motile at some stage of the life cycle – having two different flagella. One flagellum is situated in a girdle groove around the middle of the cell (for rotation) and the other projects from the sulcus groove (at one end) for propulsion. Careful use of a microscope is required to see these flagella. Dinoflagellates may be armoured (thecate – with cellulose cell walls made of plates) or unarmoured (non-thecate or naked). Armoured dinoflagellates are usually irregular in shape, bearing horns, ridges and wings, which aid in identification.

Over 80 species of marine dinoflagellates produce cysts and at least 16 of these species are known to cause red tides and seven species to be toxic. Cysts can be of two types: either temporary cysts (whereby the cell quickly re-established itself after a brief encystment) or resting cysts, which sink from the water column and often remain in the sediment anywhere from 6 weeks to many years, depending on the species. The purpose of cyst formation is probably a survival strategy, which is regulated by both physiological and environmental factors such as:

- protection from adverse conditions (such as temperature or nutrient availability)
- a refuge from predation
- alternation between planktonic and benthic habitats
- as part of the reproductive process
- aid in dispersion/seed population for the subsequent bloom.

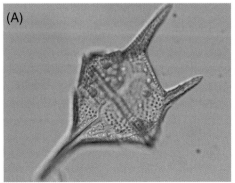 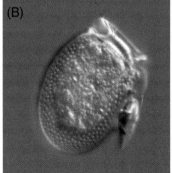 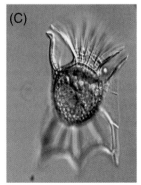

Fig. 6.2. Light microscopy images of some common dinoflagellate species found in temperate coastal waters of New South Wales, Australia: **(A)** *Ceratium pentagonum;* **(B)** *Dinophysis acuminata;* and **(C)** *Ornithocerus magnificus* (photos: P. Ajani).

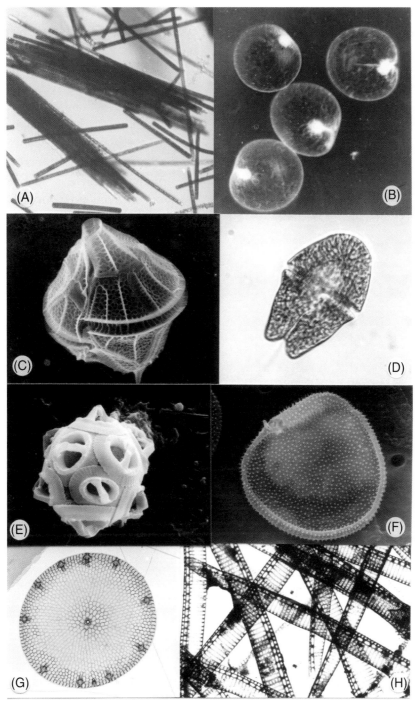

Fig. 6.3. Common bloom species in New South Wales marine and estuarine waters: **(A)** light microscopy (LM) of the filamentous cyanobacterium *Trichodesmium erythraeum* producing raft-like bundles, up to 750 µm long; **(B)** LM of the balloon-shaped, colourless dinoflagellate *Noctiluca scintillans*, 200–500 µm diameter; **(C)** scanning electron microscopy (SEM) of the dinoflagellate *Gonyaulax polygramma*, showing ornamented cellulose plates with longitudinal ridges, 29–66 µm long; **(D)** LM of the large, unarmoured dinoflagellate *Akashiwo sanguinea*, 50–80 mm long; **(E)** SEM of the calcareous nanoplankton *Gephyrocapsa oceanica*, 15 µm diameter; **(F)** SEM of the triangular, armoured dinoflagellate cells of *Prorocentrum cordatum*, 10–15 µm wide and covered with minute spinules; **(G)** transmission electron microscopy (TEM) of the weakly silicified cell of the centric diatom *Thalassiosira partheneia*, 10 µm diameter; **(H)** TEM of a bloom of the pennate diatom *Pseudo-nitzschia pseudodelicatissima*, 57–150 µm long (photos courtesy of Ajani *et al*. 2001a).

Many dinoflagellates make daily diurnal migrations up and down the water column. During the day, they migrate towards the surface of the water (for better light availability) and at night, they move down to a depth of several metres (for better access to nutrients). This vertical migration is an important consideration when sampling or when analysing the results of sampling activities.

A regularly occurring red tide on the south-east Australian coast is caused by the dinoflagellate *Noctiluca scintillans* (Fig. 6.3B) (Ajani *et al.* 2001a). *Noctiluca* are large (up to 1 mm diameter)

Box 6.4 Species belonging to the *Dinophysis* genus

Even at very low cell densities (<10^3 cells/L), species belonging to the dinoflagellate genus *Dinophysis* can produce phycotoxins that cause diarrhetic shellfish poisoning (DSP), a type of gastroenteritis in consumers of seafood. *Dinophysis* is widespread in Australian waters, with 36 species reported thus far. Toxic representatives include *D. acuminata, D. acuta, D. caudata, D. fortii* and *D. hastata.* There have been three major DSP events in Australia to date. In 1997 *D. acuminata* and *D. tripos* were implicated in the contamination of pipis (*Plebidonax deltoides*) in New South Wales (NSW) in which 102 people were affected, and 56 cases of gastroenteritis reported. In March 1998 a second outbreak was reported in which 20 cases of DSP poisoning were reported. In March 2000, a third outbreak occurred in Queensland and was again linked to the consumption of pipis, when only one person was affected.

Dinophysis species are common in Australian coastal waters, but rarely abundant. Although the highest abundance of *D. acuminata* has been observed in spring in the Hawkesbury River estuary (maximum abundance 4500 cells/L), the highest abundance of *D. caudata* was in summer to autumn (maximum 12 000 cells/L), highlighting the species-specific seasonality of this toxic group.

Being a cosmopolitan genus, 'diarrhetic shellfish poisoning' caused by *Dinophysis* occurs in many parts of the world (Ajani *et al.* 2016b and references therein) so this genus is the focus of many harmful algal monitoring programs.

---

balloon-shaped, heterotrophic dinoflagellates that consume other algae, some zooplankton and even fish eggs. They have no photosynthetic pigments, although in tropical waters they may appear green due to endosymbiotic, green-pigmented prasinophytes living in their vacuoles. *Noctiluca* are positively buoyant (due to ammonia production) causing dense red slicks to form in surface waters. As blooms die off the ammonia is released into the environment, which is potentially dangerous to fish. *Noctiluca* are bioluminescent at night (i.e. they glow), especially around a moving boat or breaking wave. Interestingly, the frequency of this species off south-eastern Australia has increased in recent years. Using continuous plankton recorder data, McLeod *et al.* (2012) reported on the climate-driven range expansion of this species into the Southern Ocean. This species appeared to be transported south by an East Australian Current warm-core eddy, most likely because of the increased poleward penetration of this western boundary current.

Dinoflagellates have the largest number of known harmful species (around 40 species), with new discoveries still being made. Similar to diatoms, they can produce toxic compounds that accumulate in filter-feeding bivalves and the larvae of commercially important crustaceans and finfish. Consumption of these seafoods by humans can result in a range of symptoms including gastroenteritis, headaches, muscle and joint pain. In extreme cases, paralysis and respiratory failure can occur, caused by several major poisoning syndromes: paralytic shellfish poisoning (PSP); diarrhetic shellfish poisoning (DSP) (Box 6.4); neurotoxic shellfish poisoning (NSP); azaspiracid shellfish poisoning (AZP); and ciguatera fish poisoning (CFP). On a global scale, ~60 000 human intoxications occur per year worldwide, with an overall mortality of ~1.5% (Kantiani *et al.* 2010).

6.4 Cyanobacteria (blue-green algae)

Cyanobacteria are primitive algae characterised by the absence of membrane-bound cell components (i.e. they are prokaryotes; see also Section 5.2). Cyanobacteria are often blue-green in colour. They have unicellular, colonial and filamentous forms, and do not have flagellate cells at any stage in their life cycle.

Cyanobacteria include benthic and planktonic forms. Many species have adaptations to aid survival in extreme and diverse habitats, such as gas vacuoles for buoyancy control, akinetes (resting stages) and heterocysts (specialised cells that can fix atmospheric nitrogen) for survival in waters where the nitrate and ammonia levels are relatively low. Not all taxa have these features. In marine and brackish waters, cyanobacteria have produced toxins that have resulted in neuromuscular and organs distress as well as external contact irritation.

Six genera of cyanobacteria have been implicated in blooms in Australian coastal waters: *Dolichospernum, Microcystis, Amphizomenon, Nodularia, Trichodesmium* and *Lyngbya,* several which produce phycotoxins (see Chapter 3). *Trichodesmium* is a most common cyanobacterium in most waters worldwide, and is able to 'fix' atmospheric nitrogen into nitrate (i.e. a diazotroph) and therefore important in tropical marine ecosystems. This tropical/subtropical species produces episodic 'red tides' that were historically reported as 'sea sawdust' during Captain Cook's voyage through the Coral Sea in 1778. The filaments of this cyanobacterium are united (parallel) into small raft-like bundles that are just visible to the naked eye (around 1 mm). The filaments are generally cylindrical, uniformly broad or slightly tapering at the tips, and are straight or slightly curved. *Trichodesmium* filaments do not have any specialised cells such as heterocysts or akinetes (Fig. 6.3A).

Blooms of *Trichodesmium* have been linked to respiratory distress and contact dermatitis in humans, as well as toxicity in zooplankton and higher trophic levels in coral reef and tropical environments (Golubic *et al.* 2010). Blooms of *Trichodesmium* are most commonly seen in subtropical and tropical waters most commonly around northern Australia and the Red Sea. They occur in northern NSW waters in spring, summer and early autumn when the East Australian Current (EAC) transports these algal masses into NSW from Queensland waters. These blooms appear yellow-grey in their early stages, while they become a reddish-brown later.

6.5 Other marine phytoplankton

6.5.1 Raphidophyceae (raphidophytes)

Raphidophytes are unicellular flagellates that have two unequal, heterodynamic flagella arising from a sub-apical shallow groove. The forward-directed flagellum has two rows of fine hairs, while the trailing flagellum is smooth and lies close to the surface of the cells. Their cells are unarmoured, dorsoventrally flattened (potato-shaped) and contain numerous ejectosomes, trichocysts and/or mucocysts that readily discharge upon stimulation. They have a characteristic 'raspberry-like' appearance upon preservation, which can make identification difficult (Fig. 6.4E). Many raphidophytes can be toxic to fish and bloom events have been reported throughout the world in coastal and estuarine waters (Box 6.5). *Heterosigma, Chattonella* and *Fibrocapsa* commonly bloom in summer. Toxic species in this group are a particular concern for aquaculture, where large-scale fish kills may occur due to the inability of caged fish to escape from deteriorating conditions.

Box 6.5 Toxic raphidiophyte blooms

A toxic raphidiophyte, *Chattonella* cf. *globosa*, bloomed sporadically in Sydney Harbour, Australia on a few occasions. Blooms of related species have caused significant mortality of cultured yellowtail and red sea bream in Japanese inland seas and implicated in the mass mortality of farmed blue-fin tuna in South Australia (Marshall and Hallegraeff 1999 and references therein). The production of superoxide radicals as the major mechanism of fish mortality is also hypothesised for this genus. Evidence for brevetoxin-like production is still being investigated.

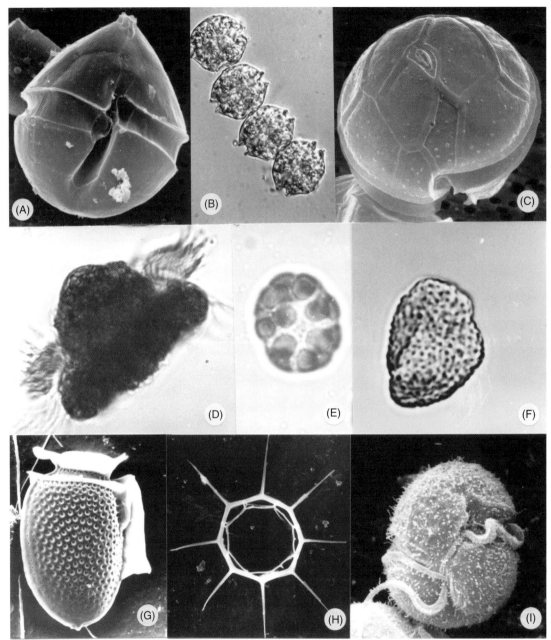

Fig. 6.4. Common bloom species in marine and estuarine waters of eastern Australia: **(A)** scanning electron microscopy (SEM) of the red-water dinoflagellate *Scripsiella trochoidea*, 16–36 μm long. Note tube-shaped apical pore on top of the cell and nearly equatorial (not displaced) girdle groove; **(B)** light microscopy (LM) of the chain-forming dinoflagellate *Alexandrium catenella* – the causative organism of paralytic shellfish poisoning. Individual cells 20–22 μm long; **(C)** SEM of the red water dinoflagellate *Alexandrium minutum* – the causative organism of paralytic shellfish poisoning. Individual cells 24–29 μm diameter. Note the hook-shaped apical pore on top of the cell and characteristic shape of the first apical plate; **(D)** LM of the ciliate *Mesodinium rubrum*, with two systems of cilia arising from the waist region, 30 μm diameter; **(E)** LM of the 'raspberry-like' cell of the fish-killing flagellate *Heterosigma akashiwo* ('Akashiwo' = red tide), containing numerous disc-shaped chloroplasts, cell 11–25 μm long; **(F)** LM of an undescribed flagellate resembling *Haramonas*. The cell surface is covered by numerous mucus-producing vesicles, cells 30–40 μm long; **(G)** SEM of the small armoured dinoflagellate *Dinophysis acuminata* – the causative organism of diarrhetic shellfish poisoning, cells 38–58 μm long; **(H)** SEM of the siliceous skeleton of the silicoflagellate *Octactis octonaria*, 10–12 μm diameter; **(I)** SEM of the small unarmoured, fish-killing dinoflagellate *Karlodinium micrum*, 15 μm diameter (photos courtesy of Ajani *et al.* 2001a).

6.5.2 Dictyochophyceae (silicoflagellates)

Silicoflagellates are unicellular cells with a single flagellum and a life-cycle that includes a siliceous skeleton. Identification to species level is based on the shape of this silica skeleton. *Octactis* (formerly *Dictyocha*) is the most common genus found in our waters and is perhaps toxic to fish (Box 6.6). In addition to diatoms, these silica skeletons can provide detailed information on past environmental conditions through analysis of their accumulation patterns in sediments.

6.5.3 Prymnesiophyceae = Haptophyta (coccolithophorids, prymnesiophytes, golden brown flagellates)

Prymnesiophytes, also known as haptophytes, are either unicellular or colony-forming flagellates that

have two equal or unequal flagella, as well as a 'third flagellum' – a haptonema – a thin filamentous organelle sometimes used for anchoring the cell and sometimes in food uptake. Most species are small and belong to the nanoplankton (2–20 μm). The cell surface is covered with tiny scales or granules of organic material (cellulose), which is used extensively in taxonomy but requires powerful techniques such as scanning electron microscopy for definitive identification (Fig. 6.5). In addition, there may be spectacular calcified scales called coccoliths, which are characteristic of the coccolithophorids (Box 6.7). Coccolithophorids have formed geological deposits, such as the White Cliffs of Dover in the UK, yet are vulnerable to future ocean acidification by virtue of their calcifying scales.

6.5.4 Cryptophyceae (cryptomonads)

Cryptomonads are very small, ovoid phytoplankton (6–20 μm) with a rigid protein coat and two flagella protruding from a 'gullet' at one end (two equal or unequal in length, with one or two rows of tubular hairs).

6.5.5 Euglenophyceae (euglenids)

Euglenids (or euglenoids) are large (15–500 μm), green, single-celled flagellates that have a deep fold or gullet where the flagellum is attached. The cell has a spiral construction and is surrounded by a pellicle that is composed of proteinaceous

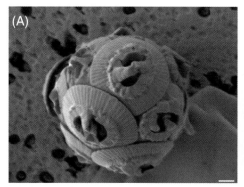

Fig. 6.5.   Scanning electron micrograph showing the 'coccoliths' or scales covering the haptophytes: **(A)** *Gephyrocapsa* sp. and **(B)** *Emiliania huxleyi*, both from the Southern Ocean (photos: Ruth Eriksen, courtesy of Australian Antarctic Division Electron Microscopy Unit © Commonwealth of Australia). Scale bar 1 μm.

Box 6.7 Blooms of prymnesiophytes and coccolithophorids

One prymnesiophyte (*Phaeocystis*) is the cause of extensive mounds of sea foam 1–2 m deep around the coasts of the north Atlantic, and the east coast of Australia, usually occurring after blooms coincide with major storms. An unprecedented, bloom of the small (<10 μm) cosmopolitan coccolithophorid *Gephyrocapsa oceanica* (Fig. 6.3E) occurred for 4 weeks in Jervis Bay on the south east Australian coast (Blackburn and Cresswell 1993). The bloom turned the waters milky green, which caused economic hardship for SCUBA diving businesses during the peak tourist season. Upwelling of cool nutrient-rich slope water and an influx of warm East Australian Current waters from an adjacent eddy may have enhanced the nutrients and upper layer temperatures. The maximum cell density of 2×10^7 cells/L was greater than any previously recorded of this species in Australian waters.

interlocking strips that wind helically around the cell (giving the cells a striped pattern). A conspicuous eyespot located in the cytoplasm can also usually be seen. Most of the Euglenophyta are freshwater species (Section 5.6), with only a few marine species reported – mainly belonging to the genera *Eutreptiella*.

6.5.6 Chlorophyceae (green algae – prasinophytes, chlorophytes)

The chlorophytes (green flagellate algae) and the prasinophytes (scaly green flagellate algae) are the two main groups of the Chlorophyceae represented in coastal waters. The prasinophytes are generally small flagellates that are covered in organic scales. From one up to 16 flagella (covered in minute scales and simple hairs) may be present and are used in many species to produce the characteristic stop and start swimming movement. The presence or absence and number of layers of scales covering the cell are used in the taxonomy of the group:

- scales absent (*Micromonas*)
- one layer of scales (*Mantoniella*)
- two or three layers (*Pyramimonas*)
- fused scales (*Tetraselmis*).

The chlorophytes represent a great variety of levels of organisation and include the macroalgae such as *Ulva* (sea lettuce), *Enteromorpha*, *Cladophora* and *Caulerpa*. Marine microalgae are mainly represented by the genera *Dunaliella* and *Chlamydomonas*. These phytoplankton are distinguished by their bright green appearance, flagella and naked cell wall.

6.6 References

Ajani P, Hallegraeff GM, Pritchard T (2001a) Historic overview of algal blooms in marine and estuarine waters of New South Wales, Australia. *Proceedings of the Linnean Society of New South Wales* **123**, 1–22.

Ajani P, Lee R, Pritchard T, Krogh M (2001b) Phytoplankton dynamics at a long-term coastal station off Sydney, Australia. *Journal of Coastal Research* **34**, 60–73.

Ajani P, Murray S, Hallegraeff G, Brett S, Armand L (2013a) First reports of *Pseudo-nitzschia micropora* and *P. hasleana* (Bacillariaceae) from the Southern Hemisphere: morphological, molecular and toxicological characterization. *Phycological Research* **61**(3), 237–248. doi:10.1111/pre.12020

Ajani P, Murray S, Hallegraeff G, Lundholm N, Gillings M, Brett S, *et al.* (2013b) The diatom genus *Pseudo-nitzschia* (Bacillariophyceae) in New South Wales, Australia: morphotaxonomy, molecular phylogeny, toxicity, and distribution. *Journal of Phycology* **49**(4), 765–785. doi:10.1111/jpy.12087

Ajani P, Kim JH, Han MS, Murray SA (2016a) The first report of the potentially harmful diatom *Pseudo-nitzschia caciantha* from Australian coastal waters. *Phycological Research* **64**(4), 312–317. doi:10.1111/pre.12142

Ajani P, Larsson ME, Rubio A, Brett S, Bush S, Farrell H (2016b) Modelling bloom formation of the toxic dinoflagellates *Dinophysis acuminata* and *Dinophysis caudata* in a highly modified drowned river valley, south eastern Australia. *Estuarine Coastal Shelf Science* **183**, 95–106.

Blackburn SI, Cresswell G (1993) A coccolithophorid bloom in Jervis Bay. *Australian Journal of Marine and Freshwater Research* **44**, 253–260. doi:10.1071/MF9930253

Golubic S, Abed RMM, Palinska K, Pauillac S, Chinain M, Laurent D (2010) Marine toxic cyanobacteria: diversity, environmental responses and hazards. *Toxicon* **56**(5), 836–841. doi:10.1016/j.toxicon.2009.07.023

Hallegraeff GM (1991) *Aquaculturists' Guide to Harmful Marine Microalgae.* Fishing Industry Training Board of Tasmania/CSIRO, Division of Fisheries, Hobart, Australia.

Hallegraeff GM (1994) Species of the diatom genus *Pseudo-nitzschia* in Australian Waters. *Botanica Marina* **37**, 397–411. doi:10.1515/botm.1994.37.5.397

Horner RA (2002) *A Taxonomic Guide to Some Common Marine Phytoplankton.* Biopress, Bristol, UK.

Kantiani L, Llorca M, Sanchis J, Farre M, Barcelo D (2010) Emerging food contaminants: a review. *Analytical and Bioanalytical Chemistry* **398**(6), 2413–2427. doi:10.1007/s00216-010-3944-9

Marshall JA, Hallegraeff GM (1999) Comparative ecophysiology of the harmful alga *Chattonella marina* (Raphidophyceae) from South Australian and Japanese waters. *Journal of Plankton Research* **21**, 1809–1822.

McCabe RM, Hickey BM, Kudela RM, Lefebvre KA, Adams NG, Bill BD *et al.* (2016) An unprecedented coastwide toxic algal bloom linked to anomalous ocean conditions. *Geophysical Research Letters.* **43**(19):10366–10376. doi:10.1002/2016GL070023

McLeod DJ, Hallegraeff GM, Hosie GW, Richardson AJ (2012) Climate-driven range expansion of the red-tide dinoflagellate *Noctiluca scintillans* into the Southern Ocean. *Journal of Plankton Research* **34**(4), 332–337. doi:10.1093/plankt/fbr112

Saunders KM, Hodgson DA, Harrison J, McMinn A (2008) Palaeoecological tools for improving the management of coastal ecosystems: a case study from Lake King (Gippsland Lakes) Australia. *Journal of Paleolimnology* **40**(1), 33–47. doi: 10.1007/s10933-007-9132-z

6.7 Further reading

Ajani P, Hallegraeff G, Allen D, Coughlan A, Richardson AJ, Armand L, *et al.* (2016) Establishing baselines: a review of eighty years of phytoplankton diversity and biomass in southeastern Australia. *Oceanography and Marine Biology - an Annual Review* **54**, 387–412.

Hallegraeff GM (2015) *Aquaculturist's Guide to Harmful Australian Microalgae.* 3rd edn. School of Plant Science, University of Tasmania, Hobart, Australia.

Hallegraeff GM (2010) Ocean climate change, phytoplankton community responses, and harmful algal blooms: a formidable predictive challenge. *Journal of Phycology* **46**(2), 220–235. doi:10.1111/j.1529-8817.2010.00815.x

Hallegraeff GM, Bolch CJS, Hill DRA, Jameson I, LeRoi JM, McMinn A, *et al.* (2010) *Algae of Australia: Phytoplankton of Temperate Coastal Waters.* CSIRO Publishing, Melbourne, Australia.

Hoppenrath M, Murray SA, Chomérat N, Horiguchi T (2014) *Marine Benthic Dinoflagellates – Unveiling Their Worldwide Biodiversity.* Kleine Senckenberg-Reihe, Band 54, Schweizerbart, Stuttgart, Germany.

Jeffrey SW, Hallegraeff GM (1990). Phytoplankton ecology of Australasian waters. In *Biology of Marine Plants.* (Eds MN Clayton and RJ King) pp. 310–348. Longman Cheshire, Melbourne, Australia.

Tomas CR (Ed.) (1997) *Identifying Marine Diatoms and Dinoflagellates.* Academic Press, London, UK.

UNESCO (1995) *Manual on Harmful Marine Microalgae.* (Eds GM Hallegraeff, DM Anderson, AD Cembella and HO Enevoldsen). UNESCO, Paris, France.

van den Hoek C, Mann DG, Jahns HM (1995) *Algae: An Introduction to Phycology.* Cambridge Scientific Press, London, UK.

7

Freshwater zooplankton: diversity and biology

Tsuyoshi Kobayashi, Ian A.E. Bayly, Russell J. Shiel and Anthony G. Miskiewicz

7.1 Identifying freshwater zooplankton

The important groups of freshwater zooplankton are rotifers, cladocerans, copepods, protozoans and larval fish. Occasionally other essentially non-planktonic groups of animals appear in the plankton, particularly in areas inhabited by the abundance of rooted water plants. These animals are collectively termed either 'accidental plankton' or tychoplankton or neuston, and may include water mites, worms, water beetles, dragonfly larvae, ostracods and even snails (Fig. 7.1) (see also Gooderham and Tsyrlin 2009). This tychoplankton appears in the water column as plankton only when they swim among their habitats such as the surface of submerged logs or water plants and are suspended due to disturbance. For example, towing a plankton net among rooted water plants can dislodge the nearby animals, transporting them into the water column.

Zooplankton are found in freshwater habitats of almost all sizes, ranging from phytotelmata (small bodies of water held by terrestrial plants such as pitcher plants), to small pools such as panholes or gnammas, ponds, dams and reservoirs, to the largest and deepest lakes on Earth. Zooplankton also occur in extreme environments: calanoid copepods (Bayly *et al.* 2003), cyclopoid copepods (Karanovic *et al.* 2014) and rotifers (Hansson *et al.* 2012) are all found in Antarctica lakes. Cladocerans and copepods occur near Mount Everest (Manca *et al.* 1994), and even in groundwaters (Brancelj and Dumont 2007). Salt lakes present another form of extreme habitat confronting freshwater organisms and requiring complex physiological solutions (Bayly and Boxshall 2009). In Australia and South America, many inland waters are saline, rather than fresh, and contain dense populations of zooplankton. In Australia, Lake Eyre and Lake Corangamite, and in South America Lake Poopo (Bolivia) and Mar Chiquita (Argentina), are prime examples. In the last two lakes, brine shrimps and calanoid copepods are so abundant that they constitute a major food for the Chilean Flamingo (Bayly 1993).

Rotifers are distinctive little animals (many <1 mm long), with most species occurring only in fresh water. Planktonic cladocerans and copepods are tiny crustaceans, and are mostly 0.5–5.0 mm long. Protozoans are single-celled organisms and most are smaller than the other three groups. Larval fish in freshwater systems are typically 2–20 mm in total length, and can therefore be seen with the naked eye. Many zooplankters are fairly transparent, but in clear-water alpine lakes and shallow transparent pools they may be bright red or blackish due to photo-protective pigments (Persaud *et al.* 2007).

Self-sustaining zooplankton communities are usually thought of as largely restricted to standing-water bodies such as lakes and ponds (lentic environments). In fast-flowing streams and rivers (lotic environments), zooplankton density is often low. However, when rivers cease to flow or flow slowly they resemble lentic systems, and pseudo-lentic zooplankton communities may become established, if only for a limited time (Shiel *et al.* 1982). In fresh water, various prime habitats support different species (Table 7.1). Limnetic species are found in open water (such as in the centre of a lake or pond) and are fully adapted to planktonic life.

Littoral species are those occurring among water plants near the shore or bank. Littoral species are thus not truly planktonic, but constitute an important part of the aquatic biota. The littoral fauna may be more species rich than those in the limnetic zone.

7.2 Rotifers

Rotifers are found primarily in fresh or inland saline water habitats. They are particularly diverse in the littoral zone of stagnant water bodies with soft, slightly acidic water and under oligo- to

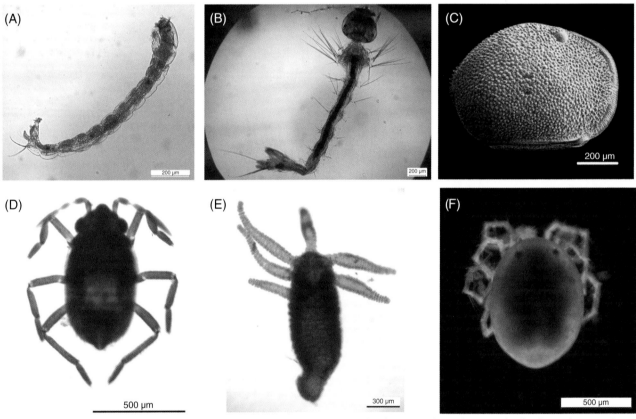

Fig. 7.1. Tychoplankton (accidental plankton): **(A)** Chironominae (midge larvae) – lateral view. The body is elongated, cylindrical with a pair of short legs just behind the head and on the terminal segment; **(B)** Culicidae (mosquito larvae) – dorsal view. The body is elongated and covered with bristles. It has a large head and thick thoracic segments; **(C)** *Newhamia fenestrata* (ostracods) – lateral view. The ventral region consists of a flattened, grid-like plate that helps stick to the water surface face down. The eye lenses are located on the top of each valve and used to spot predators as the animal swims upside down; **(D)** Veliidae (small water striders) – dorsal view. The body has relatively short legs that are used for running and walking; **(E)** Hydridae (hydras) – lateral view. The body is elongated, cylindrical with tentacles encircling the mouth. These are essentially sessile animals but also capable of moving slowly on the substrate. They are carnivorous; **(F)** Acarina (water mites) – dorsal view. The body is rounded with eight legs. Two pigmented eyes are located near the anterior region of the body. They are carnivorous.

Table 7.1.   Typical freshwater zooplankton in Australia and elsewhere

Limnetic taxa are those occurring in open water (such as the centre of a lake or pond). Littoral taxa are those occurring among water plants near the shore or bank. The taxa marked by an asterisk * are illustrated in this chapter.

Limnetic copepods	Limnetic cladocerans	Limnetic rotifers

Calanoids
  *Calamoecia**
  *Boeckella**
  Eodiaptomus
  *Gladioferens**
Cyclopoids
  Acanthocyclops
  Australocyclops
  Eucyclops
  *Mesocyclops**
  Metacyclops
  Thermocyclops
  Tropocyclops
	*Bosmina**
	*Ceriodaphnia**
	*Daphnia**
	*Diaphanosoma**
	Moina
	*Chydorus**
		*Asplanchna**
		*Brachionus**
		Conochilus
		*Filinia**
		Hexarthra
		*Keratella**
		Polyarthra
		Synchaeta
		*Trichocerca**

Littoral copepods	**Littoral cladocerans**	**Littoral rotifers**

Calanoids
  *Gladioferens**
  Tropodiaptomus
Cyclopoids
  Diacyclops
  Ectocyclops
  Eucyclops
  Macrocyclops
  *Mesocyclops**
  Paracyclops
	*Acroperus**
	Alona
	Camptocercus
	*Chydorus**
	Ilyocryptus
	*Macrothrix**
	Neothrix
	Scapholeberis
	Simocephalus
		Euchlanis
		Lecane
		Lepadella
		Notommata
		Bdelloids

mesotrophic conditions (low to moderate nutrient conditions; Segers 2008). Most rotifers are 0.1–0.5 mm long. Their body shape varies widely between groups: they can be spherical, cylindrical or elongated. The body can be soft or may have a firm covering called a lorica. Some rotifers are enclosed in a gelatinous case. Many have different types of spines and a foot, and some even have 'toes'. The structure of the jaw (or trophi) is distinctive for each species and is used for identification. To observe the jaws as a wet mount under a compound microscope (Fig. 4.7B), it is necessary to dissolve body tissues with a chemical such as bleach. The cilia surrounding the mouth of a rotifer form a circle, called a corona or wheel organ. Rapid movements of the cilia create water currents for swimming and feeding. A key to the orders and families

of freshwater rotifers is shown in Key 7.1 (see also Fig. 7.2, and Table 7.2).

Rotifer populations consist only of females under normal environmental conditions. They produce eggs that hatch into females without the need for male fertilisation: a process known as parthenogenesis. Eggs are relatively large compared with the body size of females, and are normally attached to the posterior part of their bodies before being released into the water. It may take less than a week for juveniles of many rotifers to become mature.

However, under certain conditions, females produce eggs that hatch into males. Fertilised female rotifers then produce special resting eggs. Resting eggs can withstand extreme temperatures, drought and other adverse conditions. Eggs can remain viable long after the female rotifers that produced them have died. Resting eggs remain dormant – buried in the sediments for many years. New populations of female rotifers can establish from resting eggs when environmental conditions become favourable.

Rotifers eat bacteria, including cyanobacteria, and phytoplankton. Some are carnivorous and eat other rotifers. Rotifers may be abundant in both standing and running waters. A maximum of 3500 rotifers have been recorded from 1 L of water in an Australian river (Kobayashi *et al.* 1998). It is common to find more than 20 000 rotifers per litre in some reservoirs and billabongs.

7.3 Cladocerans

Cladocerans are primarily a freshwater group of invertebrates, with more than 600 species currently known. A high diversity of cladocerans may be found in the littoral zone of stagnant waters, as well as in temporary water bodies (Forró *et al.* 2008). Most cladocerans are less than 1–2 mm long, but there are some notable exceptions: specimens 5–6 mm in length have been found in some water bodies. Females are usually larger than males. The body consists of a rigid, clam-like shell – called a carapace – which is transparent, but can be yellowish or brownish in colour. Pairs of appendages called thoracic limbs are inside the carapace and are important for collecting and transferring food particles to the mouth. The head of a cladoceran is usually compact, with prominent eyes and large antennae used for swimming. Some cladocerans develop conspicuous head and tail spines, helmet or 'neck-teeth' (Fig. 7.3). A key to the families of freshwater cladocerans is shown in Key 7.2 (see also Fig. 7.4). Cladoceran taxonomy is constantly being reviewed, including the additions of new families (e.g. Santos-Flores and Dodson 2003; Sinev 2015). See Section 8.3.4 for information on marine cladocerans.

Like rotifers, there are female-only populations of cladocerans under normal environmental conditions. They produce female eggs inside a chamber on the dorsal side of the body within which the eggs hatch. Newly hatched young – which look like

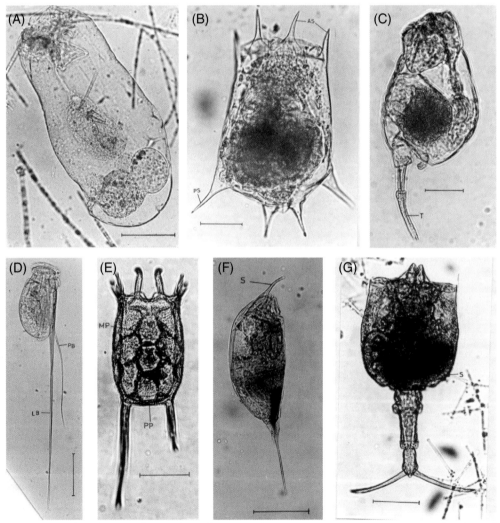

Fig. 7.2. Rotifers: **(A)** *Asplanchna priodonta* – foot absent. Body transparent. Specimen preserved in formalin often strongly contracts. Jaw (trophus) needs to be examined for identification of species. Scale bar 100 μm; **(B)** *Brachionus calyciflorus amphiceros* – four anterior spines (AS) on dorsal side of lorica. Long posterior spines (PS). Scale bar 50 μm; **(C)** *Cephalodella gibba* – body fusiform, with slender toes (T); **(D)** *Filinia longiseta* – body shape oval. Body with two long lateral bristles (LB) and one short posterior bristle (PB). Scale bar 100 μm; **(E)** *Keratella tropica* – three six-sided median plaques (MP) on dorsal side of lorica. Single small four-sided posterior plaque (PP). Scale bar 50 μm; **(F)** *Trichocera chattoni* – body cylindrical, more or less squat. Single long curved spine (S) at margin of head opening. Scale bar 100 μm; **(G)** *Trichotria truncata* – head, body and foot segments distinctive and rigid. Lorica margin with small spines (S). Scale bar 50 μm.

Key 7.1 Key to orders of freshwater rotifers (modified from Shiel 1995) (Fig. 7.2)

Phylum Class	Rotifera Monogononta/Bdelloidea	
1a Body with a single ovary; body often with a lorica or tube		**Class Monogononta 2**
1b Body with paired ovary; body without a lorica or tube Orders Adinetida, Philodinida Philodinavida (fresh to brackish) and others		**Class Bdelloidea**
2a Mastax malleoramate Family Conochilidae: *Conochilopsis* and *Conochilus* Family Floscularidae: *Floscularia, Lacinularia, Sinantherina* and others Family Testudinellidae: *Pompholyx, Testudinella* and others Family Trochosphaeridae: *Filinia* (Fig. 7.2D) and others		**Order Flosculariacea**
2b Mastax not malleoramate		⟶ 3
3a Mastax uncinate Family Collothecidae: *Collotheca* and others		**Order Collothecacea**
3b Mastax not uncinate Family Asplanchnidae: *Asplanchna* (Fig. 7.2A) and others Family Brachionidae: *Anuraeopsis, Brachionus* (Fig. 7.2B), *Keratella* (Fig. 7.2E), *Notholca, Platyias* and others Family Gastropodidae: *Ascomorpha* and *Gastropus* Family Lecanidae: *Lecane* Family Lepadellidae: *Colurella, Lepadella* and *Squatinella* Family Mytilinidae: *Mytilina* Family Notommatidae: *Cephalodella* (Fig. 7.2C), *Monommata* and others Family Synchaetidae: *Polyarthra, Synchaeta* and others Family Trichocercidae: *Ascomorphella, Elosa* and *Trichocerca* (Fig. 7.2F) Family Trichotriidae: *Trichotria* (Fig. 7.2G) and others		**Order Ploima**

Table 7.2. **A summary of specialised terms used in this chapter**

Structures	Definition
Cilia	Fine hair-like structures; ciliates are a group of protozoa having cilia in rows
Corona	The circle of cilia surrounding a rotifer's mouth, also called a wheel organ
Cyst	A capsule-like covering that encloses a small organism such as a protozoan or rotifer in a dormant state
Ephippium (plural ephippia)	A saddle-shaped formation enclosing resting eggs in daphniids and other members of Cladocera; becomes detached from body on death of parent
Flexion	Stage of larval fish development when the notochord (precursor to the backbone) bends slightly upwards (dorsally) to form the tail fin
Gas bladder	Gas-filled sac located above the gut which helps in regulating buoyancy
Incudate	A type of rotifer mastax (or trophi) with a seizing, pincer-like shape, characteristic of carnivorous animals
Lorica	A firm shell covering the body of a rotifer or protozoan and some algae
Malleoramate	A type of rotifer mastax (or trophi) with many teeth attached to the base structure
Mastax	Mouthparts of rotifers (*see also* malleoramate, uncinate)
Melanophores	Nucleated cells containing the brown and back pigment melanin; melanophores can expand and contract thus changing in size and shape, and remain in larval fishes after fixation
Myomeres	Muscle bands aligned sequentially and in transverse series along the trunk and tail; total number of myomeres is approximately equal to the number of vertebrae
Resting eggs	Fertilised eggs with thick shell in rotifers and cladocerans
Swim bladder	*See* Gas bladder
Uncinate	A type of rotifer mastax (or trophi) with three to five pairs of tearing-type teeth attached to the base structure, found only in the Family Collothecidae

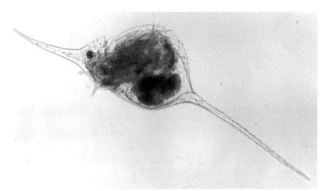

Fig. 7.3. A species such as *Daphnia lumholtzi* can produce conspicuously long head and tail spines, resulting in the extension of an overall body length. Long head and tail spines can make it more difficult for fish to eat *Daphnia*, thus reducing the level of predation by fish.

small adults – remain there until they are ready to swim.

When environmental conditions deteriorate (through a lack of food or drying of the water body), the females produce eggs that hatch into males. Fertilised females then produce one or two special resting eggs encased in a thick protective covering to form an ephippium, which is released into the water (Fig. 7.5). Ephippia can withstand a wide range of environmental conditions, surviving for many years in dry sediments. Cladocerans can establish new populations from ephippia when environmental conditions once again become favourable.

Cladocerans moult several times as they grow into adults. A new carapace is formed inside the old, which is then discarded as the body grows bigger. The discarded carapaces made of chitin are called exuviae (singular: exuvium). Collections of plankton samples may contain exuviae as well as live animals. Exuviae can also be used to identify species that have occupied a habitat in the past. Those preserved in sediments can also be used to identify past cladocerans in habitats up to 10 000 years ago. The science of studying such remains is called palaeolimnology, and is helpful in understanding past environmental conditions and climate change.

Cladocerans normally occur from spring to early summer, reaching densities of 10–30 animals per litre in ponds, lakes and reservoirs. In a special

case, a high density of 500 cladocerans per litre has been reported from a waste stabilisation pond (Mitchell and Williams 1982). Cladocerans, especially large *Daphnia*, eat a wide variety of phytoplankton and other suspended matter, such as decayed plant material and clay particles. They may greatly reduce phytoplankton abundance. There are globally several genera of carnivorous cladocerans and all cladocerans are important prey for small fish in lakes and ponds (Box 7.1).

7.4 Copepods

The greatest diversity of copepods is found in the marine environment. About 2800 species of copepods occur in fresh or inland water habitats (~20%, Section 8.3.1). Most freshwater copepods are planktonic, but some (~330 species) are parasitic, with fish and molluscs as hosts (Boxshall and Defaye 2008). Freshwater planktonic copepods comprise two major groups: calanoids and cyclopoids. The calanoid copepods have an elongated body and long first antennae (Fig. 7.6A,B), while the cyclopoid copepods have a stout body and short first antennae (Fig. 7.6C,D). A third group, the harpacticoids, have cylindrical bodies and very short first antennae. Freshwater harpacticoids are almost entirely benthic, living in or on the bottom mud or sand but occasionally appear in plankton (see also Fig. 8.10). A key to the orders of freshwater copepods is shown in Key 7.3 (see also Fig. 7.7).

The bodies of calanoids are often 1–2 mm long and cyclopoids and harpacticoids are usually less than 1 mm long. The body of a copepod is clearly segmented and females are larger than males. Females and males are also distinguished by the shape of the first antennae that are attached near the anterior end of the body and by other features (see Key 7.3 for details, see Fig. 8.10).

Copepods have pairs of different appendages on the ventral side of the body. For calanoid copepods, the appendages under the head are used for creating water currents to collect, filter and/or capture food particles. The appendages along the mid

Box 7.1 Range extension of freshwater zooplankton

Many species of freshwater zooplankton naturally occur within a certain range of geographical areas. However, we find species in areas that are very remote from the natural range, due to various transport and dispersal mechanisms. They become new records in those areas and may subsequently establish their populations (i.e. they are invasive species). In recent decades, certain zooplankton have been introduced to North America's Great Lakes via ballast water (see Box 3.2) from northern Europe and Asia, including the carnivorous cladoceran *Bythotrephes*, the spiny water flea. It feeds on other zooplankton including *Daphnia*, and has far-reaching impacts on the Great Lakes planktonic ecosystem, including the prey of larval fish.

Key 7.2 Key to families of freshwater cladocerans (modified from Smirnov and Timms 1983) (Fig. 7.4)

Phylum	Arthropoda	
Subphylum	Crustacea	
Class	Branchiopoda	
Order	Diplostraca	
Suborder	Cladocera	
1a Body and swimming legs not covered with a carapace		⟶ 2
1b Body and swimming legs covered with a carapace		⟶ 3
2a Body short with four pairs of swimming legs Family Polyphemidae: *Polyphemus* 2b Body long with six pairs of swimming legs Family Leptodoridae: *Leptodora*		
3a Six pairs of swimming legs inside the carapace all similar		⟶ 4
3b Five or six pairs of swimming legs inside the carapace not similar		⟶ 5
4 Body length much greater than body height; second antennae with large branch-like appendages Family Sididae: *Diaphanosoma* (Fig. 7.4E) and others		
5a First antennae long and slender, like an elephant's trunk Family Bosminidae: *Bosmina* (Fig. 7.4A) and *Bosminopsis*		
5b First antennae usually short		⟶ 6
6a Second antennae two-branched, both with three segments; mostly small body length, hemispherical or circular in lateral view Family Chydoridae: *Acroperus* (Fig. 7.4B), *Alona*, *Chydorus* (Fig. 7.4D), *Graptoleberis*, *Pleuroxus* and others		
6b Second antennae two-branched, one with three segments and the other with four segments		⟶ 7
7a First antennae not flexible and short Family Daphniidae: *Ceriodaphnia* (Fig. 7.4C), *Daphnia* (Figs 7.3 and 7.5), *Simocephalus* and others		
7b First antennae flexible and long relative to body length		⟶ 8
8a First antennae on mid-abdominal side of head; oval body Family Moinidae: *Moina* and *Moinodaphnia*		
8b First antennae on frontal side of head		⟶ 9
9a Postabdomen with distal, terminal claw Family Macrotrichidae: *Macrothrix* (Fig. 7.4F) and others 9b Postabdomen lacks terminal claw Family Neotrichidae: *Neothrix*		

to lower body are used for swimming. Cyclopoid copepods use their mouth parts for capturing animal prey – most species are carnivorous. The legs along the mid to posterior body of copepods are mainly used for swimming. Calanoids and cyclopoids have five pairs of swimming legs and harpacticoids have five or six pairs. The detailed structure of fifth legs in the male is useful in

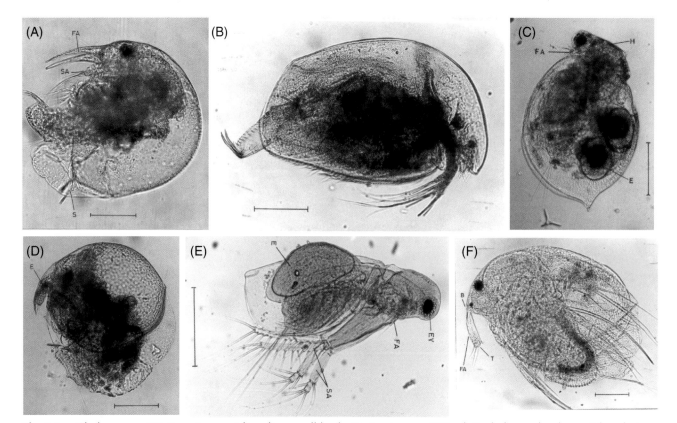

Fig. 7.4. Cladocerans: **(A)** *Bosmina meridionalis* – small body. First antennae (FA) relatively long, slender, not fused at their bases. Second antennae (SA) relatively small. Often a pair of spine-like elongation (S) at ventro-posterior corner of body. Scale bar 100 µm; **(B)** *Acroperus* sp. – body flattened laterally. Bottom dwelling. Normally found among water plants. Scale bar 100 µm; **(C)** *Ceriodaphnia* sp. – body shape broadly oval. Head (H) small. Short first antennae (FA). Normal eggs (E). Scale bar 200 µm; **(D)** *Chydorus* sp. – body small, spherical. Small eyes (E). Bottom dwelling, but also appears in plankton. Scale bar 100 µm; **(E)** *Diaphanosoma excisum* – body without a tail spine. Head relatively large, rectangular with large eye (EY). First antennae small (FA) small. Second antennae (SA) large and well developed. Large normal egg (E). Scale bar 300 µm; **(F)** *Macrothrix spinosa* – body flattened laterally, without tail spine. First antennae (FA) situated frontal side of head. Tip of first antennae (T) wider than its base (B). Bottom dwelling. Normally found among water plants. Scale bar 100 µm.

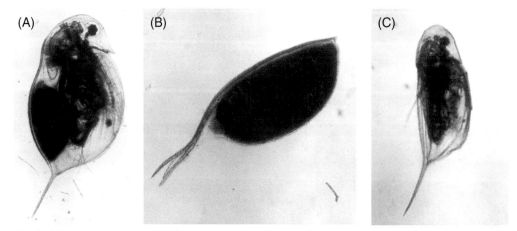

Fig. 7.5. *Daphnia*'s resting eggs in an ephippium can survive in adverse environmental conditions even after the females that produced the ephippium die. **(A)** An ephippium is formed on the dorsal side of a female; **(B)** the ephippium usually detaches after the female dies; **(C)** young *Daphnia* will hatch from the resting eggs when the environmental conditions become favourable again.

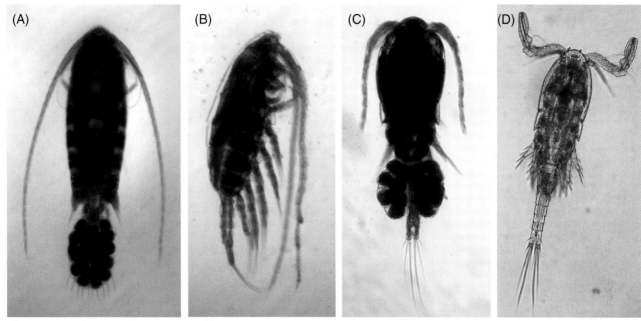

Fig. 7.6.   Freshwater calanoid and cyclopoid copepods: **(A)** calanoid (egg-carrying female, dorsal view); **(B)** calanoid (male, side view); **(C)** cyclopoid (egg-carrying female, dorsal view); **(D)** cyclopoid (male, dorsal view).

identifying calanoid species. Fourth and fifth legs in the female are important in identifying cyclopoid species. All swimming legs are important in identifying harpacticoid species.

Copepods moult up to 11 times before becoming adults, with body shape and size changing after each moult (see Fig. 2.5). There are two distinct young stages: nauplius larvae and copepodites.

Key 7.3 Key to orders of freshwater copepods (Fig. 7.6)

Phylum	Arthropoda
Subphylum	Crustacea
Class	Copepoda

1a First antennae long, slender body   **Order Calanoida** *Calamoecia* (Figs 7.6A,B), *Boeckella* (Fig. 7.6C), *Diaptomus*, *Eodiaptomus*, *Gladioferens* (Fig. 7.6D), *Pseudodiaptomus* and others Key to sexes Right and left first antennae similar in shape = female Right and left first antennae dissimilar; right antenna geniculate (with an elbow-knee-like hinge) = male
1b First antennae short; head often much wider than lower body when seen from above   **Order Cyclopoida** *Australocyclops*, *Diacyclops*, *Macrocyclops*, *Mesocyclops* (Fig. 7.6E) *Thermocyclops* and others Key to sexes Right and left first antennae similar in shape = female Right and left first antennae similar in shape but geniculate and often strongly curved = male
1c First antennae short; cylindrical body   **Order Harpacticoida** *Canthocamptus*, *Fibulacamptus*, *Parastenocaris* (Fig. 7.6F) and others Key to sexes Right and left first antennae similar in shape = female Right and left first antennae similar in shape, but geniculate = male

A copepodite has fewer body segments and appendages, but looks like a small adult.

Female copepods produce eggs that always need to be fertilised by males. Females carry the eggs in one or two sacs attached to the ventral side of the body. The egg sacs and eggs are easily observed under a microscope. Some copepods produce resting eggs that withstand drought and other adverse environmental conditions. The resting eggs of certain calanoid copepods can live in lake sediments for centuries (Hairston *et al.* 1995), as can many other eggs and resting stages of small crustaceans.

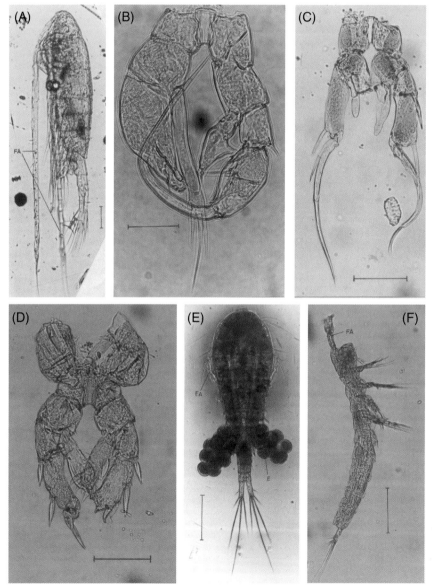

Fig. 7.7.   Copepods: **(A)** *Calamoecia ampulla* – body elongated, with long first antennae (FA). Small calanoid copepod. Male fifth legs need to be examined for identification of species. Scale bar 100 μm; **(B)** *Calamoecia ampulla* – Male fifth legs (posterior aspect). Scale bar 50 μm; **(C)** *Boeckella fluvialis* – male fifth legs (posterior aspect). Body elongated, with long first antennae. Relatively large calanoid copepod. Scale bar 100 μm; **(D)** *Gladioferens pectinatus* – male fifth legs (anterior aspect). Body elongated, with long first antennae. Relatively large calanoid copepod in fresh and salt water. Scale bar 100 μm; **(E)** *Mesocyclops* sp. – body relatively stout, with short first antennae (FA). Scale bar 200 μm; **(F)** *Parastenocaris* sp. – body cylindrical, with very short first antennae (FA). Bottom dwelling, but may also appear in plankton.

Copepods may occur in the plankton all year round, usually reaching densities of 5–20 animals per litre in ponds, lakes, reservoirs and slow-flowing rivers.

Calanoids eat a wide variety of phytoplankton species and other suspended matter such as decayed plant material and clay particles. Some eat other small zooplankton, such as rotifers and ciliated protozoans. Cyclopoids are primarily carnivorous – eating other zooplankton.

For the Australasian copepods, a recent comprehensive key to the *Boeckella* species is provided by Quinlan and Bayly (2017). Morton (1985, 1990) provides keys to the species of *Acanthocyclops*, *Diacyclops*, *Australocyclops*, *Eucyclops* and *Ectocyclops*.

7.5 Protozoans

Protozoans are mainly microscopic (much less than 1 mm long) and need to be observed with a wet mount (see Fig. 4.7B). They have various body shapes (spherical, oval or elongate) and often have one or more long, fine, whip-like appendages, called flagella, or many short hair-like structures, called cilia. Some produce temporary foot-like protrusions called pseudopodia. These body parts are important for locomotion and feeding. A key to the phyla of protozoans is shown in Key 7.4 (see also Fig. 7.8).

Protozoans are efficient at eating bacteria, including cyanobacteria, and small phytoplankton. Some are carnivorous and eat other zooplankton (e.g. the ciliate *Bursaria* may include rotifers in their diet). Protozoans grow quickly and increase in numbers by means of cell duplication. They are abundant in many types of water bodies, from fish tanks and sewage ponds to lakes and reservoirs. In running waters, such as streams and rivers, protozoans found in the plankton are often those that have been swept from the surfaces of submerged rocks, water plants or sediments. Many freshwater protozoans can live in extremely low oxygen environments and may persist locally for long periods as cysts (Finlay and Esteban 1998).

7.6 Larval fish

Larval fish (or ichthyoplankton – which includes pelagic fish eggs as well as larval fish) are a common, seasonal and potentially diverse component of the zooplankton of most freshwater habitats. Compared with estuarine and marine fish species, only a limited number of identification guides is available for freshwater larvae (e.g. Snyder 1983; Moser *et al.* 1984; Neira *et al.* 1998; Serafini and Humphries 2004, Snyder *et al.* 2004). Identification of eggs and fish larvae to species is difficult, even for experienced workers, as species descriptions are based on adult characteristics, which are not apparent in early developmental stages. In addition, the eggs and larval development of many species are undescribed. A new technique used for identification of adult fish and now being applied

Key 7.4 Key to phyla of protozoans (modified from Jahn *et al.* 1979) (Fig. 7.8)

1a Body with cilia or tentacles	Phylum Ciliophora (often called ciliates)
	Epistylis (Fig. 7.8C), *Frontonia*, *Paramecium*, *Paradileptus* (Fig. 7.8E), *Vorticella* and others
1b Body without cilia or tentacles	⟶ 2
2a Body with other structures for locomotion	⟶ 3
2b Body without obvious structures for locomotion	⟶ 4
3a Body with one or more flagella	Phylum Mastigophora (often called flagellates)
	Ceratium, *Euglena*, *Peridinium* and others
3b Body with pseudopodia	Phylum Sarcodina (often called amoebae)
	Arcella (Fig. 7.8A), *Cyphoderia* (Fig. 7.8B), *Euglypha* (Fig. 7.8D), *Difflugia*, amoebae without a rigid test (Fig. 7.8F) and others
4 Movement by body flexions; all parasitic	Phylum Sporozoa
	Plasmodium (the causative organism of malaria) and others

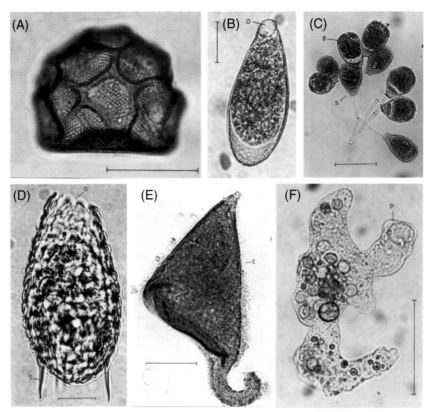

Fig. 7.8. Protozoans: **(A)** *Arcella mitrata* – body with test. Test circular from above, dome-like on top. Small central opening. Scale bar 50 μm; **(B)** *Cyphoderia* sp. – body with test. Test oval, short cylindrical neck. Round opening (O) oblique to body of test. Test with a yellow-brown matrix. Scale bar 30 μm; **(C)** *Epistylis* sp. – Bell-shaped body (B), with a stalk (S). Stalk splits into two branches and cannot contract. Scale bar 100 μm; **(D)** *Euglypha* sp. – body with oval test, made of scales of equal sizes. Opening (O) terminal. Some with spines (S) on test. Scale bar 20 μm; **(E)** *Paradileptus* sp. – body with cilia. Relatively large protozoans. Scale bar 50 μm; **(F)** Amoeba (unidentified) – body with no test and no cilia. Note pseudopodia (P). Scale bar 100 μm.

to eggs and larval fish is DNA barcoding, using the cytochrome c oxidase gene subunit I (COI or CO1) (Becker *et al.* 2015; Frantine-Silva *et al.* 2015; Almeida *et al.* 2018).

As for estuarine and marine fish (Section 8.3.10.2), the most common method for identifying larvae is the series method (or using existing keys and descriptions where they are available). The series method involves identifying the largest available larval or juvenile specimen, based on adult characteristics such as fin meristics and vertebral number (equivalent to the number of myomeres or muscle blocks – in larvae). The largest specimen is linked to smaller specimens in the series by using morphological and pigment characteristics. A variety of characters can be used to

identify fish larvae including their general morphology such as the body shape and gut length and degree of coiling, the number of myomeres, pigmentation patterns (melanophores), the sequence of development of fins and the pattern of head spination (Table 7.3, Fig. 7.9). The length and stage of development are important features in identification of larvae. For example, the stage of flexion is when the notochord begins to grow upwards (dorsally) and the bony structures of the tail begin to form on the ventral surface. Compared with the larvae of estuarine and marine fish, larvae of many freshwater species have a large yolk sac and morphological changes such as notochord flexion and development of the fin elements occurs at a larger size.

Most freshwater fishes have seasonal reproduction, with peaks in reproduction, and therefore larval abundance, generally occurring in spring and summer (Wootton 1998; King *et al.* 2013). Some species spawn over a relatively long-time period (months), while for others the spawning period is brief (a few days) (Matthews 1998; Pusey *et al.* 2004). Therefore, the potential species composition of the ichthyoplankton is likely to change considerably from one sampling time to the next.

7.6.1 Life history strategies

Larval fish can be found in rivers, creeks, lakes, reservoirs, off-channel habitats such as billabongs (ox bow lakes), wetlands and even in temporarily inundated habitats such as floodplains and ephemeral creeks. Larvae use a variety of habitat patches in freshwater systems, such as open water (pelagic) habitats, complex submerged macrophytes and woody debris, interstitial spaces of gravels, littoral habitats and backwaters. Some species also have fairly specific requirements at certain developmental stages; for example, larval carp have a downstream drifting dispersal phase, while other species require parental care in protected nest areas, such as in hollow logs.

Freshwater fish species use a large diversity of habits, with a variety of selective pressures. They have been grouped into three categories related to their spawning style, association or lack of association with flooding and significance for recruitment (King *et al.* 2013). This model was originally proposed for North American freshwater fish and applies broadly to North American (Winemiller and Rose 1992), European (Blanck *et al.* 2007) and south-east Australian (Humphries *et al.* 1999; King *et al.* 2013) freshwater fish communities. The three categories are:

- **opportunistic species** – small short-lived species with early maturation, low fecundity,

Table 7.3.   Freshwater fish larval characteristics (modified from Neira *et al.* 1998 and Serafini and Humphries 2004)

Family	Features
Retropinnidae (smelt)	45–53 myomeres, body very elongate, lightly pigmented, gut very long and straight, demersal eggs (Fig. 7.9A)
Galaxiidae (whitebait)	36–64 myomeres, body very elongate, lightly to heavily pigmented, gut long to very long and straight, demersal eggs (Fig. 7.9B)
Atherinidae (hardyhead)	34–36 myomeres, body very elongate, moderately pigmented, gut coiled and compact, demersal eggs (Fig. 7.9C)
Melanotaenidae (rainbowfish)	34 myomeres, body elongate, moderately pigmented, gut short and coiled, demersal eggs (Fig. 7.9D)
Poeciliidae (mosquitofish)	31–33 myomeres, body moderate, moderately pigmented, gut short and coiled, live bearer (Fig. 7.9E)
Cyprinidae (carp)	36–40 myomeres, body elongate, moderate to heavily pigmented, gut long and straight (Fig. 7.9F)
Percichthyidae (cods/pigmy perch)	27–36 myomeres, body elongate to moderate, moderate to heavily pigmented, gut moderate to long and loosely coiled, large yolk sac in some genera, weak pre-opercular spines, demersal eggs (Fig. 7.9G,H)
Terapontidae (silver perch/grunter)	25 myomeres, body elongate, lightly pigmented, gut coiled and moderate, small pre-opercular spines (Fig. 7.9I)
Eleotridae (gudgeons)	28–34 myomeres, body elongate, lightly pigmented, gut moderate and slightly coiled, conspicuous gas bladder, demersal eggs (Fig. 7.9J,K)
Percidae (redfin)	39–41 myomeres, body elongate, lightly pigmented, gut moderate and loosely coiled, conspicuous gas bladder demersal eggs (Fig. 7.9L)
Plotosidae/Ariidae (catfish)	>77 myomeres, body elongate, moderately to heavily pigmented, gut moderate and loosely coiled, mouth barbells, large yolk sac, demersal eggs (Fig. 7.9M)

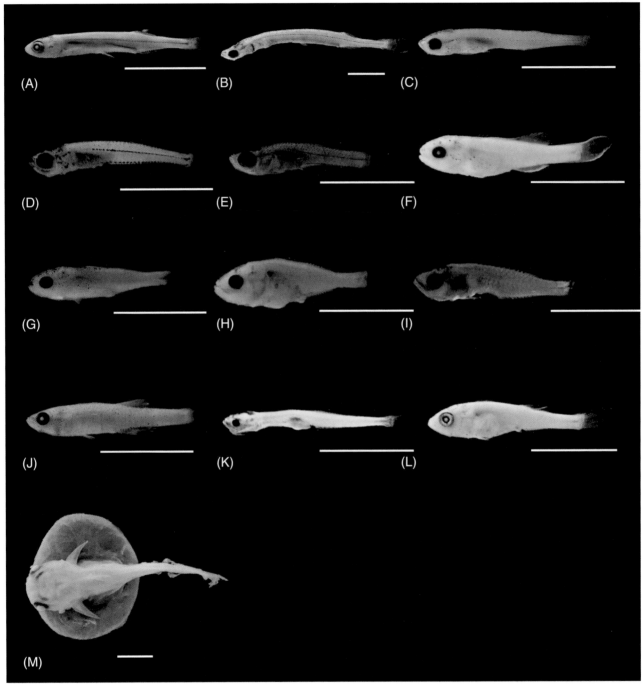

Fig. 7.9. The post flexion stages of some dominant families of fish typically occurring in freshwater plankton samples: **(A)** Retropinnidae – *Retropinna semoni* (smelt); **(B)** Galaxiidae – *Galaxias occidentalis* (whitebait), **(C)** Atherinidae – *Craterocephalus honariae* (hardyhead); **(D)** Melanotaeniidae – *Melanotaenia splendida australis* (rainbowfish); **(E)** Poeciliidae – *Gambusia holbrooki* (mosquitofish); **(F)** Cyprinidae – *Cyprinus carpio* (carp), **(G)** Percichtyidae – *Macquaria ambigua* (yellowbelly); **(H)** Percichtyidae – *Macquaria ikei* (eastern cod); **(I)** Terapontidae – *Amniataba caudavittata* (trumpeter); **(J)** Eleotridae – *Hypseleotris galii* (firetail gudgeon); **(K)** Eleotridae – *Philypnodon grandiceps* (flathead gudgeon); **(L)** Percidae – *Perca fluviatalis* (redfin); **(M)** Ariidae – A*rius graeffei* (catfish). Note the 5 mm scale bar beside each image.

fast growth and small energy investment per offspring

- **periodic species** – relatively large long-lived species, late maturity, high fecundity per batch and small energetic investment per offspring
- **equilibrium species** – medium to large species, late maturing, low fecundity per batch and large energetic input per offspring.

The early life of fishes – from embryo to larvae to juveniles – is marked by rapid changes in morphology, ecology, growth and behaviour (Fuiman and Higgs 1997, Trippel and Chambers 1997). These changes often result in dramatic changes in habitat and diet use within a species. For example, some riverine fishes are exclusively found in shallow, still, off-channel habitats as newly hatched larvae, but then move to a variety of mid-channel habitats as older larvae and juveniles (e.g. Schiemer and Spindler 1989). Similarly, in lakes, many species occur in structurally dense, shallow, littoral habitats as small larvae and then move to mid water, deeper habitats as larger individuals. Movements of larval fish can also occur vertically, with diel migrations between surface waters and benthic habitats being common, particularly in deeper environments.

Larval fishes are a useful and sensitive tool for monitoring the effects of various anthropogenic influences on the system. For example, the presence of fish larvae clearly indicates that fish have spawned recently, and this can be used to elucidate the success of particular rehabilitation strategies targeted to enhance spawning, such as environmental flows (Humphries and Lake 2000; Rayner *et al.* 2015).

7.7 Specific issues in sampling and monitoring freshwater zooplankton

Temporal and spatial scales of zooplankton sampling and monitoring in fresh water depend on the type and extent of ecological concern, issues and hypotheses that are going to be examined. The general framework of ecological sampling and monitoring and statistical considerations are applicable to zooplankton sampling and monitoring (such as the original BACI design or its modifications and trend analyses; Chapter 4). A pilot sampling and monitoring program is always helpful in determining the methods of sampling (e.g. plankton net versus plankton trap) and in providing basic data on species composition, density, biomass and their variability (Bottrell *et al.* 1976).

There is a large diversity of types of gear currently available for the collection of larval fish in freshwater habitats. The most commonly used types of gear are designed to filter volumes of water through fine mesh, including drift nets, trawl nets, seines and pumps with fitted mesh nets (Kelso and Rutherford 1996). Electrofishing gear modified for sampling small-bodied fish has also recently been used increasingly in freshwater habitats (Copp 1989; King and Crook 2002). There is also a range of more passive collection gears, such as light traps, baited traps and activity traps – where fish are either attracted into the trap or are captured while moving through the habitat. However, knowledge of the target fish reproductive life history and larval behaviour and ecology is required in the choice of collection methods, gear types, sampling periodicity and sampling habitat.

For other types of zooplankton, a conical plankton net is often useful in collecting pelagic species (Table 7.4). Depending on the mesh size and specifications of the plankton net used, the net may clog partially or fully after towing certain distances and its filtering efficiency may drop dramatically. The clogging of a net is primarily due to collection of phytoplankton and detrital particles that are larger than the mesh size. This problem is often encountered in eutrophic waters, as well as highly turbid waters. The volume of water filtered by the net needs to be calibrated with a flow meter if zooplankton are needed to be collected quantitatively (see Chapter 4). Plankton traps or pumps are generally more efficient quantitative zooplankton samplers than plankton nets.

Zooplankton are seldom distributed uniformly within a water body. Some species exhibit a diurnal vertical migration – often concentrating in deep waters during the day and in surface waters during the night (Chapter 2). Zooplankton samples should be collected in a depth-integrated manner from the bottom to the surface or from multiple discrete depths.

It is difficult to properly tow a plankton net in the littoral zone – often resulting in the collection of large amounts of aquatic-plant debris that clog the net. Specialised sampling devices and techniques are recommended to use in collecting littoral zooplankton (Campbell *et al.* 1982; Sakuma *et al.* 2002).

7.8 Conclusions

Zooplankton are diverse and ubiquitous organisms in fresh water. Zooplankton occupy an intermediate trophic level – functioning as an important food source for a variety of animals, including juvenile and larger fish. In turn, they can be important in the control of bacterial and algal abundances and quickly increase in number following increased bacterial and algal numbers.

Zooplankton are also sensitive to various substances that enrich or pollute water, and have often been used as indicators to monitor and assess the condition and changes in the freshwater environment, particularly in the Northern Hemisphere (see Section 3.8). They display fairly consistent, measurable changes in response to water quality and various forms of pollution. These findings provide a basis for 'where to look' when zooplankton are used as indicators in freshwater ecosystems.

As a general trend, microzooplankton are more tolerant than macrozooplankton to different forms of pollution. Possible mechanisms to explain this trend include:

- reduced food availability for large zooplankton in acidified systems (Havens 1991)
- the short generation time and ability to recover quickly after stress shown by small zooplankton in agricultural pollution (Havens and Hanazato 1993)

Table 7.4. Sampling devices for freshwater zooplankton

Type	Comments	References
Conical or cylindrical-conical plankton nets	Widely used, different type of nets available, easy to deploy, very suitable for depth-integrated as well as horizontally integrated samples. The filtration efficiency of a net must be determined for more quantitative sampling of zooplankton.	Evans and Sell (1985), Wetzel and Likens (1991), McQueen and Yan (1993)
Bottles (e.g. Van Dorn and Niskin samplers)	Suitable for fixed volume sampling and discrete depth sampling. Light weight allowing samples to be taken easily from a small boat. Effective in collecting small organisms such as protozoans and rotifers.	Rice *et al.* (2012)
Traps (e.g. Schindler-Patalas trap)	Suitable for fixed volume sampling, and discrete depth sampling. Light weight allowing samples to be taken easily from a small boat. Suitable for collecting larger organisms, such as adult copepods and cladocerans, as well as small rotifers and protozoans.	Schindler (1969), Haney (1971), Shiel *et al.* (1982), Wetzel and Likens (1991)
Pumps	Easy to deploy; suitable for collecting littoral organisms, such from the surface of submerged aquatic plants.	Campbell *et al.* (1982), Malone and McQueen (1983), Sollberger and Paulson (1992)

- the predation of large zooplankton (particularly *Daphnia*) by fish in eutrophication (Brooks and Dodson 1965).

Zooplankton have been frequently used as ecotoxicological test organisms to assess the acute and chronic effects of various toxic substances that are found in the freshwater environment. Importantly, the lethal and effective values obtained from these bioassays are not necessarily applied to the evaluation of ecosystem impact of a toxicant. For example, Lampert *et al.* (1989) reported that *Daphnia* showed low sensitivity to the herbicide atrazine when direct effects (i.e. acute toxicity) were measured, but became very sensitive to the chemical in the moderately complex 'food chain' mesocosm experiment. Clearly biological interactions play a significant, and unexpected, role in the modified response of *Daphnia*.

Pollution management and monitoring programs that depend on a small number of indicators may fail to consider the full complexity of ecosystems. It may be necessary to use a suite of indicators representative of the structure, function and composition of ecosystems (Dale and Beyeler 2001). The useful application of zooplankton as indicators in freshwater ecosystems can only be realised by understanding the characteristics and dynamics of the ecosystems that are subject to various water resource management activities. In addition, the design of any monitoring program needs to consider the importance of temporal and spatial variability in sampling for zooplankton, to allow for meaningful conclusions from the data.

7.9 References

Almeida FS, Frantine-Silva W, Lima SC, Garcia DAZ, Orsi ML (2018) DNA barcoding as a useful tool for identifying non-native species of ichthyoplankton in the neotropics. *Hydrobiologia* **817**, 111–119. doi:10.1007/s10750-017-3443-5

Bayly IAE (1993) The fauna of athalassic saline waters in Australia and the Altiplano of South America: comparisons and historical perspectives. *Hydrobiologia* **267**, 225–231. doi:10.1007/BF00018804

Bayly IAE, Boxshall GA (2009) An all-conquering ecological journey: from the sea, calanoid copepods mastered brackish, fresh, and athalassic saline waters. *Hydrobiologia* **630**, 39–47. doi:10.1007/s10750-009-9797-6

Bayly IAE, Gibson JAE, Wagner B, Swadling KM (2003) Taxonomy, ecology and zoogeography of two East Antarctic freshwater species: *Boeckella poppei* and *Gladioferens antarcticus*. *Antarctic Science* **15**, 439–448. doi:10.1017/S0954102003001548

Becker RA, Sales NG, Santos GM, Santos GB, Carvalho DC (2015) DNA barcoding and morphological identification of neotropical ichthyoplankton from the upper Parana and Sao Francisco. *Journal of Fish Biology* **87**, 159–168. doi:10.1111/jfb.12707

Blanck A, Tedesco PA, Lamouroux N (2007) Relationships between life-history strategies of European freshwater fish species and their habitat preferences. *Freshwater Biology* **52**, 843–859. doi:10.1111/j.1365-2427.2007.01736.x

Bottrell HH, Duncan A, Gliwicz ZM, Grygierek E, Herzig A, Hillbricht-Ilkowska A, Kurasawa A, Larson P, Weglenska T (1976) A review of some problems in zooplankton production studies. *Norwegian Journal of Zoology* **24**, 419–456.

Boxshall GA, Defaye D (2008) Global diversity of copepods (Crustacea: Copepoda) in freshwater. *Hydrobiologia* **595**, 195–207. doi:10.1007/s10750-007-9014-4

Brancelj A, Dumont HJ (2007) A review of the diversity, adaptations and groundwater colonization pathways in Cladocera and Calanoida (Crustacea), two rare and contrasting groups of stygobionts. *Fundamental and Applied Limnology* **168**, 3–17. doi:10.1127/1863-9135/2007/0168-0003

Brooks JL, Dodson SI (1965) Predation, body size and composition of plankton. *Science* **150**, 28–35. doi:10.1126/science.150.3692.28

Campbell JM, William JC, Kosinski R (1982) A technique for examining microspatial distribution of Cladocera associated with shallow water macrophytes. *Hydrobiologia* **97**, 225–232. doi:10.1007/BF00007110

Copp GH (1989) Electrofishing for fish larvae and 0+ juveniles: equipment modifications for increased efficiency with short fishes. *Aquaculture and Fisheries Management* **20**, 453–462.

Dale VH, Beyeler SC (2001) Challenges in the development and use of ecological indicators. *Ecological Indicators* **1**, 3–10. doi:10.1016/S1470-160X(01)00003-6

Evans MS, Sell DW (1985) Mesh size and collection characteristics of 50-cm diameter conical plankton nets. *Hydrobiologia* **122**, 97–104. doi:10.1007/BF00032095

Finlay BJ, Esteban GF (1998) Freshwater protozoa: biodiversity and ecological function. *Biodiversity and Conservation* **7**, 1163–1186. doi:10.1023/A:1008879616066

Forró L, Korovchinsky NM, Kotov AA, Petrusek A (2008) Global diversity of cladocerans (Cladocera: Crustacea) in freshwater. *Hydrobiologia* **595**, 177–184. doi:10.1007/s10750-007-9013-5

Frantine-Silva W, Sofia SH, Orsi ML, Almeida FS (2015) DNA barcoding of freshwater ichthyoplankton in the Neotropics as a tool for ecological monitoring. *Molecular Ecology Resources* **15**, 1226–1237. doi:10.1111/1755-0998.12385

Fuiman LA, Higgs DA (1997) Ontogeny, growth and the recruitment process. In *Early Life History and Recruitment in Fish Populations*. (Eds RC Chambers and EA Trippel) pp. 225–250. Chambers and Hall, London, UK.

Gooderham J, Tsyrlin E (2009) *The Waterbug Book. A Guide to the Freshwater Macroinvertebrates of Temperate Australia*. CSIRO Publishing, Melbourne, Australia.

Hairston NG, Jr, Van Brunt RA, Kearns CM (1995) Age and survivorship of diapausing eggs in sediment egg bank. *Ecology* **76**, 1706–1711. doi:10.2307/1940704

Haney JF (1971) An *in situ* method for the measurement of zooplankton grazing rates. *Limnology and Oceanography* **16**, 970–977. doi:10.4319/lo.1971.16.6.0970

Hansson LA, Hylander S, Dartnall HJG, Lidström S, Svensson JE (2012) High zooplankton diversity in the extreme environments of the McMurdo Dry Valley lakes, Antarctica. *Antarctic Science* **24**, 131–138. doi:10.1017/S095410201100071X

Havens KE (1991) Crustacean zooplankton food web structure in lakes of varying acidity. *Canadian Journal of Fisheries and Aquatic Sciences* **48**, 1846–1852. doi:10.1139/f91-218

Havens KE, Hanazato T (1993) Zooplankton community responses to chemical stressors: a comparison of results from acidification and pesticide contamination research. *Environmental Pollution* **82**, 277–288. doi:10.1016/0269-7491(93)90130-G

Humphries P, King AJ, Koehn JD (1999) Fish, flows and flood plains: links between freshwater fishes and their environment in the Murray-Darling River system, Australia. *Environmental Biology of Fishes* **56**, 129–151. doi:10.1023/A:1007536009916

Humphries P, Lake PS (2000) Fish larvae and the management of regulated rivers. *Regulated Rivers: Research and Management* **16**, 421–432. doi:10.1002/1099-1646(200009/10)16:5<421::AID-RRR594>3.0.CO;2-4

Jahn TL, Bovee EC, Jahn FF (1979) *How to Know the Protozoa*. 2nd edn. Wm. C. Brown Publishers, Dubuque IA, USA.

Karanovic T, Gibson JAE, Hawes I, Anderson DT, Stevens MI (2014) *Diacyclops* (Copepoda: Cyclopoida) in continental Antarctica, including three new species. *Antarctic Science* **26**, 250–260. doi:10.1017/S0954102013000643

Kelso WE, Rutherford DA (1996) Collection, preservation and identification of fish eggs and larvae. In *Fisheries Techniques*. (Eds BR Murphy and DW Willis) pp. 255–302. American Fisheries Society, Bethesda MD, USA.

King AJ, Crook DA (2002) Evaluation of a sweep net electrofishing method for the collection of small fish and shrimp in lotic freshwater environments. *Hydrobiologia* **472**, 223–233. doi:10.1023/A:1016307602735

King AJ, Humphries P, McCasker NG (2013) Reproduction and early life history. In *Ecology of Australian Freshwater Fishes*. (Eds P Humphries and

K Walker) pp. 159–194. CSIRO Publishing, Melbourne, Australia.

Kobayashi T, Shiel RJ, Gibbs P, Dixon PI (1998) Freshwater zooplankton in the Hawkesbury-Nepean River: comparison of community structure with other rivers. *Hydrobiologia* **377**, 133–145. doi:10.1023/A:1003240511366

Lampert W, Fleckner W, Pott E, Schober U, Storkel KU (1989) Herbicide effects on planktonic systems of different complexity. *Hydrobiologia* **188/189**, 415–424. doi:10.1007/BF00027809

Malone BJ, McQueen DJ (1983) Horizontal patchiness in zooplankton populations in two Ontario kettle lakes. *Hydrobiologia* **99**, 101–124. doi:10.1007/BF00015039

Manca M, Cammarano P, Spagnuolo T (1994) Notes on Cladocera and Copepoda from high altitude lakes in the Mount Everest Region (Nepal). *Hydrobiologia* **287**, 225–231. doi:10.1007/BF00006371

Matthews WJ (1998) *Patterns in Freshwater Fish Ecology*. Chapman and Hall, New York, USA.

McQueen DJ, Yan ND (1993) Metering filtration efficiency of freshwater zooplankton hauls: remainders from the past. *Journal of Plankton Research* **15**, 57–65. doi:10.1093/plankt/15.1.57

Mitchell BD, Williams WD (1982) Population dynamics and production of *Daphnia carinata* (King) and *Simocephalus exspinosa* (Koch) in waste stabilisation ponds. *Australian Journal of Marine and Freshwater Research* **33**, 837–864. doi:10.1071/MF9820837

Morton DW (1985) Revision of the Australian Cyclopidae (Copepoda: Cyclopoida). I. *Acanthocyclops* Kiefer, *Diacyclops* Kiefer and *Australocyclops*, gen. nov. *Australian Journal of Marine and Freshwater Research* **36**, 615–634. doi:10.1071/MF9850615

Morton DW (1990) Revision of the Australian Cyclopidae (Copepoda: Cyclopoida). II. *Eucyclops* Claus and *Ectocyclops* Brady. *Australian Journal of Marine and Freshwater Research* **41**, 657–675. doi:10.1071/MF9900657

Moser HG, Richards WJ, Cohen DM, Fahay MP, Kendall AW Jr, Richardson SL (Eds) (1984) *Ontogeny and Systematics of Fishes*. Special Publication 1. American Society of Ichthyologists and Herpetologists, Lawrence KS, USA.

Neira FJ, Miskiewicz AG, Trnski T (1998) *Larvae of Temperate Australian Fishes. Laboratory Guide for Larval Fish Identification*. University of Western Australia Press, Perth, Australia.

Persaud AD, Moeller RE, Williamson GE, Burns CW (2007) Photoprotective compounds in weakly and strongly pigmented copepods and co-occurring cladocerans. *Freshwater Biology* **52**, 2121–2133. doi:10.1111/j.1365-2427.2007.01833.x

Pusey BJ, Kennard MJ, Arthington AH (2004) *Freshwater Fishes of North-eastern Australia*. CSIRO Publishing, Melbourne, Australia.

Quinlan K, Bayly IAE (2017) A new species of *Boeckella* (Copepoda: Calanoida) from arid Western Australia, an updated key, and aspects of claypan ecology. *Records of the Western Australian Museum* **32**, 191–206. doi:10.18195/issn.0312-3162.32(2).2017.191-206

Rayner TS, Kingsford RT, Suthers IM, Cruz DO (2015) Regulated recruitment: native and alien fish responses to widespread floodplain inundation in the Macquarie Marshes, arid-Australia. *Ecohydrology* **8**, 148–159. doi:10.1002/eco.1496

Rice EW, Baird RB, Eaton AD, Clesceri LS (Eds) (2012) *Standard Methods for the Examination of Water and Wastewater*. 22nd edn. American Public Health Association, Washington DC, USA.

Sakuma M, Hanazato T, Nakazato R, Haga H (2002) Methods for quantitative sampling of epiphytic microinvertebrates in lake vegetation. *Limnology* **3**, 115–119. doi:10.1007/s102010200013

Santos-Flores CJ, Dodson SI (2003) *Dumontia oregonensis* n. fam., n. gen., n. sp., a cladoceran representing a new family of 'Water-fleas' (Crustacea, Anomopoda) from USA, with notes on the classification of the Anomopoda. *Hydrobiologia* **500**, 145–155. doi:10.1023/A:1024638620460

Schiemer F, Spindler T (1989) Endangered fish species of the Danube River in Austria. *Regulated Rivers: Research and Management* **4**, 397–407. doi:10.1002/rrr.3450040407

Schindler DW (1969) Two useful devices for vertical plankton and water sampling. *Journal of the Fisheries Research Board of Canada* **26**, 1948–1955. doi:10.1139/f69-181

Segers S (2008) Global diversity of rotifers (Rotifera) in freshwater. *Hydrobiologia* **595**, 49–59. doi:10.1007/s10750-007-9003-7

Serafini LG, Humphries P (2004) *Preliminary Guide to the Identification of Larvae of Fish, with a Bibliography of their Studies, from the Murray-Darling Basin.* CRC for Freshwater Ecology. Identification Guide No. 48. Murray-Darling Freshwater Research Centre, Albury, Australia.

Shiel RJ, Walker KF, Williams WD (1982) Plankton of the lower River Murray, South Australia. *Australian Journal of Marine and Freshwater Research* **33**, 301–327. doi:10.1071/MF9820301

Shiel RJ (1995) *A Guide to Identification of Rotifers, Cladocerans and Copepods from Australian Inland Waters.* CRC for Freshwater Ecology. Identification Guide No. 3. Murray-Darling Freshwater Research Centre, Albury, Australia.

Sinev AY (2015) Revision of the *pulchella*-group of *Alona* s. lato leads to its translocation to *Ovalona* Van Damme et Dumont, 2008 (Branchiopoda: Anomopoda: Chydoridae). *Zootaxa* **4044**, 451–492. doi:10.11646/zootaxa.4044.4.1

Smirnov NN, Timms BV (1983) Revision of the Australian Cladocera (Crustacea). *Records of the Australian Museum* **1**(Supplement), 1–132. doi:10.3853/j.0812-7387.1.1983.103

Snyder DE (1983) Fish eggs and larvae. In *Fisheries Techniques.* (Eds LA Nielsen and DL Johnson) pp. 165–179. American Fisheries Society, Bethesda MD, USA.

Snyder DE, Muth RT, Bjork CL (2004) *Catostomid Fish Larvae and Early Juveniles of the Upper Colorado River Basin – Morphological Descriptions, Comparisons and Computer-Interactive Key.* Technical Publication No. 42, Colorado Division of Wildlife, Denver CO, USA.

Sollberger PJ, Paulson LJ (1992) Littoral and limnetic zooplankton communities in Lake Mead, Nevada-Arizona, USA. *Hydrobiologia* **237**, 175–184. doi:10.1007/BF00005849

Trippel EA, Chambers RC (1997) The early life history of fishes and its role in recruitment processes. In *Early Life History and Recruitment in Fish Populations.* (Eds RC Chambers and EA Trippel) pp. 21–32. Chambers and Hall, London, UK.

Wetzel RG, Likens GE (1991) *Limnological Analyses.* 2nd edn. Springer, New York, USA.

Winemiller KO, Rose KA (1992) Patterns of life-history diversification in North American fishes: implications for population regulation. *Canadian Journal of Fisheries and Aquatic Sciences* **49**, 2196–2218. doi:10.1139/f92-242

Wootton RJ (1998) *Ecology of Teleost Fishes.* 2nd edn. Kluwer Academic Publishers, Dordrecht, The Netherlands.

7.10 Further reading

7.10.1 Taxonomy and general biology

Dumont HJ (Ed.) (1992–2006) *Guides to the Identification of the Microinvertebrates of the Continental Waters of the World.* 23 vols. SPB Academic Publishing, The Hague, The Netherlands and Backhuys Publishers BV, Leiden, The Netherlands.

Foissner W, Helmut B (1996) A user-friendly guide to the ciliates (Protozoa, Ciliophora) commonly used by hydrobiologist as bioindicators in rivers, lakes, and waste waters, with notes on their ecology. *Freshwater Biology* **35**, 375–482.

Patterson DJ (1996) *Free-living Freshwater Protozoa: A Colour Guide.* John Wiley & Sons, New York, USA.

Romanowski N (2013) *Living Waters: Ecology of Animals in Swamps, Rivers, Lakes and Dams.* CSIRO Publishing, Melbourne, Australia.

7.10.2 Environmental issues

Gyllström M, Hansson LA (2004) Dormancy in freshwater zooplankton: induction, termination and the importance of benthic-pelagic coupling. *Aquatic Sciences* **66**, 274–295. doi:10.1007/s00027-004-0712-y

Mimouni EA, Pinel-Alloul B, Beisner BE (2015) Assessing aquatic biodiversity of zooplankton communities in an urban landscape. *Urban Ecosystems* **18**, 1353–1372. doi:10.1007/s11252-015-0457-5

Pace ML, Carpenter SR, Johnson RA, Kurtzweil JT (2013) Zooplankton provide early warnings of a regime shift in a whole lake manipulation. *Limnology and Oceanography* **58**, 525–532. doi:10.4319/lo.2013.58.2.0525

Rehse S, Kloas W, Zarfl C (2016) Short-term exposure with high concentrations of pristine microplastic particles leads to immobilisation of *Daphnia magna*. *Chemosphere* **153**, 91–99. doi:10.1016/j.chemosphere.2016.02.133

Winder M, Schindler DE (2004) Climate change uncouples trophic interactions in an aquatic ecosystem. *Ecology* **85**, 2100–2106. doi:10.1890/04-0151

8

Coastal and marine zooplankton: identification, biology and ecology

Anthony J. Richardson, Julian Uribe-Palomino, Anita Slotwinski, Frank Coman, Anthony G. Miskiewicz, Peter C. Rothlisberg, Jock W. Young and Iain M. Suthers

I often towed astern a net made of bunting and thus caught many curious animals.

Charles Darwin, 1833, during the voyage of the *Beagle* off Patagonia (Darwin 1845)

8.1 Identifying coastal and marine zooplankton

Charles Darwin was intrigued with the creatures he captured towing a fine-mesh net through a patch of water. Today, with modern microscopes with vastly improved optics over those in Darwin's era, it is even more enthralling to peer down a microscope at zooplankton. One is struck by the dazzling diversity of sizes, shapes, motions, colours, textures and degrees of transparency (Fig. 8.1 shows images of typically diverse zooplankton samples at different magnifications). Your amazement, however, can quickly turn to bewilderment when faced with the seemingly overwhelming task of identifying what is there! Where do you begin?

It is not as daunting as it might first seem to identify zooplankton. Even when they are dead and their colours have faded, zooplankton are relatively easily identified to major groups, and even sometimes to species for those with distinctive features. In this chapter, we will show you how.

8.1.1 The structure of this chapter

This chapter is designed for readers who are either new to zooplankton identification or have an intermediate level of experience; we expect experienced zooplankton ecologists will go directly to particular taxonomic resources for guidance. The chapter therefore focuses on features of specimens that are easy-to-see using a stereo microscope, rather than taxonomic keys or diagnostic characters that require a compound microscope and specialist training. Part of the reason it can be difficult to identify zooplankton is the associated nomenclature, which can seem impenetrable, especially when you are beginning. We explain key terms as simply as possible and leave out taxonomic jargon, although a minimum level is needed to be proficient in zooplankton identification. To help, Tables 8.1 and 8.2 explain commonly used terminology to describe key structures used in identification and also the names of larval stages in different zooplankton groups. We also use several comparative figures to separate difficult groups: for example, calanoid, cyclopoid, poecilostomatoid and harpacticoid copepods from each other; decapod larvae from euphausiids and mysids; and salps from doliolids. We include images captured using a high-resolution camera to showcase the different zooplankton taxa, because it can be easier to start with an overall feeling of the form of different groups with images rather than line drawings. Note that in images where we have identified the species we include the species name. However, if a genus name is followed by spp. we are referring to multiple images of multiple species in that genus, and when a genus name is

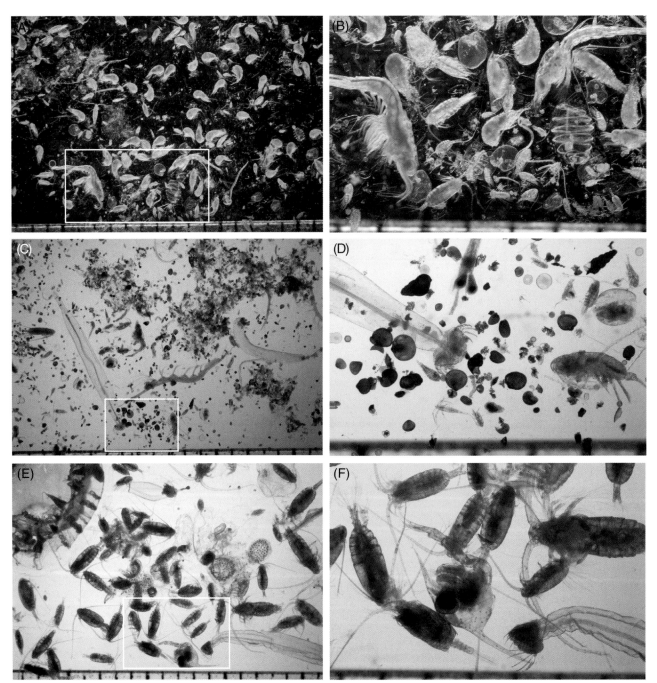

Fig. 8.1.   Typical mixed zooplankton assemblage viewed with a stereo microscope. Each row in the figure represents an individual sample, and the magnification for the photos in the left column is 7.3× and for the right column is 20×. **(A, B)** Sample collected with 100 μm (0.1 mm) mesh net in Darwin Harbour, Australia. **(C, D)** A different sample collected with a 100 μm mesh net in Darwin Harbour. **(E, F)** Sample collected with a 500 μm (0.5 mm) mesh net off North Stradbroke Island, Australia. The scale bar along the bottom of each image is in 1 mm intervals. All samples collected as part of the Australian Integrated Marine Observing System (IMOS). Note the amazing diversity of shapes in the samples.

Table 8.1.   **Commonly used terms to describe structures in different zooplankton groups**

Structures	Group	Definition
Abdomen	Crustaceans	The rear most part of the body, segmented and usually supporting swimming legs (pleopods), except in copepods
Antennae	Crustaceans	A pair of thin sensory appendages on the head
Appendix	Doliolids	Projection where asexual clones bud
Carapace	Decapods, euphausiids, mysids, stomatopods, cumaceans	Hard upper shell covering the cephalothorax
Cephalothorax	Crustaceans	Fused head and thorax
Chaetae	Polychaetes	Bristles made of chitin
Chelae	Crustaceans, particularly decapods	A pincer-like claw
Cilia	Larvae of annelids, echinoderms, molluscs, adult tintinnids	Hair-like projections from a cell
Cnidoblasts	Cnidarians	Stinging cells that are the most complex cell in any animal
Colloblasts	Ctenophores	Sticky cells that can capture prey (not stinging cells as in cnidarians)
Combs	Ctenophores	Ctenophores have eight longitudinal rows of cilia, known as combs (or ctenes or comb plates), which produce a rainbow effect known as iridescence
Exopodites (or exopods)	Crustaceans	Crustacean appendages are biramous (divided in two): the outer branch (exopodite or exopod) may be smaller in comparison to the inner branch (endopodite or endopod)
Gas bladder	Larval fish	Gas-filled sac located above the gut, which helps in regulating buoyancy
Lorica	Tintinnids	Outer covering made from protein and often reinforced with sand grains
Marsupium	Mysids, amphipods, isopods, cumaceans	A brood pouch, formed by appendages attached to the base of the pereiopods, that protects eggs
Maxilliped	Crustaceans	Often highly modified pair of appendages that pass food to the maxilla (pair of mouthparts used for chewing)
Melanophores	Larval fish	Nucleated cells containing the brown and back pigment melanin; melanophores can expand and contract thus changing in size and shape, and remain in larval fishes after fixation
Myomeres	Larval fish	Muscle bands aligned sequentially and in transverse series along the trunk and tail; total number of myomeres is approximately equal to the number of vertebrae
Nectophore	Siphonophores	A specialised individual in a colony
Parapodia	Polychaetes	Paired fleshy limbs on each segment, with bundles of chaetae extending out
Pereiopods	Crustaceans	Thoracic appendages, often for walking (except euphausiids)
Photophores	Euphausiids	Light organs on the abdomen at the base of the pleopods
Pincers	Decapods	Small claws or chelae
Pleopods	Decapods, euphausiids, mysids, cumaceans, amphipods	Abdominal appendages often used for swimming
Polyp	Cnidarians	One of two body forms, typically but not always benthic stage characterised by a sac-like body with mouth and tentacles
Pseudopodia	Forams, radiozoans	A temporary extension of the cell surface for feeding and movement
Rostrum	Decapods, copepods, euphausiids, mysids	Extension of the cephalothorax between the eyes
Setae	Most invertebrates	Stiff hairs or bristles (similar to chaetae)

(Continued)

Table 8.1. (Continued)

Structures	Group	Definition
Statocyst	Mysids, cnidarians, ctenophores, molluscs	Organ for balance and orientation. Consists of a hollow sphere with a statolith inside that brushes against tiny bristles
Swim bladder	Larval fish	*See* Gas bladder
Tail	Appendicularians	Contains nerve cord and muscle bands. The tail beats to concentrate food within the appendicularian house
Telson	Decapods, euphausiids, mysids	Posterior (last) section of the abdomen
Thorax	Arthropods	Middle part of body between head and abdomen. In crustaceans often fused with head as cephalothorax
Trunk	Appendicularians	Head and thorax
Trunk	Chaetognaths	Body, excluding head
Uropods	Crustaceans	The last pair of abdominal appendages forming part of the tail fan

Table 8.2. Commonly used names to describe larval stages of different zooplankton groups

Stages	Group	Definition
Antizoea/pseudozoea/ alima/ericthus	Stomatopods	Larval names differ among stomatopod groups. Antizoea and pseudozoea are early stages, and alima and ericthus are late stages
Bipinnaria	Sea stars (Asteroidea)	Ciliated and folded first larval stage
Calyptopis	Euphausiids	Stage after naupliar stage, when the abdomen has developed, but eyes remain within carapace
Copepodite	Copepods	Stages that look like the adult and are after the nauplius stage. Increasing segmentations with limbs following each moult. The last copepodite stage is the adult
Cyphonautes	Bryozoan	The only larval stage
Cyprid/cypris	Barnacle	After the nauplius stage. This is a non-feeding stage that settles on the substrate
Echinopluteus	Echinoid echinoderms (sea urchins)	Larval stage with two to six pairs of ciliated arms supported by calcite rods
Epitokes	Polychaetes	Posterior segments containing gametes that break off, and then swim autonomously to the surface. When epitokes reach the surface, gametes (eggs and sperm) are released
Furcilia	Euphausiids	Stage after calyptopis, when stalked eyes emerge from carapace
Larva/larvae	Most marine zooplankton	Young stage/s of animals that undergo true metamorphosis (i.e. they look different to adults)
Medusa	Cnidaria	The generally bell-shaped, free-swimming, sexually reproducing stage of some cnidarian groups. Alternates in life cycle with benthic stages (polyps)
Megalopa	Decapods	Last larval stage that resembles adults with developed adult appendages
Mysis	Decapods	Intermediate larval stage between zoea/protozoea and postlarvae
Nauplius/nauplii	Crustaceans	First larval stage with rudimentary appendages
Ophiopluteus	Ophiuroid echinoderms (brittle stars)	Larval stage with four pairs of ciliated arms
Paralarva	Cephalopods	Young cephalopod in the planktonic stage between hatchling and subadult. This stage differs from larval stages of animals that undergo true metamorphosis
Pediveliger	Bivalves	Following on from the veliger stage, characterised by development of the foot; this stage can swim and crawl

(Continued)

Table 8.2.   (Continued)

Stages	Group	Definition
Phyllosoma	Achelata (spiny and slipper lobsters)	Spider-like, flattened and transparent planktonic stage with long legs
Pluteus	Ophiurids (brittle stars), echinoids (urchins)	A-shaped spiny larvae
Postlarva		After the typical larval stage and similar to adult in appearance. Uses abdominal appendages (pleopods) for locomotion. A post-larval crab is a megalopa. In fish, the stage following the larval stage when the yolk sac is absorbed and juvenile characters appear
Protozoea	Decapods	Early larval stages with little development of appendages
Trochophore	Molluscs, polychaetes	A roughly spherical-shaped larva, with a band of cilia, and a spinning motion
Veliger	Gastropods, bivalves	The final larval stage before settling to the sea floor
Zoea	Decapods	Intermediate larval stages with developed cephalothorax appendages, but no abdominal appendages

followed by sp. we are referring to a single species, but one that we are not sure of.

If you are new to zooplankton identification, reading this chapter in the order it is presented will benefit you. The chapter starts with some definitions and explanations that are needed for zooplankton identification. Section 8.1.2 'What is zooplankton?' deals with several common misconceptions in the definition of zooplankton and helps to frame the problem of identifying zooplankton. Section 8.1.3 'What is meroplankton?' explains one of the major reasons why zooplankton are so diverse in form, and how this makes identifying zooplankton more challenging than many other animals, but also more interesting!

Section 8.2 'Identifying zooplankton using body types' then introduces 14 representative body types that cover the range of zooplankton forms. This section uses a visual approach to define zooplankton body types based on easy-to-identify features. We believe that this approach is the one everyone uses consciously or subconsciously when they begin identifying zooplankton. The 14 body types are based on the shape, texture, size, and degree of transparency of a specimen, and are only loosely based on taxonomy. This section has two subsections to aid identification of body types: Section 8.2.1 'Size matters' and Section 8.2.2 'Abundance matters'. Using an analogy with bird identification field guides, both the size of birds and whether

they are common or a vagrant in a region are key pieces of information used in identification.

The main part of the chapter is Section 8.3 'From body types to taxonomic groups'. Once a specimen is placed into a body type, one can then identify its taxonomic group based on a limited number of options. Most sub-sections start with a 'Quick guide' that provides a concise synopsis of the main distinguishing features of the different taxonomic groups within each body type. There is then a detailed description of each taxonomic group, with a consistent structure, describing: identification, diversity, some common species, lifespan, feeding, reproduction, locomotion, ecology and human interactions. The more important groups have longer sections: for example, we have an extensive section on identification of fish larvae to the family level because this is a key group in fisheries research. It is worth remembering that taxonomy is constantly evolving – we have endeavoured to present the most up-to-date taxonomic classification at the time of publication, based on the World Register of Marine Species (WoRMS, http://www.marinespecies.org). The chapter finishes with Section 8.4. 'Top tips for identifying zooplankton', which outlines ways that you can accelerate your learning.

If you are new to identifying zooplankton, then the body type classification is a valuable approach for you. Once you have gained confidence and

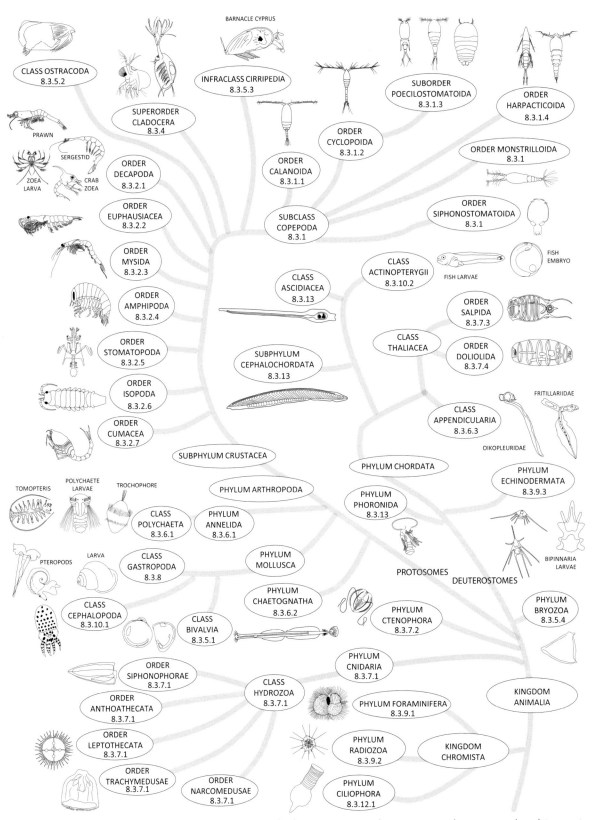

Fig. 8.2. Simplified taxonomic tree of life of major zooplankton groups, with section numbers. Note that this tree is not comprehensive because it covers only the main groups, and there is some debate about the exact position of different groups in the tree of life.

experience in zooplankton identification (or if you already have it), then we suggest you might prefer to use the taxonomic tree in Fig. 8.2 to navigate directly to specific taxonomic sections in the chapter. This simplified tree of life is not comprehensive but does cover almost all zooplankton you will see in a sample. This taxonomic tree short-circuits the need for using zooplankton body types to decide upon the taxonomic group a specimen belongs to. For example, molluscs have very varied appearances and thus appear in several different body types (e.g. bivalves or snails or squid or trochophore larvae), but if you already know that your specimen is a snail and not sure whether it is a pteropod or proso-branch, then the taxonomic tree allows you to go directly to the relevant section to decide.

8.1.2 What is zooplankton?

Zooplankton (from the Ancient Greek *zôion* meaning 'animal' and *planktós* meaning 'to drift') is defined as any animal that floats in the water but cannot swim against a current. A common misconception about zooplankton is that they are microscopic animals, passively floating in the water without the ability to swim. This is not true.

Most zooplankton are microscopic, but some are huge. Scyphozoan jellyfish, the typical large bell-shaped jellyfish we are familiar with, can be up to 2 m in diameter, and colonies of siphonophore jellyfish can reach 50 m in length and are the longest marine animals. Another common misconception is that zooplankton cannot swim. Almost all zooplankton can swim, but they cannot swim strongly enough to progress against a current and thus are defined as plankton.

This book separates the zooplankton from the phytoplankton, which are plankton that photosynthesise. Even this seemingly clear distinction is not absolute: many small zooplankton have photosynthetic symbionts, and many phytoplankton feed on other organisms. We also separate the zooplankton from the nekton (from the Ancient Greek *nēktón* meaning 'swimming'): animals that are large enough to swim against currents. The importance of the distinction between zooplankton and nekton will become clearer in the next section.

8.1.3 Meroplankton

Young stages (eggs and larvae) of marine animals often look very different to their adults (Fig. 8.3). In fact, Linnaeus used the Latin word *larvae*, meaning 'ghosts' or 'masks', because immature forms look very different to their adults, masking their true identity and making identification challenging.

Zooplankton can be divided into holoplankton (from the Ancient Greek *hólos* meaning 'whole') – permanent members of the plankton that spend their entire lives in the water column; and mero-plankton (from the Ancient Greek *méros* meaning 'portion') – temporary members that spend only a portion of their life as plankton (see Section 2.4). The holoplankton includes krill, copepods, chae-tognaths, appendicularians (larvaceans), salps and some jellyfish (i.e. those without a bottom-dwelling stage), because they remain drifting as zooplankton their entire life.

There are various reasons why meroplankton only spend a portion of their lives in the plankton:

- they have young stages (eggs and larvae) that float in the plankton, but they leave the plankton when they settle out to the sea floor or shoreline as adults (e.g. crabs, lobsters or barnacles)
- they have young stages that float in the plankton until they become large enough to swim against currents (e.g. fish, squid and octopus)
- they have alternation of generations, where the pelagic stage is planktonic and the polyp stage lives on the sea floor (e.g. many jellyfish)
- they live as part of the plankton on a near-daily basis to feed, mate or move, but live much of their time on the sea floor (e.g. stomatopods, cumaceans).

The diversity of zooplankton forms reflects the reality that most living animal phyla evolved in the oceans and that almost all marine animals have a meroplanktonic stage for dispersal.

Like a caterpillar metamorphosing into a butterfly, many meroplankton look nothing like their adults (Fig. 8.3). In fact, some larvae have peculiar names, unrelated to their adults, because they were

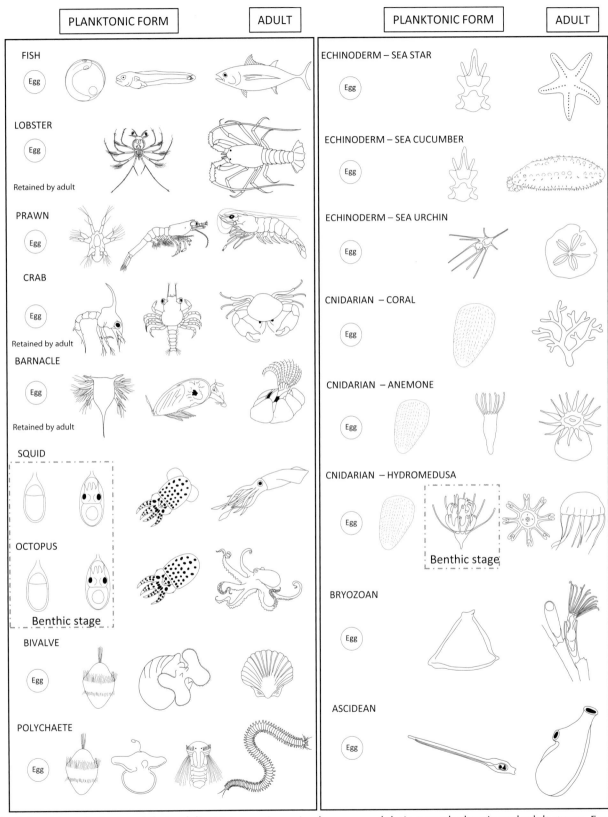

Fig. 8.3. Meroplankton and their adults. Major marine animal groups and their meroplanktonic and adult stages. Eggs characteristic of each group are not shown because they are generally difficult to distinguish (but see Section 8.3.11 for separating some common types).

identified in the plankton before they were grown in laboratories from eggs to adult and were not recognised as the same species. Most marine animals have egg stages that are free floating, but they are difficult to distinguish, and they are represented generically here in this diagram (but see Section 8.3.11 for identification of some common types). Probably the easiest type of meroplankton to identify is fish larvae, because they resemble miniature adult fish. You will likely notice the distinctive fish-like eye first. Spiny and slipper lobsters have a stunning planktonic phase, known as a phyllosoma (from the Ancient Greek *fýllo* meaning 'leaf'), which is flat and looks like a spider. These lobsters can moult through a dozen or so phyllosoma stages over many months before they settle on the sea floor as juveniles and are no longer planktonic. Some crustacean (from Latin *crusta* meaning 'shell') groups such as copepods and barnacles have a nauplius stage that looks like a tick. It is easy to forget that barnacles, common on the rocky shore, are crustaceans, and thus have the typical crustacean nauplius phase. Crab larvae often have a spiky head, probably for protection against predation, and then leave the plankton as they settle to the sea floor. Squid and octopus have benthic eggs, but planktonic paralarval stages that look like miniature adults, and these are no longer part of the zooplankton once they are large enough to swim against currents. Bivalve (clams and scallops) larvae are common in coastal samples and look like hairy spinning tops – they leave the plankton when they settle out on a hard substrate (e.g. mussels) or in the sediment (e.g. clams). There are a few holoplanktonic polychaetes (worms), but most polychaete larvae settle on the sea floor. Many echinoderms have ciliated larvae without many distinguishing features (e.g. bipinnaria larvae of sea stars and sea cucumbers), but sea urchins and brittle stars have distinctive spiky larvae. Cnidarian (corals and jellyfish) larvae are rarely seen, but they settle out (e.g. corals or jellyfish polyp stage) or remain as holoplankton.

The abundance of meroplankton in samples fluctuates more than holoplankton because marine animals often spawn seasonally, particularly in spring and summer, releasing vast numbers of eggs that develop into larvae. Meroplankton are also more likely to be found in coastal areas where many of their adults live, particularly at night when many organisms spawn and when some species move up into the water column.

8.2 Identifying zooplankton using body types

To identify zooplankton, you will need some basic equipment – at least a microscope and a dish (Box 8.1). Once set up, you are ready to go!

Birdwatchers use the term jizz (gestalt) for the overall impression of a bird's appearance, based on features including shape, posture, size, colouration and location. The human brain is good at rapidly processing these overall features and coming up with an identification. We apply the concept of using the jizz of a specimen to identify zooplankton body types: that way you are not considering the full range of taxonomic groups from the outset. As practicing ecologists identifying zooplankton, we find that we subconsciously use this approach of first identifying body types, before placing specimens into specific taxonomic groups.

We rarely use taxonomic keys and will not do so here. The focus of this chapter is on identification based on features that are readily distinguishable under a stereo microscope.

The first step in identifying zooplankton is to place a specimen in its body type, based on its shape (spherical, round, vase-like, worm-like, bullet-like), size (tiny: <0.5 mm, small: 0.5–5 mm, large: >5 mm), degree of transparency (clear, semi-transparent, opaque) and texture (soft, hard). To focus your attention on the basic shapes in your sample, it might help to pick a few of the different forms and sketch them, which helps to focus your attention on key features. For identification, it is best to look at a suite of features to place a specimen into a body type rather than rely on any one feature. We have split typical specimens you see in a zooplankton sample into 14 different body types (Fig. 8.5), based on a suite of general diagnostic features and rough relative sizes.

Box 8.1 Equipment for zooplankton identification in the laboratory

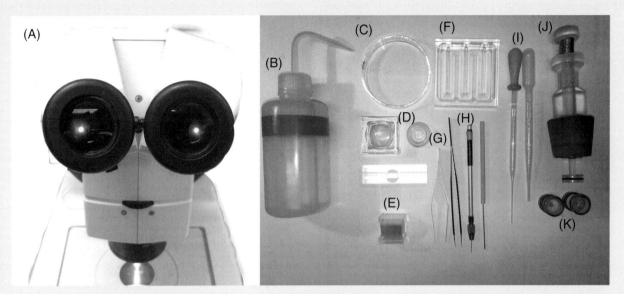

Fig. 8.4. Equipment for identifying zooplankton in the laboratory: **(A)** stereo (dissecting) microscope; **(B)** squeeze bottle; **(C)** Petri dish; **(D)** watch glasses/embryo dishes; **(E)** coverslips; **(F)** Bogorov tray; **(G)** forceps; **(H)** fine metal probe; **(I)** eye droppers; **(J)** Stempel pipette; and **(K)** inserts of different volumes for Stempel pipette.

To look at the zooplankton in a sample, you will need some equipment. Fig. 8.4 shows some of these items, but you will not need everything here when starting off. The most important piece of equipment is a stereo (dissecting) microscope with a selection of lenses. To see the hidden world of zooplankton, you will need a microscope with at least 5× magnification, and preferably 10×, with the higher magnification allowing you to see more features. Microscopes can range from cheap digital ones available over the internet that open your eyes to the world of zooplankton, to traditional versions with sharp optics and high magnifications for research or commercial purposes. Illumination of the sample is best from the bottom. Although live specimens are beautiful and transparent, their constant motion makes them difficult to identify and count, and they decompose quickly after several hours (you will never forget the stench!). Therefore, if zooplankton is to be taken to the laboratory, the sample is typically preserved in dilute buffered formalin solution (5%, Section 4.8.3, Box 4.8) immediately after collection and placed in a plastic jar with a screw top lid. Buffered formalin contains ~5% formaldehyde buffered with sodium borate. To view zooplankton under a microscope in the laboratory, it is easiest to use an eye dropper to transfer a small number of individuals from the jar to a flat-bottomed Petri dish, preferably unlined on the bottom for better clarity. A fine metal probe is useful for moving and flipping zooplankton to see diagnostic features. A set of fine forceps is handy so you can gently pick up specimens and place them in smaller dishes (known as watch glasses or embryo dishes) for more detailed investigation.
A plastic squeeze bottle filled with tap water is useful to top up dishes so there is sufficient water to fully immerse specimens. When counting zooplankton routinely, you will need a way of measuring the exact volume of a sub-sample, and one method is to use a Stempel pipette with inserts of different volumes. The sub-sample from a Stempel pipette is then usually transferred to a Bogorov tray, which has a channel in a zig-zag pattern so you can systematically work through a sample. For identifying smaller species, sometimes the specimen is placed on a glass slide with a coverslip and identified using a compound microscope with higher magnification.

Copepods (see Section 8.3.1): Small, bullet or tear-drop shaped crustaceans that are semi-transparent, with a moderately hard jointed outer skeleton. Almost always the most abundant zooplankton.

Shrimp-like (8.3.2): Large, elongate crustaceans with stalked eyes and large limbs, semi-transparent, and a moderately hard jointed outer skeleton. Common, particularly in coastal waters, but not as abundant as copepods. Decapods such as prawns, krill (euphausiids) and mysids are reminiscent of shrimp, while others such as stomatopods, amphipods, cumaceans and isopods are more loosely bug or shrimp-like.

Tick-like (8.3.3): Small, often hairy, and look like ticks or mites. These are the young nauplius stages of copepods and other crustaceans. They can be abundant if you use a fine mesh net.

Water fleas (8.3.4): Also known as cladocerans, and similar in appearance to water fleas in fresh water. Sometimes abundant in coastal waters.

Clam-like (8.3.5): Small and look like a clam or bivalve – that is, they have two shells (valves), usually hard or moderately hard. These include the true bivalves (e.g. mussels, clams), bryozoans, and small crustaceans such as ostracods and the cyprid stage of barnacles. They are sometimes abundant in coastal waters.

Worm-like (8.3.6): Small to medium, long, semi-transparent, soft, and often abundant. This includes the true worms (polychaetes, which are annelids with bristles), the arrow worms (chaetognaths), and the appendicularians (larvaceans, within the chordates).

Jelly-like (8.3.7): Small to large, often round, transparent, soft, and periodically abundant. A few are large and conspicuous but most are small. These are the jelly-like (gelatinous) zooplankton comprising the true jellyfish (cnidarians), comb jellies (ctenophores), and salps and doliolids (thaliaceans, within the chordates).

Snails (8.3.8): Medium to large, often with a prominent hard shell, but containing a soft animal that is transparent and jelly-like. Periodically abundant. These molluscs include the pteropods, heteropods and prosobranchs. Pteropods and heteropods are holoplanktonic, and prosobranchs are meroplanktonic.

Spiky (8.3.9): Tiny to small, with spines and spikes, often hard, and only sometimes abundant. These include forams and radiozoans (radiolarians), echinoderm larvae, and polychaete larvae (young of true worms).

Fish and squid (8.3.10): Large, usually elongate, prominent eyes, semi-transparent, soft. These are larvae of fish and squid. Squid closely resemble their adults, but fish less so.

Eggs (8.3.11): Small, spherical, transparent and soft, they rarely have spines. Most marine animals shed their eggs into the water, and the most common are those of copepods, krill, decapods and fish.

Vase-like (8.3.12): Tiny, vase or condom-shaped, semi-transparent, soft and sometimes abundant.

Oddities (8.3.13): Other zooplankton that either have few distinguishing features or are distinctive but rare. These include brachiolaria larvae, hemichordate larvae, trochophore larvae, insects and cephalochordates, among others.

Phytoplankton (8.3.14): Larger phytoplankton (photosynthetic plankton) that usually has a greenish tinge. Phytoplankton is spherical, in chains, or in filaments, with little internal structure visible. Phytoplankton is sometimes abundant in zooplankton samples.

A good way to learn to recognise these 14 different body types commonly found in zooplankton samples is to separate the specimens in a sample into these body types. This is best done by using a set of fine forceps and gently picking up and placing similar specimens in separate dishes (with a little tap water from a squeeze bottle). It takes a bit of practice so that you can see the end of the forceps while looking down the microscope. Once you have placed similar specimens in individual dishes, compare the specimens across dishes, paying careful attention to differences in body type (i.e. their jizz). This approach will train you to be more observant of key distinguishing features,

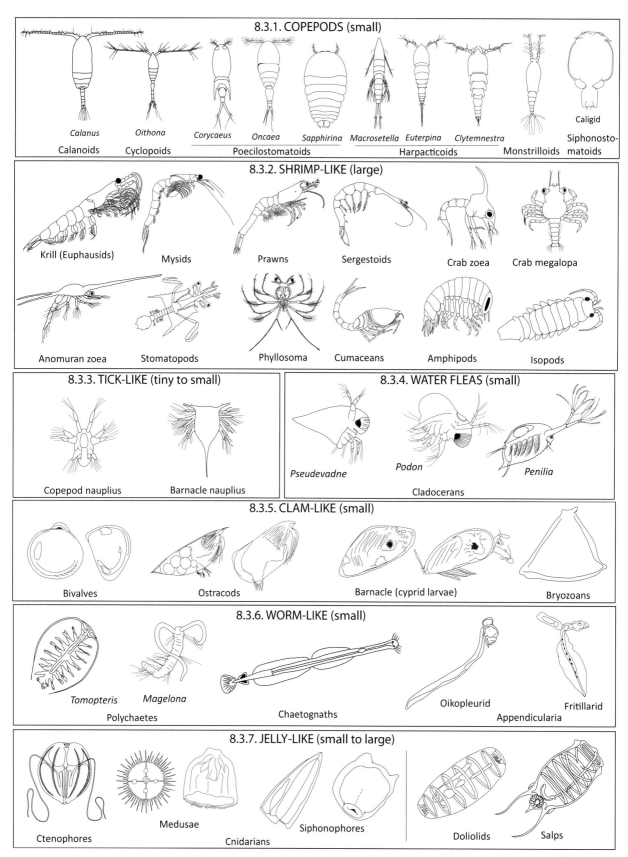

8.3.1. COPEPODS (small)

Calanus — Calanoids
Oithona — Cyclopoids
Corycaeus *Oncaea* *Sapphirina* — Poecilostomatoids
Macrosetella *Euterpina* *Clytemnestra* — Harpacticoids
Monstrilloids
Caligid — Siphonostomatoids

8.3.2. SHRIMP-LIKE (large)

Krill (Euphausids)
Mysids
Prawns
Sergestoids
Crab zoea
Crab megalopa
Anomuran zoea
Stomatopods
Phyllosoma
Cumaceans
Amphipods
Isopods

8.3.3. TICK-LIKE (tiny to small)

Copepod nauplius
Barnacle nauplius

8.3.4. WATER FLEAS (small)

Pseudevadne
Podon
Penilia
Cladocerans

8.3.5. CLAM-LIKE (small)

Bivalves
Ostracods
Barnacle (cyprid larvae)
Bryozoans

8.3.6. WORM-LIKE (small)

Tomopteris *Magelona*
Polychaetes
Chaetognaths
Oikopleurid
Fritillarid
Appendicularia

8.3.7. JELLY-LIKE (small to large)

Ctenophores
Medusae
Cnidarians
Siphonophores
Doliolids
Salps

(Continued)

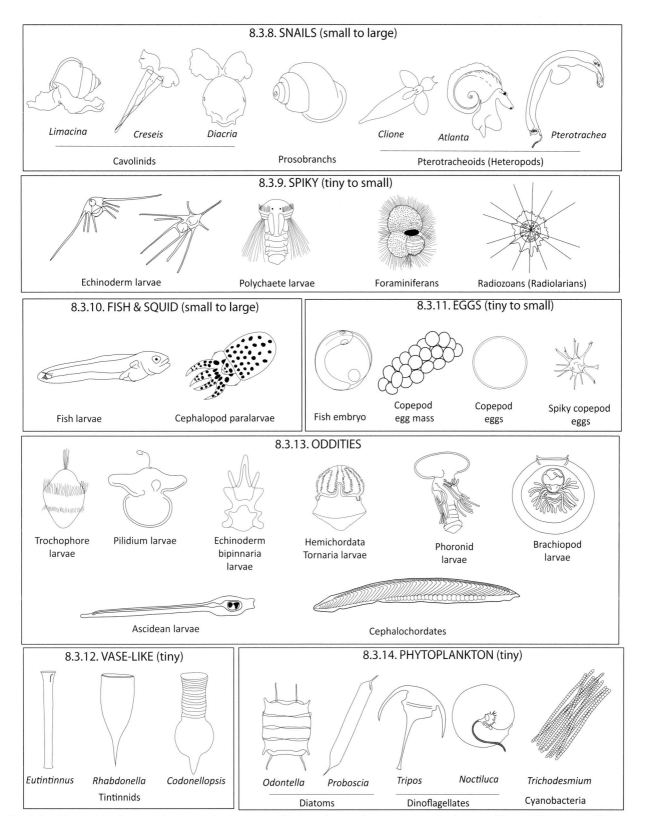

Fig. 8.5. Typical body types in a zooplankton sample. Numbers refer to sections in the text. Tiny <0.5 mm, small 0.5–5 mm (size range of most copepods), large >5 mm.

increasing your identification skills and confidence. The next two sections will mention briefly how we can use both body size (Section 8.2.1) and abundance (Section 8.2.2) to assist in the identification of specimens, before we describe the 14 body types and their taxonomic groups in detail (Section 8.3).

8.2.1 Size matters

Body size can help us identify zooplankton because different body types and their taxonomic groups have characteristic sizes. Although most zooplankton are microscopic, they span five orders of magnitude in size, from ~0.1 mm to greater than 10 000 mm (10 m) (Fig. 8.6). Size not only provides

taxonomic insights but is often used instead of taxonomic identification to understand and model marine ecosystems because metabolism, growth, swimming speed, prey size, and predator size are all related to the size of an organism: a phenomenon known as allometry.

Size can be used to help narrow down the possible body types a specimen belongs to. For example, if you have a crustacean greater than ~5 mm in size, then it is very unlikely to be a copepod because most are smaller than 3 mm. It would also not be a water flea (cladoceran) because none are over 2 mm in size. Although Fig. 8.5 provides the relative sizes of the different body types from tiny to large, a better sense of their relative sizes is provided in

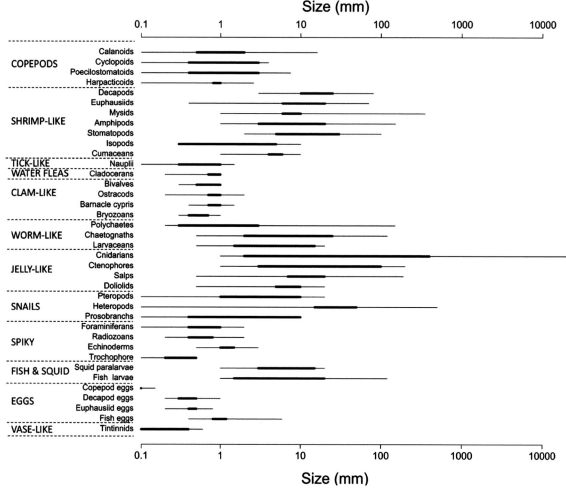

Fig. 8.6. Zooplankton sizes. Dark lines represent the size range of common members of each zooplankton group, and light lines the extreme size ranges. Note the use of a log size scale because zooplankton vary in size by five orders of magnitude. Sizes sourced from books on the reference list.

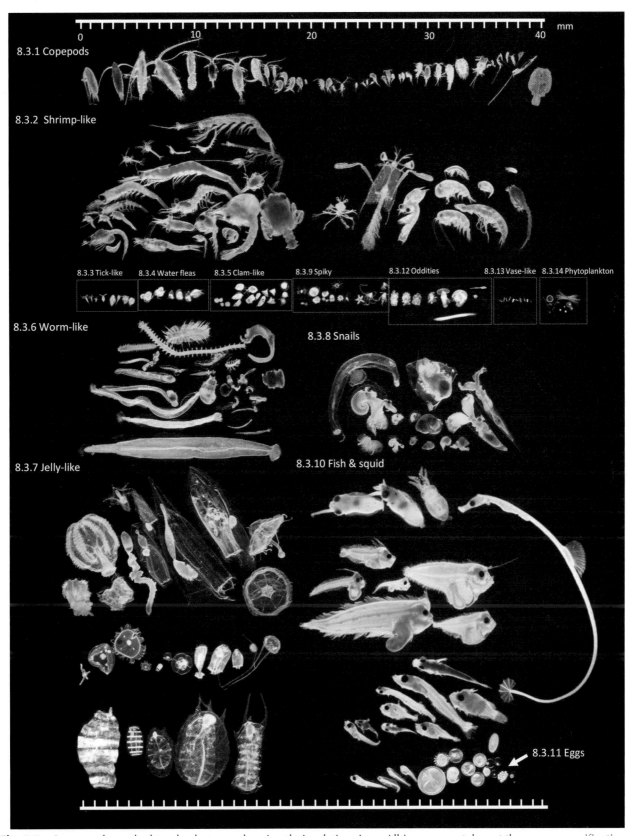

Fig. 8.7. Images of zooplankton body types showing their relative sizes. All images are taken at the same magnification. Numbers refer to sections in the text. The scale bar is 40 mm long and divided into 1 mm increments.

Fig. 8.7 where images of all 14 body types were taken at the same magnification. The range of sizes is striking. When looking down a microscope, assessing size relative to the small yet super-abundant copepods is the easiest approach. It is clear how tiny the vase-like body type is, and how much smaller body types such as water fleas, clam-like, tick-like and spiky are than copepods, and how much smaller copepods are than zooplankton that are shrimp-like, worm-like, jelly-like, and fish and squid.

8.2.2 Abundance matters

The relative abundance of different zooplankton groups can inform your identification. This is akin to bird identification guides that provide information on whether a species is abundant, common, rare or a vagrant – this is useful information because an unknown bird that you see is more likely to be an abundant species than a vagrant. Fig. 8.8 provides examples of what might be expected to be seen most frequently in a zooplankton sample. This illustration was based on 659 daytime samples collected monthly with a 100 µm mesh net hauled vertically in coastal waters from nine locations around Australia. The relative abundance of zooplankton groups in your samples is likely to be somewhat similar but not the same. It depends on: the mesh size of the net used (greater proportion of larger groups using a larger mesh size); the type of tow (oblique tows at 2 knots capture relatively more larger groups than vertical tows); the time of day of sampling (relatively more meroplankton are captured at night); distance from coast (more coastal and meroplanktonic species close to the coast); depth of sampling (some groups are mainly herbivorous and are found predominantly in the top 100 m), among other factors. Nevertheless, Fig. 8.8 provides a useful guide to the typical relative abundance of different body types and taxonomic groups.

Although the zooplankton in your sample are likely to cover a range of sizes, copepods almost always dominate numerically. This is why we have

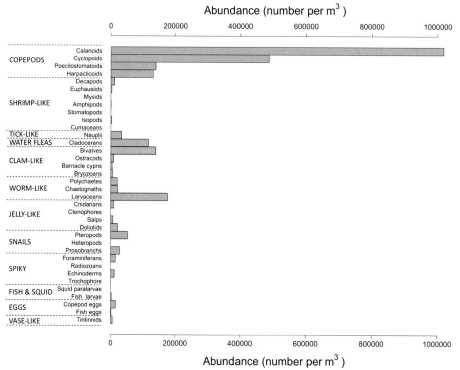

Fig. 8.8.   Relative abundances of different zooplankton body types and taxonomic groups. Based on 659 samples collected with a 100 µm-mesh net from nine National Reference Stations that are part of the Australian Integrated Marine Observing System (IMOS).

defined copepods as small (0.5–5 mm), and all other zooplankton sizes (tiny: <0.5 mm and large: >5 mm) relative to copepods. Note that when you look down a microscope at plankton, you do not always know the size of the organisms exactly (it is dependent upon magnification and field of view), but you always see copepods.

8.3 From body types to taxonomic groups

Once you have identified a specimen to body type in Fig. 8.5, you can go to the relevant section on that body type and then select the taxonomic group from a limited number of options. Almost all specimens are identifiable to body type and taxonomic group, but going further can be challenging. It not only requires experience, but some specimens may not be identifiable to species or even genus because many young stages do not possess the diagnostic features of adults, or specimens are damaged. So, this chapter focuses on identification to major taxonomic groups.

8.3.1 Copepods (Subclass Copepoda)

Marine copepods (from Ancient Greek *kōpē* meaning 'oar' and *podós* meaning 'foot') are small crustaceans 0.2–10 mm in length, but more typically 0.5–3 mm long, with a teardrop-shaped body and a pair of antennae, although there are many variations on this typical body form (Fig. 8.9). Like other crustaceans, they have an outer exoskeleton, but

this is so thin that the entire body is almost transparent. Copepods have no carapace (i.e. no hard upper shell like a crab) and a single compound, sessile eye. There are no limbs on the abdomen, which is a distinctively thinner 'tail' compared with the thorax. Many copepods have feathery appendages and long caudal setae ('hairs' at the end of the abdomen) that slow their rate of sinking.

About 15 000 species of free-living and parasitic copepods have been described, but there is considerable uncertainty in the total number of copepod species, with some estimates up to 450 000 species. The good news is that almost all the common species have been described!

Marine copepods are pelagic, hyperbenthic, benthic, commensal or parasitic. They are found in all depth and biogeographical zones of the ocean. As copepods can constitute 90% of the abundance of marine zooplankton, they are an important trophic link in marine food webs. Copepods grow from an egg through six larval (nauplius) stages and through a further six juvenile or copepodite (looks like an adult copepod) stages, with the last stage being a sexually reproducing adult (see Chapter 2).

Quick guide

The four main groups of pelagic copepods – calanoids, cyclopoids, poecilostomatoids and harpacticoids – are strictly distinguished based on their articulation or division of their body into major

Fig. 8.9.   Copepods. CALANOIDA: **(A)** *Temora turbinata*, **(B)** *Acrocalanus gibber*, **(C)** *Parvocalanus crassirostris*; CYCLOPOIDA: **(D)** *Oithona* spp. (large one on left is *O. plumifera*); POECILOSTOMATOIDA: **(E)** *Oncaea* spp. (mainly *O. venusta*), **(F)** *Corycaeus* spp., **(G)** *Sapphirina* spp., **(H)** *Copilia mirabilis*; HARPACTICOIDA: **(I)** *Euterpina acutifrons*, **(J)** *Macrosetella gracilis*, **(K)** *Microsetella norvegica*, **(L)** *Clytemnestra* spp., **(M)** *Metis holothuriae*; MONSTRILLOIDA: **(N)** monstrilloid; SIPHONOSTOMATOIDA: **(O)** juvenile caligid.

parts – the head and body (prosome) and their 'tail' (urosome) – although this articulation is difficult to see using a stereo microscope and is not used in routine identification. The main distinguishing features are the size, body shape, length of antennae, the relative length of their body (prosome) and their 'tail' (urosome), and the position of the fifth leg relative to the prosome and urosome (Fig. 8.10).

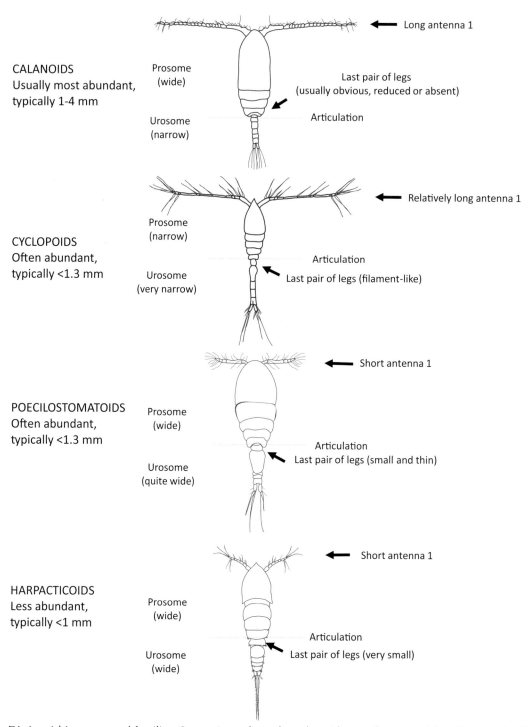

CALANOIDS
Usually most abundant, typically 1-4 mm

Long antenna 1

Prosome (wide)

Last pair of legs (usually obvious, reduced or absent)

Articulation

Urosome (narrow)

CYCLOPOIDS
Often abundant, typically <1.3 mm

Relatively long antenna 1

Prosome (narrow)

Articulation

Urosome (very narrow)

Last pair of legs (filament-like)

POECILOSTOMATOIDS
Often abundant, typically <1.3 mm

Short antenna 1

Prosome (wide)

Articulation

Last pair of legs (small and thin)

Urosome (quite wide)

HARPACTICOIDS
Less abundant, typically <1 mm

Short antenna 1

Prosome (wide)

Articulation

Last pair of legs (very small)

Urosome (wide)

Fig. 8.10. Distinguishing copepod families. Separating calanoid, cyclopoid, poecilostomatoid and harpacticoid copepods under a stereo microscope.

Calanoids: Typically, the largest (1–4 mm) and most numerous copepods, calanoids are bullet-shaped, have long antennae about the length of the body, and have a wide prosome and narrow urosome. The prosome is longer than their urosome, and the last pair of legs when present is clearly on the prosome.

Cyclopoids: Usually smaller than calanoids (<1.3 mm) and often abundant, cyclopoids are delicate, have relatively long antennae, and a narrow prosome and a very narrow urosome. The prosome is about the same length as the urosome, and the last pair of legs is filament-like and on the urosome.

Poecilostomatoids: Smaller than calanoids (<1.3 mm) and often abundant, poecilostomatoids have short antennae, and a wide prosome and a quite wide urosome. The prosome is longer than urosome, and the last pair of legs is on the urosome and not easily visible. Poecilostomatoids are generally sturdier than cyclopoids and their urosome is not as narrow.

Harpacticoids: Small copepods (<1 mm) less abundant in the water column than other copepod orders and most are benthic. Harpacticoids have short antennae, and a wide prosome and quite wide urosome, resulting in little distinction between the two and making them appear a little 'worm-like'. The prosome is only a bit longer than the urosome, and the last pair of legs is on the urosome and not easily visible.

With a bit of practice, you will be able to distinguish the four main groups of copepods based on these diagnostic characteristics. There are two other groups – the monstrilloids and siphonostomatoids – that are occasionally seen in zooplankton samples (Fig. 8.9). Monstrilloida is an order of copepods with a worldwide distribution, but relatively rare. The life cycle is unusual because the larvae are parasites of benthic gastropods and polychaetes, whereas the adults are pelagic and cannot feed and their sole purpose is for reproduction. Siphonostomatoida is an order of copepods that contains 75% of all copepods parasitic on fish. Only their young stages are seen in the zooplankton, whereas their adults have found a host fish. These copepod orders will not be dealt with further.

8.3.1.1 Calanoids (Order Calanoida)

The Order Calanoida is the most numerous and diverse group of pelagic copepods.

Identification: Large (1–4 mm), bullet-shaped copepods, with their body (prosome) broader and longer than their tail (urosome), and antennae about the length of the body (Fig. 8.11). They are often the most common zooplankton present in a sample (Fig. 8.8). The sexual dimorphism in calanoid copepods, where males and females have

Fig. 8.11.   CALANOIDA: **(A)** *Cosmocalanus darwinii*, **(B)** *Calanoides* sp., **(C)** *Pareucalanus sewelli*, **(D)** *Subeucalanus crassus*, **(E)** *Euchaeta concinna*, **(F)** *Candacia ethiopica*, **(G)** *Labidocera acuta*, **(H)** *Undinula vulgaris*, **(I)** *Centropages furcatus*, **(J)** *Acartia pacifica*, **(K)** *Temora turbinata*, male (bottom) and female (top), **(L)** *Acrocalanus gibber* (male), **(M)** *Acrocalanus gibber* (female), **(N)** *Parvocalanus crassirostris*.

different morphology, is used for identifying species and separating the sexes (Box 8.2).

Diversity: There are ~40 families of calanoid copepods, with ~1800 species in both marine and freshwater systems.

Some common species: Tropical species of calanoid copepods are often small, such as *Acrocalanus* (Figs 8.9B, 8.11L,M) and *Parvocalanus crassirostris* (Figs 8.9C, 8.11N), the smallest species in the world. *Temora turbinata* (Figs 8.9A, 8.11K) is a widespread medium-sized calanoid species. *Acartia* (Fig. 8.11J) is a global genus that has a prominent eye spot on its head and is often common in both estuarine and oceanic waters. In productive upwelling areas, *Calanoides* spp. are common (Fig. 8.11B) because they have lipid reserves to survive diapause ('hibernation') in cold bottom-waters until they are brought to the surface during an upwelling event. Large copepods from the genera *Cosmocalanus* (Fig. 8.11A), *Pareucalanus* (Fig. 8.11C) and *Subeucalanus* (Fig. 8.11D) mainly feed on phytoplankton. Carnivorous genera such as *Eucheata* (Fig. 8.11E), *Candacia* (Fig. 8.11F) and *Labidocera* (Fig. 8.11G) are large and have prominent raptorial feeding appendages. The copepod *Undinula vulgaris* (Fig. 8.11H) is a large copepod that often has a blue tinge when alive (lost when fixed) and forms dense blooms in tropical coastal regions.

Lifespan: Calanoid copepods generally live from a couple of weeks to months, depending on species and water temperature. Diapause can extend life to years for cold-water species.

Feeding: Calanoid copepods can be detritivores, herbivores, omnivores or carnivores, or a combination of these, depending on the food environment. For herbivorous species, feeding currents are created by moving the antennae and maxillipeds (some of the mouth parts) to create a feeding current that draws water and phytoplankton towards the copepod. For carnivorous species, sensors on the antennae detect prey based on size, movement or 'smell', and then they use large raptorial feeding appendages to capture their prey.

Reproduction: Generally, females produce pheromones, and males search, find and follow the signal. Males thus have more chemo-sensors on their antennae than females. Some species form swarms, allowing males to find females more easily. Males attach a spermatophore to the urosome of the female. In calanoid copepods, the highly fecund broadcast spawners shed their eggs into the water and have no parental care, whereas the less fecund sac spawners carry their eggs and invest more in parental care. Most species produce more than one clutch of eggs.

Locomotion: The five pairs of thoracic limbs are well developed for swimming using comb-like rows of setae. Sinking is controlled by their fat content and extension of the antennae.

Ecology: Calanoid copepods are often the dominant zooplankton, commonly constituting 55–95% of the organisms in plankton samples (Fig. 8.8). As calanoids are also the largest copepods, they frequently dominate the biomass. They are therefore

Box 8.2. Using sexual dimorphism in calanoid copepods to separate sexes and species

If you look carefully with a stereo microscope, you will be able determine the sex of adult copepods. Sexual dimorphism is the morphological difference between males and females. Female and male calanoid copepods differ in several ways (for images, see *Australian Marine Zooplankton: Taxonomic Guide and Atlas* www.imas.utas.edu.au/zooplankton). First, females are usually 20% larger than males. Second, females typically have fewer segments in their urosome than males. Third, females have a bulging gonadal segment (where the eggs are released) on their urosome; this is absent in males. Fourth, females have symmetrical antennae, whereas males often have one geniculate antenna (it appears to have an 'elbow') that is used to grab the female during mating. Last, the fifth pair of swimming legs (i.e. the one closest to the abdomen) of females is symmetrical, whereas the fifth pair of legs for males is often asymmetrical, with a longer one modified for grabbing females during mating. The fifth legs of males, and to a lesser extent of females, are used for species-level identification.

important in energy transfer from phytoplankton to higher trophic levels. Many commercial fish are dependent on calanoid copepods as food during their larval, juvenile or adult stages. They are also important food items for baleen whales and some seabirds. Many calanoid copepods perform diel vertical migration, where they typically migrate to phytoplankton-rich surface waters at night to feed and then descend to deeper layers during the day where they are not as detectible to visual predators. The excretion, egestion and death of calanoid copepods at depth while vertically migrating is an important component of the biological pump, where carbon is transferred from surface layers to the deeper ocean.

Human interactions: The calanoid copepod *Calanus finmarchicus* is reported to be the largest renewable and harvestable resource in the North Atlantic and is now the basis for a bio-industry harvesting 1000 tonnes per year for omega-3 fatty acids. Calanoid copepods, particularly those of the genus *Acartia* that have resting eggs, are cultured for use as a live feed in marine aquaculture production to supplement feeding of juvenile fish and prawns. *Acartia* are also used in toxicity testing because they are extremely sensitive to chemicals, they can be mass cultured, they grow quickly, and they have distinct stages that provide endpoints to determine toxicity of contaminants. Zooplankton with chitinous exoskeletons, particularly copepods, are hosts for bacterial pathogens such as *Vibrio cholerae*, which is responsible for ~5 million cases and 120 000 deaths per year. Other *Vibrio* species found in zooplankton include *Vibrio parahaemolyticus* and *Vibrio vulnificus*, which were responsible for significant bacterial pandemics. The amazing diversity of calanoid copepod forms have been popularised by Ernst Haeckel's classic 'Art forms in Nature' in 1904 (https://en.wikipedia.org/wiki/Kunstformen_der_Natur#/media/File:Haeckel_Copepoda.jpg).

8.3.1.2 Cyclopoids (Order Cyclopoida)

Identification: Cyclopoids (from Ancient Greek *Kúklōps* meaning 'Cyclops', a one-eyed giant from Greek mythology) are generally delicate, have short antennae, and a much broader body (prosome) than their very thin 'tail' (urosome), which is relatively long (about the same length as the prosome) (Figs 8.9D, 8.10). Cyclopoids are small (<1.3 mm), with many benthic and some pelagic species.

Diversity: The taxonomic position of cyclopoids and poecilostomatoids has been in a state of flux in recent years. Before 1991, cyclopoids and poecilostomatoids were grouped together in the Order Cyclopoida. From 1991 to 2017, poecilostomatoids were elevated to their own order, distinct from the cyclopoids. Since 2017, poecilostomatoids have been once again placed in the Order Cyclopoida. In this book we treat cyclopoids and poecilostomatoids as distinct groups. The most common marine genus of cyclopoids is *Oithona*.

Some common species: Most species of cyclopoids are found in fresh water, although *Oithona* is cosmopolitan in marine waters (Fig. 8.9D). The species *Oithona similis* could be the most abundant multicellular animal on Earth. *Oithona* species are difficult to discriminate, even for experts, although the large and setose *O. plumifera* is distinctive and common in oceanic waters (Fig. 8.9D). In coastal waters *O. simplex* and *O. brevicornis* are common.

Lifespan: Cyclopoids typically live several weeks, although deeper water and polar species are likely to live longer.

Feeding: Cyclopoid copepods are omnivorous, often feeding on 'marine snow' that forms from remains of phytoplankton, small zooplankton, and appendicularian houses, which may be colonised by bacteria.

Reproduction: As it can be difficult to find a mate in the vast marine environment, many cyclopoid copepods exhibit pre-copulatory mate guarding, where the male attaches to the female while still she is still immature. This means that he can fertilise the female as soon as she matures. Once fertilised, cyclopoid copepods carry up to 100 eggs (60–100 µm in size) in two egg sacs attached on each side of the urosome on females. Most species produce more than one clutch of eggs.

Locomotion: Male *Oithona* spp. make forward spiralling movements looking for females, and females making circular swimming motions searching for food. Nauplii, copepodites and females move forward by leaping, whereas males move forward by paddling.

Ecology: Due to their abundance, cyclopoid copepods play an important role in aquatic food webs as either primary consumers or predators. They can be an important food source for larval, juvenile and adult fish. A conventional 200 μm (0.2 mm) mesh net probably captures <10% of meso-zooplankton numbers, primarily because of under-estimating *Oithona* spp. and poecilostomatoids.

Human interactions: Cyclopoid copepods, including *Oithona* spp., are cultured for use as a live feed in marine aquaculture production to supplement feeding of juvenile fish and prawns.

8.3.1.3 Poecilostomatoids (Suborder Poecilostomatoida)

Identification: Small (<1.3 mm) and often very abundant, poecilostomatoids have short antennae (much shorter than calanoids), and a stockier body (Fig. 8.9E–H, 8.10) than the cyclopoid *Oithona*. The classification of poecilostomatoid copepods has been established on their mouth structure, although this is not easily visible using a stereo microscope.

Diversity: Although poecilostomatoids were placed in the Order Cyclopoida in 2017, here we treat poecilostomatoids and cyclopoids as distinct groups, as is common practice currently. There are currently 1103 species of poecilostomatoids described. The most common genera are *Oncaea*, *Corycaeus* and *Sapphirina*.

Some common species: The most common genus of pelagic poecilostomatoids is *Oncaea*, of which four species are commonly encountered in near surface plankton tows (Fig. 8.9E): *O. venusta*, *O. media*, *O. mediterranea* and *O. clevei*. The genus *Corycaeus* is also commonly encountered and is distinguished by its distinctive eye lenses for detecting prey (Fig. 8.9F). The genus *Copilia* is large and has a squarish prosome with large eye lenses (Fig. 8.9H). When alive, the most beautiful copepod in the

world is probably *Sapphirina* (Fig. 8.9G). Sometimes looking over the side of a boat on a still day you will see the red and purple iridescent glint off the flattened body of *Sapphirina*. Even more of a treat is to see the iridescence of *Sapphirina* when you shine a light on it at night. The sparkle in *Sapphirina* is only seen in males, which live free in the water column. Females live inside salps and are parasitic on their hosts. Males 'spiral swim' as they shimmer, competing for the attention of females. The iridescence in *Sapphirina* is a consequence of light shining on microscopic layers of crystal plates inside their cells, akin to interference colours in an oil sheen.

Lifespan: No information is available about the longevity of poecilostomatoids in the natural environment. In the laboratory, a long period without feeding was recorded for deep-sea *Triconia* species from the subarctic Pacific Ocean (minimum longevity: 208 days for females and 158 days for males).

Feeding: Many poecilostomatoids have a more pseudo-pelagic lifestyle than calanoids or cyclopoids, and are often associated with gelatinous organisms such as salps or appendicularians upon which they also feed. They have also been observed attacking organisms much larger than themselves, such as chaetognaths or larger copepods. The structure of their mouthparts shows that they are not filter-feeders, but their feeding habits are opportunistic, feeding on prey ranging from particle-loaded mucus and small algae to crustaceans larger than themselves.

Reproduction: Some poecilostomatoids exhibit pre-copulatory mate guarding, where the male attaches to the female while she is still immature, and must detach and reattach when she moults. This means that he can fertilise the female as soon as she matures. Although a few species carry single egg sacs, most females have paired sacs attached laterally or dorsally, never on the ventral surface as in harpacticoids.

Locomotion: There are four pairs of swimming legs; the fifth pair is rudimentary and not used for swimming. The swimming behaviour of poecilostomatoids is variable; they are able to swim freely (diurnal vertical movements of several hundred metres have been reported for some larger species such as *Triconia conifera*), but they are usually attached to or crawling on substrates or other material/organisms in the water column.

Ecology: Despite living in the water column, *Oncaea* spp. are pseudo-benthic, as they find sinking aggregates by following their chemical trail and feed on their surface. Because poecilostomatoids are small, they are massively underestimated using a 200 µm mesh net. Many poecilostomatoids are ectoparasites in the mouth or gills of marine fish or invertebrates. *Oncaea* spp. are food for fish larvae and particularly mesopelagic lanternfishes (Myctophidae) in the mesopelagic zone.

Human interactions: Some species of poecilostomatoids are used in marine aquaculture, especially *Oncaea*.

8.3.1.4 Harpacticoids (Order Harpacticoida)

Identification: Harpacticoids (from Ancient Greek *harpacticon* meaning 'rapacious predator') are small (<1 mm), have short antennae, and have little distinction between the prosome and the urosome (Fig. 8.9I–M). Harpacticoids are not as abundant in the pelagic environment as other copepod orders. To reduce sinking rates, some harpacticoids have distinctively long tail setae almost as long as the animal (Fig. 8.9K,M).

Diversity: Although there are ~3400 species of harpacticoids described, most are benthic or epibenthic, a few are ectoparasites of corals, tunicates, crabs and baleen whales, ~700 are found in freshwater, and there are only 17 truly pelagic species. Non-pelagic species are occasionally taken if a plankton net is towed too close to the sea floor, but their identification requires knowledge of benthic species.

Some common species: Common harpacticoid species are *Euterpina acutifrons* (Fig. 8.9I), *Macrosetella gracilis* (Fig. 8.9J) and *Microsetella* spp. (Fig. 8.9K). Other genera likely to be encountered are *Clytemnestra* (Fig. 8.9L) and *Metis* (Fig. 8.9M).

Lifespan: Harpacticoids are likely to live for many months.

Feeding: Harpacticoids feed inefficiently on suspended food, but feed well on food attached to surfaces, such as those offered by marine aggregates such as 'marine snow' that forms from remains of phytoplankton, small zooplankton and appendicularian houses.

Reproduction: Many harpacticoid copepods exhibit pre-copulatory mate guarding, where the male finds a female through a pheromone trail and then attaches to her while still immature – he will release her while she moults between stages, and then re-grasp her. This means that he can fertilise the female as soon as she matures. Female harpacticoids carry their eggs in paired sacs on the abdomen.

Locomotion: Most harpacticoids are not well adapted for swimming but can crawl on surfaces.

Ecology: Because of their poor swimming ability, most pelagic harpacticoids live a pseudo-benthic existence in the pelagic environment. This pseudo-pelagic existence offers an opportunity for life in the open ocean, possible shelter from predation, and greater dispersal compared with benthic copepods. For example, various life stages of the harpacticoid copepod *Macrosetella gracilis* (Fig. 8.9J) live and feed on colonies of the nitrogen-fixing cyanobacterium *Trichodesmium* spp., which is common in nutrient-poor tropical and subtropical waters. The presence of *M. gracilis* helps support a mini-community on the *Trichodesmium* spp., making them more nutritious for higher trophic levels. Another common pelagic harpacticoid, *Microsetella norvegica*, is unable to feed efficiently on suspended food, but can graze well on the surface of attached food such as marine aggregates. During bloom conditions, *M. norvegica* can feed on three aggregates per day.

Human interactions: Because harpacticoid copepods do not have resting eggs, they are not used in marine aquaculture as commonly as calanoid or cyclopoid copepods, although the benthic genus *Tigriopus* is used. Harpacticoids are the preferred copepod species in marine aquaria because their lifestyle of living on substrates makes them ideal for cleaning the substrate and aquarium panels, and their nauplii and copepodites provide food for invertebrates such as corals, clams and sea cucumbers.

8.3.2 Shrimp-like

Of Crustacea, there were many strange and undescribed genera.

Charles Darwin, December 1833, in his diary during the voyage of the *Beagle* (Darwin 1845)

The shrimp-like or elongate zooplankton include the larvae of commercial crustaceans that are familiar to us, although they often have different common names. For example, in Australia commercial penaeid crustaceans are called prawns, and in the US and Japan they are called shrimps (in the US big pandalid shrimp are called prawns). Here we use the term shrimp-like to not only include young stages of prawns and shrimps, but also those of other commercial species such as spiny lobsters, crabs and krill. We further include other crustaceans not of commercial importance including mysids, amphipods, stomatopods and cumaceans. This is a large group and it is challenging to describe the suite of shrimp-like organisms because different families typically have larval forms that have been given specialised names. In your zooplankton sample, depending on the group and the species present, you can capture a mix of larvae, juveniles and adults.

Quick guide

These are usually large zooplankton with big eyes and long limbs and can be identified into major groups:

Decapod larvae: Many look like miniature shrimps/prawns. Others look more crab-like with pincers on legs and spikes on the cephalothorax (combined head and thorax). Eyes have a reddish-brown ring around a darker centre.

Euphausiids (krill): Adult and juvenile krill appear shrimp-like. Presence of external gills at the base of the legs on the cephalothorax and photophores at the base of the pleopods on the abdomen

are diagnostic. Setose (hairy) thoracic legs (pereiopods) form a filter-feeding basket.

Mysids: Mysids appear shrimp-like, and have loose articulation between the cephalothorax and abdomen. Almost all species have statocysts on the uropods (they look like a tiny dartboard with a bullseye).

Amphipods: Almost all laterally flattened, so you see them on their side. Eyes can be large (hyperiids) and are always unstalked. Typically have one or two large pairs of forelimbs.

Stomatopods: Large 'arms' (maxillipeds) and large eyes for predation. Squarish carapace with two long spines at rear. Telson has polygonal shape and is serrated.

Isopods: Look like a tear drop (round at front, pointed at rear), segmented, dorso-ventrally flattened, eyes appear as spots on carapace.

Cumaceans: Bulbous head and thorax covered by a carapace with a long slender abdomen with a forked tail and telson. Most species are 1–10 mm in length and have two indistinct sessile, compound, eyes but these may be fused or absent.

When we start identifying shrimp-like animals in zooplankton samples, it can be difficult to distinguish decapods, euphausiids and mysids: Box 8.3 provides advice on how best to distinguish these groups.

8.3.2.1 Decapods (Order Decapoda)

Identification: Most decapods (from Ancient Greek *deka* meaning '10' and *podós* 'foot') captured in a zooplankton sample will be larvae rather than adults. Decapod larvae are usually relatively big (>5 mm) compared with other zooplankton, have large eyes, have segmented appendages (five pairs of legs when fully developed, with some being clawed), often have spines, normally have a distinct cephalothorax covered by a carapace, and have a 'tail' (or abdomen) (Fig. 8.12E–M). Decapods include the typical larger and well-known crustaceans, such as crabs, spiny lobster and prawns, most of which live on or close to the sea floor as juveniles and adults, but spend their early life as larvae in the plankton.

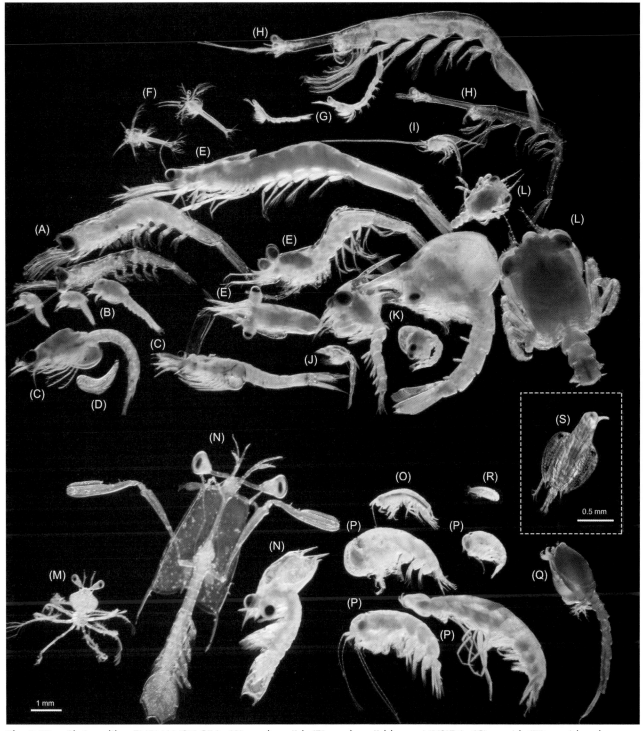

Fig. 8.12.   Shrimp-like. EUPHAUSIACEA: **(A)** euphausiid, **(B)** euphausiid larva; MYSIDA: **(C)** mysid, **(D)** mysid embryo; DECAPODA: **(E)** megalopa stage of decapod species, **(F)** sergestoid protozoea, **(G)** *Lucifer* spp. larvae, **(H)** *Lucifer* spp. adults, **(I)** anomuran zoea (porcellanid), **(J)** axiidean zoea, **(K)** brachyuran zoea, **(L)** brachyuran megalopa, **(M)** phyllosoma larva (lobster); STOMATOPODA: **(N)** stomatopod larvae; AMPHIPODA: **(O)** gammarid, **(P)** hyperiid; CUMACEA: **(Q)** cumacean adult; ISOPODA: **(R)** *Epicaridium* 1st stage isopod larva, **(S)** *Criptoniscus* 3rd stage isopod larva feeding on a calanoid copepod *Acartia pacifica* female.

Box 8.3 Distinguishing decapods, euphausiids and mysids

Several features are useful for distinguishing decapods, euphausiids and mysids. The decapods that look similar to the euphausiids and mysids are mostly the Dendrobranchiata larvae, primarily penaeid (prawn) larvae, although many carideans and axiideans also look similar. Note that crab larvae (often with prominent spines) and lobster larvae (resembling spiders) look quite different from prawns (see Fig. 8.12). Diagnostic features for separating penaeid decapods, euphausiids and mysids, include (Fig. 8.13):

1. If you see external gills at the base of the legs in the head region (the cephalothorax), setose (hairy) thoracic legs (pereiopods) that form a filter-feeding basket, and/or photophores at the base of the pleopods on the abdomen, then it is a euphausiid.
2. If you see statocysts on the uropods then it is a mysid.
3. If you see pincers on the legs then it is a decapod larvae.

   You can be confident if you see these characters, but sometimes specimens could be damaged, making diagnostic features difficult so see. There is a suite of more suggestive characters that can help build the evidence for concluding that a specimen belongs to a particular group. For example, eyes of euphausiids and mysids are typically entirely black, whereas decapod eyes have a reddish-brown ring around a darker centre. Shrimp-like decapods also typically have a more prominent rostrum than euphausiids or mysids.

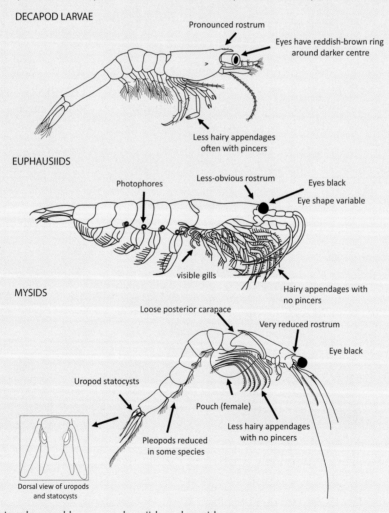

Fig. 8.13. Distinguishing decapod larvae, euphausiids and mysids.

Diversity: The largest group of decapods is the brachyurans, the true crabs, with ~5000 species worldwide, but most of their larvae remain unknown. Other groups of decapods include: the Astacidea, the clawed lobsters and freshwater crayfish; the Achelata (formerly called Palinura), the spiny and slipper lobsters; the Anomura, the hermit crabs, squat lobsters, porcelain crabs, king crabs and coconut crabs; the Caridea, the true shrimps that swim and brood their eggs; and the Dendrobranchiata, the penaeid prawns and sergestoids (e.g. *Lucifer*) that have large stalked eyes and broadcast their eggs rather than carry them on their abdomen like all other decapods. The genus *Lucifer* has recently been placed in its own family Luciferidae, but is together with the sergestids in the superfamily Sergestoidea. We thus use the term sergestoids for sergestids and *Lucifer*.

Some common species: The different types of decapods have a variety of larval forms including nauplius, protozoea, zoea, mysis, postlarva and megalopa. Some of these are illustrated here: protozoea – early larval stages with little development of appendages and use their antennae for locomotion (Fig. 8.12F); zoea – intermediate larval stages with developed cephalothorax appendages but no abdominal appendages (Fig. 8.12I–K); and postlarvae and megalopa – the last larval stages of decapods that resemble adults, with developed abdominal appendages (Fig. 8.12E,L). Crab zoeae have relatively large globular heads and thoracic bodies, often with large and distinctive dorsal and lateral spines, and quite small limbless abdomens (Fig. 8.12I–K). Crab zoea moult into a megalopa larva, taking on the appearance of a small crab (Fig. 8.12L). A distinctive decapod is the holoplanktonic sergestoid shrimp *Lucifer*, with a stalked head and eyes (the so-called ghost shrimp, Fig. 8.12 G,H). The axiidea shrimp are often used as bait and sometimes known as yabbies or mudshrimps and their larvae appear as elongate zooplankton (e.g. *Callianassa*, Fig. 8.12J). Spiny lobster phyllosoma larvae are eye-catching: they are flattened like a leaf and range in size from a few millimetres up to 20 mm (Fig. 8.12M).

Lifespan: Decapods live months to years, with some deep-sea species particularly long lived. Most larval stages metamorphose over weeks to months, a notable exception being phyllosoma larvae, which may spend a year or more in the plankton before settling.

Feeding: Decapod larvae eat a variety of phytoplankton and small zooplankton. Lobster phyllosoma specialise on soft-bodied (non-crustacean) zooplankton such as chaetognaths.

Reproduction: Eggs are fertilised internally or on release from the oviduct. In most species, eggs are carried under the abdomen until hatched then emerge at the zoeal stage. There is high mortality during larval life; for example, penaeids spawn up to 2–3 million eggs in their 1–2 year lifespan, so most offspring die.

Locomotion: Decapod larvae are classified according to the appendages they use for locomotion: nauplius (by antennae), protozoea (after metamorphosis, primarily by antennae), zoea (by exopodites on thoracic appendages) and postlarva (pleopods on abdomen and/or by flexing their abdomen). Thus, early larval stages, before the development of the pleopods, rely on the rowing action of the anterior appendages (antennae and exopodites on their thoracic appendages) to propel the animals in a jerky fashion.

Ecology: Decapods are most abundant in warm, shallow tropical waters. Decapods such as crabs and prawns in estuarine environments are important for the flow of energy from detritus, phytoplankton, benthic algae, macrobenthos and zooplankton to fish.

Human interactions: As adults, many decapod species are commercially important and form the basis of large fisheries around the world. In 2009, global wild-caught decapods (prawns, shrimps, lobsters and crabs) in the ocean totalled 3.4 million tonnes and were worth US$15.5 billion; global catches remain around 3.5 million tonnes. In 2014, a total of 4.2 million tonnes of decapods (prawns and shrimps) were produced in marine aquaculture. Apart from their use as food, chitosan is produced from decapod shells by treatment with sodium hydroxide. Chitosan is used in water treatment, cosmetics, food, beverages, agrochemicals and pharmaceuticals. The antioxidant astaxanthin is a red pigment made from decapods and has been used in the treatment of many diseases and conditions, including Alzheimer's disease, Parkinson's disease, stroke, high cholesterol, age-related macular degeneration, carpal tunnel syndrome, male infertility, symptoms of menopause and rheumatoid arthritis. Decapod larval stages of crabs and prawns are also used in short-term tests for estimating sub-lethal toxicity using moulting development variables as sensitive indicators of toxicity.

8.3.2.2 Euphausiids (Order Euphausiacea)

In the sea around Tierra del Fuego, and at no great distance from the land, I have seen narrow lines of water of a bright red colour, from the number of crustacea, which somewhat resemble in form large prawns. The sealers call them whale-food.

Charles Darwin, December 1833, made during the voyage of the *Beagle* in the Atlantic Ocean (Darwin 1845). The specimens Darwin saw were likely to be euphausiids.

Identification: Euphausiids, commonly called krill, are shrimp-like crustaceans with stalked round or bi-lobed eyes, external 'fluffy' gills, and long and setose feeding limbs (Fig. 8.12A,B). The word euphausiid derives from Ancient Greek *eû* meaning 'good' and *phausia* meaning 'light emitting', because they have bioluminescent organs (appearing as dots) on each segment of their abdomen. Technically, euphausiids are considered part of the meroplankton (up until ~5 mm), because they become part of the micronekton (i.e. small animals that can swim against currents) when larger. This distinction is frequently ignored in analysis of zooplankton net samples because they often retain substantial numbers of adults, especially if collected at night. Nevertheless, most euphausiids captured in a typical zooplankton sample will be larvae.

Diversity: There are 86 species of euphausiids worldwide, which are found in a wide range of oceanic habitats, including tropical, temperate, polar, coastal, oceanic, and deep water. A few species of euphausiids form massive blooms in polar Arctic and Antarctic seas, often discolouring surface waters.

Some common species: The most well-known species and the largest species globally is the Antarctic krill *Euphausia superba*, which supports much of the Southern Ocean ecosystem including fish, seabirds and baleen whales.

Lifespan: Tropical species live for 6 months, whereas polar species live for several years. The larval part of the life cycle is normally completed in several months.

Feeding: Euphausiids are omnivorous. Their feeding strategy varies depending on opportunity and conditions. Generally, phytoplankton is the preferred food of early stages, while older stages eat more zooplankton, fish eggs and larvae, or feed on detritus. Some euphausiids store lipids over winter as energy reserves. A general latitudinal feeding pattern has been observed for euphausiids, where true polar species rely on seasonal phytoplankton blooms and lipid stores, while further away from polar regions, some species are more carnivorous and prefer copepods.

Reproduction: Eggs are released into the water or carried in egg sacs, after internal fertilisation. Eggs of some species can sink to deeper waters, though hatched larvae return to the surface. Eggs usually hatch to a first nauplius stage. Euphausiid larvae are similar to dendrobranchiate (primitive) prawns (e.g. penaeid prawns) and these two groups are the only eucarid groups to have lecithotrophic (feed on internal yolk) nauplii. Their larval stages also have a similar appearance – calyptopis euphausiid larvae look similar to penaeid protozeae, and the furcilia euphausiid larvae look similar to the penaeid mysis. These similarities are thought to signify phylogenetic relatedness.

Locomotion: Early larval stages, before the development of the pleopods, rely on the rowing action of the anterior appendages to propel themselves in a jerky fashion. Later stage larvae swim using their pleopods on the abdomen, as the adults do.

Ecology: The term krill has become synonymous with euphausiids and was first used by Norwegian whalers who applied it to the swarming 'little fish' that signalled whale feeding grounds. Because of their large size and their strong vertical migratory ability, krill are major contributors to the biological pump. Because Antarctic krill feed on ice algae, global warming and the reduction of sea ice could lead to future declines in krill in the Southern Ocean.

Human interactions: The Antarctic krill *Euphausia superba* is fished commercially in the Southern Ocean and in 2014 reached 300 000 tonnes. Other euphausiid species are fished in waters around Japan (~70 000 t per year) and Norway. The Japanese use krill directly as a food (called okiami) and Norwegians eat krill paste. However, most of the krill harvested in the ocean is used as fish feed in aquaculture. Krill contain high levels of omega-3 essential fatty acids derived from the phytoplankton they ingest, and so are used in capsules and oils as human dietary supplements. Omega-3 fatty acids are purported to improve mental health, neurological development in children, inflammation, and reduce breast cancer.

8.3.2.3 Mysids (Order Mysida)

Identification: Mysids (commonly known as opossum shrimps) are relatively large (>5 mm) shrimp-like crustaceans and almost all have a pair of statocysts in the tail fan limbs (the uropods). Statocysts appear like a pair of translucent bulls-eyes on a dartboard (just visible in Fig. 8.12C). Statocysts are used for balance. In mysids, the carapace is relatively loose, attached only to the three anterior segments of the thorax, whereas in euphausiids and decapods the carapace is firmly attached to all eight segments of the thorax. Mysids are slender and have stalked eyes. When alive, mysids are nearly translucent and this allows you to admire their tubular heart, gut peristalsis, and beating limbs (including eight thoracic pairs). Mysid adults and juveniles are seen in plankton samples.

Diversity: Mysids are found in both shallow and deep marine environments, and can be benthic or pelagic. They are particularly abundant in estuaries. There are around 1000 species.

Some common species: Because many mysids are found in shallow coastal habitats and have restricted distributions, different regions have their own species.

Lifespan: Most mysid species probably live up to 6 months of age, although deep-water species probably survive longer.

Feeding: Mysids are omnivores that feed on algae, detritus and zooplankton.

Reproduction: Mysids are characterised by rearing their post-hatching larvae in a brood pouch (marsupium) rather than releasing them into the water column as most other crustaceans (Fig. 8.12D). Thus, mysid eggs are not seen in plankton samples.

Locomotion: Mysids swim using their pleopods.

Ecology: Many marine mysids are found in shallow coastal habitats and emerge into the water column at night. Mysids are rarely seen in daytime plankton sampling or in the open ocean. Mysids are important prey of juvenile fish and prawns.

Human interactions: Mysids are used as a food for the marine aquaculture of fish, squid and prawns. Mysids, particularly of the genus *Mysidopsis*, are used in the assessment of pollutant impacts on estuarine and marine communities. With their sensitivity to toxicants, short life cycle, ease of handling and direct larval development, mysids are ideal for both acute and chronic toxicity testing, as well as dredge spoil and effluent tests.

8.3.2.4 Amphipods (Order Amphipoda)

Identification: Amphipods (from Ancient Greek *amphi* meaning 'all over' and *podós* 'foot') are relatively large (>5 mm) and unusually among zooplankton, are laterally compressed, often with large eyes, prominent antennae, and sometimes large chelae (pincers) (Fig. 8.12O,P). Amphipods have no carapace, only a jointed exoskeleton. The body is divided into 13 segments, with the head fused to the thorax. Adult and juvenile amphipods are seen in plankton samples. When identifying amphipods in a sorting dish, they tend to float at the surface.

Diversity: Most marine amphipods are benthic, but there are 400 pelagic species that inhabit all depths and latitudes of the oceans.

Some common species: Many specimens in plankton samples are juveniles and difficult to identify to species. The most common group is the hyperiid amphipods, which are holoplanktonic, and most have large eyes and chelae (Fig. 8.12P). The other major group of amphipods is the gammarids, which have small eyes and chelae (Fig. 8.12O). Many gammarids are demersal, being found in the sediment on the sea floor during the day, but emerge at night to feed.

Lifespan: Amphipods can live up to 3 years, with females often outliving males.

Feeding: Hyperiid amphipods are all pelagic and half are parasites on gelatinous zooplankton. Gammarid amphipods are mostly benthic, with only a few pelagic species. They are mostly detritivores and scavengers.

Reproduction: Amphipods are pericarids, so females retain their eggs and larvae in a marsupium between their legs, later releasing miniature adults.

Locomotion: Amphipods swim by beating their pleopods. Some species of hyperiid amphipods live inside salps. Amphipods that settle in live salps are transported by the salp, but eventually the salp dies and the amphipod needs to provide the propulsion to find a new host.

Ecology: Many hyperiid amphipods are symbionts of gelatinous animals during part of their life cycle, including salps, cnidarian medusa, siphonophores and ctenophores. At night, demersal gammarid amphipods swim up into the plankton and return at daybreak, and thus can be important in moving matter between pelagic and benthic systems. The relative abundance of amphipods compared with other zooplankton increases with depth. Deep-sea amphipods are huge (up to 30 cm long) compared with their near-surface relatives and scavenge anything that sinks to the sea floor.

Human interactions: There are few crustaceans found on land, but beach hoppers are amphipods that are familiar to us, having successfully colonised sandy shores. Amphipods that live on the sediment interface near the coast are commonly used in bioassays for measuring toxicity. Amphipods from the deep sea have been used as bioindicators of ecosystem health because they bioaccumulate human pollutants. Contaminant levels of persistent organic pollutants such as polychlorinated biphenyls (PCBs) in amphipods that inhabit the deepest oceanic trenches (including the Mariana Trench >10 000 m deep) are higher than many polluted bays in industrialised regions. Molecules that help amphipods cope with living at high pressures are being used in bioprospecting: glycerophosphorylcholine could protect against harmful effects of waste products on our kidney, and scyllo-inositol is being tested for its ability to restore mis-folded brain proteins in patients that suffer Alzheimer's disease.

8.3.2.5 Stomatopods (Order Stomatopoda)

Identification: Stomatopod (from Ancient Greek *stóma* meaning 'mouth' and *podós* 'foot') larvae are caught in plankton samples. They are also known as mantis shrimp. They are relatively large (>5 mm), with a big rectangular flared carapace (like a transparent cloak), and two spines at the rear-facing corners (Fig. 8.12N). They have large stalked eyes and diagnostic massive raptorial mouthparts (second maxillipeds) for feeding.

Diversity: Stomatopods are marine and mostly inhabit shallow tropical and subtropical waters, but some are found in temperate seas. Adult stomatopods live on the sea floor and their larvae are planktonic.

Some common species: It is difficult to identify stomatopod larvae to species.

Lifespan: Adult stomatopods can live for several years, and have a long larval development, ranging from months to a year.

Feeding: Even as larvae, stomatopods are predatory carnivores, using their raptorial appendages to hunt other zooplankton. Adult stomatopods are the jaguars of the crustacean world. Mantis shrimps can be split into two groups based on their highly modified appendages to feed on particular prey: 'smashers' have heavy club-like maxillipeds for feeding on snails, crabs and oysters; and 'spearers' have spear-like maxillipeds for feeding on fish.

Reproduction: Some species have several partners during their life, while some species mate for life. Eggs are carried under the tail of the female or placed in a burrow. After hatching, larvae develop through several different stages (antizoea, pseudozoea, ericthus, alima, and post-larva), growing more complex appendages as they mature.

Locomotion: Larval stages swim using their pleopods. Like decapods, adults move around on both the substrate where they walk with their thoracic periopods, and in the water column where they swim with their abdominal pleopods.

Ecology: Stomatopods undergo several planktonic stages before becoming adult and settling for a benthic life.

Human interactions: Stomatopods are eaten in several countries, including in Japan as a sushi topping or in sashimi, in Vietnam, China and Philippines where they are steamed, boiled, grilled or fried, and in Italy and Spain in local seafood cuisine. The peacock mantis shrimp is often kept by saltwater aquarists because of its stunning appearance, but it can eat other marine life in the aquarium and break the glass by striking it, after seeing its reflection. Stomatopods also have one of the most advanced visual systems of any animal, and this is being used as a model for developing new technology to fight cancer. Mantis shrimp have 16 types of colour-receptive cones, compared with three in humans. Mantis shrimp thus have the ability to see polarised light – something human cannot do. As fast-growing cancer cells reflect polarised light differently, experimental work has shown that mantis shrimp can detect some types of human cancers.

8.3.2.6 Isopods (Order Isopoda)

Identification: Isopod (from Ancient Greek ísos meaning 'equal' and *podós* 'foot') larvae and juveniles are rarely seen in plankton samples because most are parasitic or benthic. Isopods tend to be large (>5 mm) and dorso-ventrally compressed with unstalked eyes and, like amphipods, have no carapace. Of all crustaceans, isopods are the most diverse in body form and the body is not always flattened. Isopods can be distinguished from other similar crustaceans because they have only one pair of uropods and lack strong clawed first thoracic legs.

Diversity: There are currently ~3200 described marine species, and nearly 10 500 species in total, including a few species from freshwater and terrestrial habitats. Terrestrial isopods (slaters, woodlice or pill bugs) are more familiar to us in our gardens. In the oceans, isopods are ubiquitous, found from littoral to abyssal depths.

Some common species: Adult isopods are rare in plankton samples, but larvae are sometimes seen. It is not possible to identify isopod larvae to species.

Lifespan: Isopods may live to be 2–3 years old (some of the giant species may live much longer) but they are only likely to spend a short period of time in the zooplankton, primarily while seeking a crustacean or fish host.

Feeding: Generally, isopods are benthic or parasitic, with only some occasionally appearing in the plankton, such as juveniles of the Family Gnathiidae that are parasitic on fish as adults, but can be found swimming around in the plankton when finding a host. Isopods are detritus feeders, carnivores, parasites and filter feeders.

Reproduction: Like amphipods, isopods are one of the peracarid crustaceans that brood their young in a marsupium under the body. Isopods brood 1–1600 eggs depending on individual species.

Locomotion: Isopods are not strong swimmers, and most adults crawl about, but some species can use their pleopods to swim.

Ecology: Generally, isopods are benthic or parasitic, with only some occasionally appearing in the plankton, such as juveniles of the Family Gnathiidae that are parasitic on fish as adults, but can be found swimming around in the plankton when sourcing a host.

Human interactions: Parasitic isopods are mostly external parasites of fish or crustaceans and feed on blood. The cleaner fish (wrasse) from coral reefs specialise on removing and eating such parasites. Isopods are problematic in aquaria and in aquaculture of sea bass, sea bream and Atlantic salmon and are usually removed manually or chemically.

8.3.2.7 Cumaceans (Order Cumacea)

Identification: Cumaceans (from Ancient Greek *kyma* meaning 'swollen') look like a tadpole, with a bulbous head and thorax, and a slender abdomen (Fig. 8.12Q). Cumaceans are peracarid crustaceans (commonly known as hooded shrimps or comma shrimps) with a large carapace that covers many thoracic segments, a narrow long abdomen, forked tail, and often a telson. Their eyes are sessile, compound, fused and sometimes absent. Their body is covered by a chitinous epidermis that can be strongly calcified, and marked with grooves, ridges, spines, teeth and setae. Cumacean adults and juveniles are rarely seen in plankton samples and there are no larval stages.

Diversity: There are currently 1636 described species of Cumacea. Cumaceans are primarily a marine group, with a few brackish and freshwater species. The majority of species are found on soft substrates, burrowing in the sediment, but they also emerge into the water column at night.

Some common species: Cumaceans are rarely seen in plankton samples and require specialist knowledge for identification.

Lifespan: Most shallow water species live for a year or less, but some deep-water species have slower metabolism and may live up to 3 years.

Feeding: Cumaceans feed on microorganisms and organic matter in the sediment. Some species browse through sand for prey, and some have specialised mandibles that pierce prey directly.

Reproduction: Adult males of many coastal species may form pelagic swarms, especially at night, where they are then joined by females for reproduction. After fertilisation, the female keeps the eggs in a brood chamber (marsupium); hatched larvae are retained there until they develop into a postlarva that resembles the adult. Most species reproduce twice in their lifetime.

Locomotion: In the water column, male cumaceans swim using their pleopods. Females have fewer pleopods (or none) than males and swim using exopods of the anterior thoracic limbs. Some species swim by beating the whole abdomen. Cumaceans use the endopods of thoracic appendages for benthic locomotion.

Ecology: At night, the normally benthic cumaceans can swarm into the water column to mate and moult. Cumaceans are usually captured in zooplankton hauls towed close to the bottom over shallow soft-substrate habitats at night. Cumaceans are eaten by fish.

Human interactions: As cumaceans can be an important component of benthic communities, they have been studied in relation to the effect of oil drilling on marine life.

8.3.3 Tick-like

Identification: The nauplius (from the Ancient Greek *naúplios* meaning 'a kind of shellfish' was thought originally to be a genus of marine animals) is the initial larval stage of crustaceans and looks like a small tick or spider. Nauplii are ~0.2–0.5 mm in length, often with a single compound eye (Fig. 8.14). Penaeid prawns have up to nine nauplius stages, copepods commonly have six stages, while barnacles have five. Early nauplius stages of most groups are similar and triangular in shape – widest at the front end of the body where the single nauplius eye is present – and are most likely to be distinguished by size. As a nauplius grows, its body lengthens (as more segments are added at the rear of the animal), it develops more limbs with setae (hairs), and its form diversifies with some becoming long and narrow and others more robust.

Diversity: As the first larval stage of crustaceans, the structure of nauplii is relatively

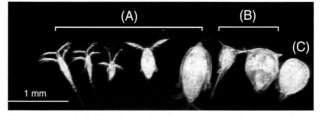

Fig. 8.14.   Tick-like. Nauplii of: **(A)** calanoid copepods, **(B)** barnacles, **(C)** Facetotecta.

conservative, with all being generally triangular, but there are differences in ornamentation, appendages, and relative proportions among groups.

Some common species: The most frequent nauplii in a zooplankton sample are those of copepods (Fig. 8.14A). The calanoid nauplius looks like a small comma when they are seen laterally. Barnacle nauplii can be easily distinguished from other nauplii by the presence of two horns, one on each corner of the anterior of the animal (Fig. 8.14B). They are also usually larger, have a posterior spine, a distinctive eye spot, and have lots of setae (i.e. they appear 'hairy'). They have five nauplius stages before moulting into a cyprid (see Section 8.3.5.3). Another distinct form is the Facetotecta (Fig. 8.14C), which is known only from its nauplius stage, with the adults presumed to be parasitic. Nauplii are not identifiable to species, although some calanoid copepod nauplii can be identified to family based on their distinctive shape (e.g. Eucalanidae, Rhincalanidae).

Lifespan: Non-feeding nauplii generally metamorphose within a few days before the yolk supply is exhausted. Feeding copepod nauplii generally take longer before they metamorphose into copepodites – often a week or a few months, depending on the species and water temperature. The duration of barnacle nauplii is variable and may be anywhere from several days to 5 months, before transforming to cyprids.

Feeding: Nauplii from different zooplankton groups have distinct feeding strategies. Krill and decapod nauplii are non-feeding larval stages relying on yolk supplies from the egg, whereas most copepod and barnacle nauplii stages feed on phytoplankton. In copepods, there is sometimes early non-feeding nauplius stages and later feeding stages. Copepod nauplii are either ambush feeders on motile prey such as dinoflagellates or produce feeding currents to draw particles such as small diatoms into their mouthparts. The nauplii of barnacle species that are sedentary feed on phytoplankton, but some parasitic barnacle species do not feed as nauplii.

Reproduction: Nauplii are larval stages and thus do not reproduce.

Locomotion: Nauplii swim using a rowing-like action of the swimming appendages at the front of their body, producing a jerky crawling motion.

Ecology: The nauplius stage is important for dispersal in many crustaceans. This is particularly true for sedentary species such as barnacles. Species that retain their eggs release fewer but more developed nauplii than their free-spawning counterparts. Because of their small size and high nutritional quality, nauplii are also an important part of the diet of many fish larvae.

Human interactions: Crustacean nauplii are commonly used as food to rear fish and crustacean larvae in commercial aquaculture operations.

8.3.4 Water fleas (Superorder Cladocera)

Quick guide

Podon: Head separated from the rounded body.

Pseudevadne: Head appears more connected to the body than in *Podon*. More elongated body (but can be more rounded when gravid with eggs or embryos), lacking spines.

Evadne: As *Pseudevadne* but has spines on the end of the body.

Penilia: Wings with serrated edges to carapace. Eye less prominent.

Identification: The best known water flea (common name for cladocerans) is the freshwater *Daphnia* (see Fig. 7.7), with its head and antennae protruding from the carapace. Marine cladocerans look similar to their freshwater counterparts; they are small (<1.3 mm) crustaceans, with an anterior, single, large compound eye (Fig. 8.15). The head usually has large, segmented and branched antennae, hence the name Cladocera from Ancient Greek

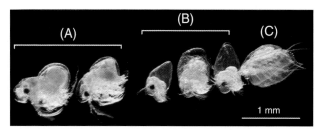

Fig. 8.15. Water fleas. CLADOCERA: **(A)** *Podon* spp.; **(B)** *Pseudevadne tergestina*; **(C)** *Penilia avirostris*.

kládos meaning 'branch' and *kéras* 'horn'. Cladocerans have a large head and their vastly reduced bodies and limbs contained within a bivalved carapace. Cladocerans are holoplanktonic.

Diversity: Although there are 400 species of cladocerans worldwide, most are found in fresh or brackish waters, and there are only eight true marine species. The genus *Pseudevadne* has only one species, *Pseudevadne tergestina*. The genus *Penilia* also has only one species, *Penilia avirostris*.

Some common species: The marine cladoceran *Penilia* (Fig. 8.15C) when viewed laterally looks similar to the freshwater cladoceran *Daphnia*. When dead in a Petri dish, *Penilia* lay on their dorsal surface and the two halves of the carapace relax open (like the wings of a small butterfly), exposing their limbs. Three other genera, *Podon* (Fig. 8.15A), *Evadne* and *Pseudevadne* (Fig. 8.15B) seem to be 'all eyes and a few limbs', showing remarkable simplification from the basic crustacean form. Cladocerans are often found floating on the surface of samples, which may be due to air trapped inside their carapace. For identification, one often needs to gently nudge them with a needle to break the surface tension and have them sink to the bottom of the Petri dish.

Lifespan: Cladocerans are short lived, but they produce resting cysts/eggs that can lay dormant for years.

Feeding: Cladocerans prey on microplankton, including phytoplankton, bacteria and microzooplankton.

Reproduction: Eggs, embryos and young stages, which are replicas of the adults, are retained in the brood chamber (in their body), so there are no free larval stages. The life cycle is dominated by parthenogenesis (asexual reproduction), with occasional periods of sexual reproduction. Under favourable conditions, cladocerans reproduce through parthenogenesis, producing only female clones, and populations can increase rapidly. These population increases are followed by production of males, and then sexual reproduction when there are unfavourable environmental conditions that stress the population. Following copulation, one or two thick-walled, dormant eggs are produced in the brood chamber. Eggs may be released by the females or reach the bottom when the female dies. They can survive for long periods in the bottom sediments, forming an egg bank for the next year or later generations, until conditions favour hatching.

Locomotion: Cladocerans swim by twirling their powerful pair of antennae.

Ecology: Cladocerans can bloom in large numbers in coastal areas. At times, they are an important food source for fish.

Human interactions: Freshwater cladocerans are commonly used in toxicity testing, but no

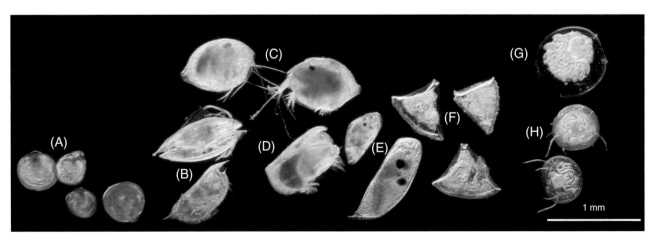

Fig. 8.16.   Clam-like. MOLLUSCA: **(A)** Bivalvia; OSTRACODA: **(B)** *Euconchoecia aculeata*; **(C)** *Cypridina* spp.; **(D)** *Conchoecia* sp.; CIRRIPEDIA: **(E)** cyprid larvae (barnacle); BRYOZOA: **(F)** cyphonautes larva; BRACHIOPODA: **(G)** lingulacean larva, **(H)** discinacean larva.

toxicity tests use marine cladocerans because they cannot be maintained in culture for longer than a few weeks due to their complex dietary requirements.

8.3.5 Clam-like

Quick guide

There are four main groups that have two shells or valves (from Latin *valva* meaning 'that which turns' and relates to the hinged shells') and appear clam-like (Fig. 8.16):

Bivalve larvae: Shells rounded like miniature clams/mussels.

Ostracods: Shells have a diagnostic small point (rostrum) at one end.

Barnacle cyprid: Has a dark eyespot and shells have no point.

Bryozoan (cyphonautes) larvae: Triangular, with ciliary band.

8.3.5.1 Bivalve larvae (Class Bivalvia)

Identification: Bivalve molluscs (from Latin *molluscus* meaning 'soft') have planktonic larvae. When sampling in coastal waters there may be many tiny bivalves (Fig. 8.16A). Once these bivalve larvae metamorphose into juveniles, they can still be swimming in the plankton for a short while before they settle. When alive, juveniles can be observed extending their slender molluscan foot between the shells and flipping themselves around. When dead, bivalve molluscs can be distinguished by the presence of concentric growth rings, visible under a microscope.

Diversity: There are ~8000 species of marine bivalves and most have pelagic larvae.

Some common species: It is difficult to distinguish species of bivalve larvae.

Lifespan: As larvae, bivalves are found in the plankton for a few hours to days. As adults attached to a substrate, some bivalves such as giant clams can live for decades; most commercial oysters live 2–4 years.

Feeding: Bivalve larvae are filter feeders on phytoplankton.

Reproduction: Most marine bivalves are free spawning and release sperm and eggs into the water where fertilisation occurs. The first larval stage is a trochophore, which is followed by a veliger, a pediveliger and then juveniles, which settle on the sea floor.

Locomotion: Locomotion in bivalve larvae changes with age. Trochophore larvae are free swimming and active at the surface and rotate. Older veliger larvae are active swimmers. Pediveligers (larvae ready to settle) crawl, swim and glide along the bottom.

Ecology: In coastal waters, bivalve larvae are extremely abundant at times and can be a periodically important food source for planktivorous fish and invertebrates. The thick shell of adult bivalves makes them difficult for predators to penetrate, although many animals have devised ingenious methods of extracting the flesh inside, including many invertebrates (sea stars, crabs, octopus) and vertebrates (fish, seabirds, otters, seals).

Human interactions: Scallops, oysters, mussels, cockles and pipis (eugaries) are all harvested commercially for their flesh. In 2014, worldwide marine aquaculture of clams, oysters and mussels was 15.8 million tonnes, worth US$18.5 billion. Certain species of oysters produce pearls (e.g. *Pinctada*) from calcium carbonate (same as their shell) and are used extensively in jewellery. Bivalve larvae, particularly from oysters, are also used in short-term toxicity testing. Because adult bivalves are filter-feeders, they can bioaccumulate toxins from toxic phytoplankton (harmful algal bloom species). When eaten by humans, these toxic bivalves cause shellfish poisoning. The most dangerous is paralytic shellfish poisoning (PSP), which involves the neurotoxin saxitoxin, and causes tingling, numbness and diarrhoea, and in severe cases paralysis and death. Harvested bivalves are regularly tested for toxicity.

8.3.5.2 Ostracods (Class Ostracoda)

Identification: Ostracods (from Ancient Greek *ostrakṓdēs* meaning 'covered with shell') are sometimes known as seed shrimps and have a clam-like shell that has a diagnostic point (or rostrum) at one end of the shell (Fig. 8.16B–D). Adults of the oceanic planktonic species range from 0.8 to 4 mm in size. The valve (hard parts) encloses the body with its appendages (soft parts).

Diversity: There are >8000 species, but the majority are benthic, with only a few holoplanktonic species, which are mostly found in the deep ocean, but some are in surface and coastal waters. Planktonic ostracods have received scant attention from ecologists.

Some common species: One of the most easily recognised and commonly encountered species is *Euconchoecia aculeata* (Fig. 8.16B). Other common genera include *Cypridina* (Fig. 8.16C) and *Conchoecia* (Fig. 8.16D).

Lifespan: Unknown, but probably only a few weeks.

Feeding: Most ostracods found in the zooplankton are probably filter feeders, consuming diatoms, bacteria and detritus.

Reproduction: Ostracod sexes are usually separate and fertilisation is internal but some species can reproduce asexually. Embryos are brooded in the postero-dorsal section of the shell or eggs are released into the water.

Locomotion: Planktonic ostracods use their antennae to swim.

Ecology: Ostracods play an important role in the recycling of material within the deep ocean. Many species are bioluminescent.

Human interactions: Benthic species are often found in fish tanks. The blue glow of ostracod bioluminescence at night in sand was used by Japanese troops during World War II to read maps, without producing enough light to alert enemy troops.

8.3.5.3 Barnacle cyprids (Order Cirripedia)

Identification: After the nauplius stage (Fig. 8.14B), barnacles (from Latin *barneca* meaning 'barnacle') metamorphose into a large dark, dense cyprid larvae (Fig. 8.16E). The clam-like carapace of barnacle cyprids lacks a notch or rostrum. Internal organs are often clearly visible through the valves, and the adductor muscle is often visible as a small dark circle. Appendages may be visible external to the valves both anteriorly and posteriorly, but are also often retracted within the valves. When they moult *en masse* during summer, the large translucent exoskeletons (exuviae) of adult barnacles can be found in plankton samples.

Diversity: There are ~1200 species of barnacles, ranging from stalked barnacles to those encased in a series of calcareous plates, and some are highly modified as parasites.

Some common species: Adult barnacles are obvious members of the intertidal zone attached to rocks and their larvae often dominate inshore plankton during their breeding season. As with other meroplanktonic animals, it is extremely difficult to identify barnacle cyprids to species.

Lifespan: As the last stage before metamorphosis to an adult, the barnacle cyprid stage survives for several days to weeks.

Feeding: Barnacle cyprids are a non-feeding stage in the barnacle life cycle.

Reproduction: Most barnacles are hermaphroditic. Sexual reproduction is difficult because adult barnacles are sessile, so males have the largest penises relative to their size in the animal kingdom. Male barnacles can also shed their sperm into the water to fertilise females further afield.

Locomotion: The appendages used to move barnacle nauplii change function or regress when the larvae metamorphose to cyprids, which develop new thoracic appendages that can propel the larvae.

Ecology: In barnacles, following the relatively long (months) nauplius stage there is a short (weeks) cyprid stage, which then settles onto a hard substrate and metamorphoses into juvenile barnacles.

Human interactions: As adults, barnacles are eaten in Japan, Spain, Portugal and Chile. Clearing fouled ships and fixed human structures of barnacles costs billions of dollars annually.

8.3.5.4 Bryozoans larvae (Phylum Bryozoa)

Identification: The term Bryozoa is from Ancient Greek *brúon* meaning 'moss' and *zôion* meaning 'animal', referring to the appearance of the adults. Bryozoan larvae (also called cyphonautes larvae, from Ancient Greek *kúphos* meaning 'bent' and *naútēs* meaning 'sailor') are distinctively triangular in shape, with a characteristic row of cilia encircling their longer convex side (Fig. 8.16F).

Diversity: There are ~4500 species.

Some common species: *Bugula neritina* is an invasive species found worldwide.

Lifespan: The cyphonautes larvae of some bryozoans are long lived.

Feeding: The beating of the cilia contributes to producing a feeding current. Cyphonautes larvae feed on phytoplankton.

Reproduction: Larval sizes range from 0.2 to 1.1 mm. Bryozoan larvae are found close to the coast and in estuaries.

Locomotion: As well as producing a feeding current, the coronal cilia propel the cyphonautes through the water as they beat.

Ecology: Adult bryozoans form an encrusting sheet of colonial, filter-feeding animals on rocks, kelp and any other firm surface.

Human interactions: Larval bryozoans provide a useful bioassay for heavy metals. Several anti-cancer agents were first derived from bryozoans. For example, *Bugula neritina* contains bryostatin, an effective anti-cancer compound, but for conservation reasons these agents are now chemically synthesised rather than extracted from wild populations.

8.3.6 Worm-like

Quick guide
There are three main groups of zooplankton that are worm-like, only one of which is a true segmented worm (Fig. 8.17):

Polychaetes: True worms, with segmented bodies that often have bristles (chaetae).

Chaetognaths: Large (>5 mm) and unsegmented, with visible hooks around the mouth and two small eyes usually visible. Fins are sometimes visible.

Appendicularians: Unsegmented and tadpole-like in appearance, with a distinct 'trunk' (looks like a head) and 'tail'.

8.3.6.1 Polychaetes (Class Polychaeta)

Identification: Polychaetes (from Ancient Greek *polus* meaning 'many' and *chaite* meaning 'hair') are the true segmented worms in Phylum Annelida and in Class Polychaeta. They are commonly called bristle worms, with the marine species having paired fleshy limbs (parapodia) with bundles of chaetae (bristles) extending outwards. Both adult and larval polychaetes are found in the plankton (Fig. 8.17A–G). Holoplanktonic polychaetes have developed special adaptations to live in the pelagic environment, including small size, long setae, enormous complex eyes, flattened or gelatinous bodies, a high degree of transparency, and sperm storage in females. Some polychaete worms are evident in the plankton sample, curled up into a ball exposing the many chaetae (like a tiny echidna).

Diversity: Most species of polychaetes are benthic, living on or in the sediment, and these are sometimes seen when they move into the water column at night. Few species are holoplanktonic.

Some common species: *Tomopteris* is a distinctive holoplanktonic polychaete with many parapodia ('paddles') (Fig. 8.17E) and unusually has yellow bioluminescence. Other holoplanktonic polychaetes include the nereids (Fig. 8.17A), alciopids (Fig. 8.17B), polynoids (Fig. 8.17C), terebellids (Fig. 8.17D) and magelonids (Fig. 8.17F). Specialist expertise is required to identify early stage polychaete larvae, even to family level.

Lifespan: Unknown, but probably weeks to months.

Feeding: Holoplanktonic species are mainly active predators that attack prey with a rapidly everted proboscis, but some filter-feed on phytoplankton.

Reproduction: Adult polychaetes may be caught at night when they swim up off the sea floor into the plankton, often for breeding. Most polychaete

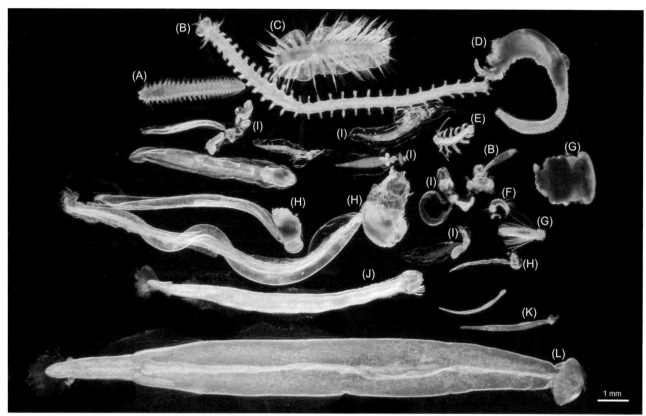

Fig. 8.17. Worm-like. POLYCHAETA: **(A)** nereid, **(B)** alciopid, **(C)** polynoid, **(D)** terebellid, **(E)** *Tomopteris* sp., **(F)** *Magelona* sp., **(G)** polychaete larval stages of different species; APPENDICULARIA (larvaceans): **(H)** oikopleurids, **(I)** fritillariids; CHAETOGNATHA: **(J)** regular size chaetognath, **(K)** small chaetognath species and juveniles, **(L)** *Flacciosagitta enflata*.

species produce eggs and larvae known as trocho-phores. Polychaete larvae may have a long or a short planktonic existence, or even remain benthic. Some polychaete species have posterior segments containing gametes that break off, and then swim autonomously to the surface as an epitoke. Gametes are released when the epitoke reaches the surface.

Locomotion: The holoplanktonic *Tomopteris* beats its parapodia rhythmically to produce a wave-like motion that propels them forward.

Ecology: Pelagic polychaetes are found in the open sea, from surface to abyssal depths, but coastal species are likely to be meroplanktonic.

Human interactions: Benthic, but not pelagic, marine polychaetes are commonly used in toxicity testing. Understanding bioluminescence in marine

organisms has led to new approaches to medical research and lighting; the bioluminescent substance in *Tomopteris* is yet to be confirmed.

8.3.6.2 Chaetognaths (Phylum Chaetognatha)

Identification: Chaetognaths (commonly called arrow worms) comes from Ancient Greek *khaítē* meaning 'flowing hair' or 'mane' and *gnáthos* meaning 'jaw' referring to their diagnostic large chitinous jaws, normally called hooks, that curve out from each side of the head (Fig. 8.17J). Chaetognaths are unsegmented, varying in length from 2 to 120 mm (Fig. 8.17J–L). Chaetognaths have elongated cylindrical bodies, are bilaterally symmetrical, and are usually transparent or slightly opaque. The body is divided into three parts by internal partitioning: head, trunk and tail. The head is slightly rounded, has a pair of pigmented eyespots, and is separated from the trunk by a constricted neck. The trunk bears one or two pairs of lateral fins, usually overlapping the septum between trunk and tail. The fins are thin, transparent and supported by fin rays and often difficult to see in preserved specimens, but their number, shape and position are diagnostic in species identification.

Diversity: Chaetognaths are in their own phylum. There are 120 species assigned to over 20 genera, with 80% of the species pelagic and the remainder benthic.

Some common species: To identify most species of chaetognaths correctly, it is necessary to examine sexually mature individuals, which can be rare. There are, however, species that can be distinguished when not completely mature including *Flaccisagitta enflata* (Fig. 8.17L).

Lifespan: The life span of chaetognaths varies from several months in tropical regions to 2 years in polar regions.

Reproduction: All chaetognath species are hermaphroditic, with a pair of testes in the tail and a pair of ovaries in the main body cavity. During mating, each individual places a sperm sac (a spermatophore) on the other individual, and fertilisation is internal. Eggs are then shed into the water, and later hatch into miniature adults.

Feeding: Chaetognaths are the tigers of the plankton. They are active carnivores and may sometimes be found grasping another animal. They have four to 14 hooks on each side of their head flanking a hollow vestibule containing the mouth. These hooks are used in hunting and are covered with a flexible hood arising from the neck region when the animal is swimming. Some species use neurotoxins to subdue prey.

Locomotion: Chaetognaths swim in short bursts using a dorso-ventral undulating body motion, where their tail fin assists with propulsion and the body fins with stabilisation and steering.

Ecology: Chaetognaths have a worldwide distribution and a wide depth range. Some species perform large diurnal vertical migrations. Chaetognaths are often abundant in the plankton and are important predators of other zooplankton.

Human interactions: Chaetognaths could negatively impact people by hosting parasites that are passed to fish that are then ingested to humans. They are important predators of fish eggs and larvae and thus influence fish recruitment and stock size. Some oceanographers use chaetognaths as indicators of particular water masses.

8.3.6.3 Appendicularians (Class Appendicularia)

Identification: Appendicularians are tadpole-like in appearance, with a distinct tiny spongy ball or trunk (looks like a head), and a longer flat fibrous tail, which seems barely attached to the trunk (Fig. 8.17H,I). Appendicularians are conspicuous and abundant holoplanktonic members of marine zooplankton in near-surface waters. Appendicularians are also known as larvaceans, named after their resemblance to the tadpole-like larvae of the related sessile ascidians. Like salps and doliolids, appendicularians are chordates (in Phylum Chordata as we are) and thus have a notochord and nerve tube as adults. Appendicularians produce a gelatinous house that is usually destroyed during sampling and thus not routinely visible.

Diversity: There are currently 69 described species of appendicularians. There are more species in warm nutrient-poor waters than colder nutrient-rich ones.

Some common species: There are two main families of appendicularians: the Oikopleuridae (Fig. 8.17H) and Fritillariidae (Fig. 8.17I), which can be distinguished based on a suite of characters (Box 8.4). *Fritillaria pellucida* is a distinctive species with four clearly visible spots on its tail. *Oikopleura longicauda* and *Fritillaria haplostoma* are abundant and widely distributed.

Lifespan: Appendicularian life spans are short; *Oikopleura dioica*, the only species to be cultured, is 10 days or less.

Reproduction: All but one species of appendicularian are hermaphrodites. Appendicularians reproduce sexually. They die after reproduction because the body wall ruptures when eggs are released (species with this type of reproductive strategy are referred to as semelparous). From the fertilised eggs (0.1 mm in diameter), larvae emerge that initially resemble adults.

Feeding: Appendicularians secrete a delicate gelatinous house of protein and cellulose around themselves like a bubble, which supports a network of filters built into its wall. The tail generates a feeding current and the fine filters concentrate food in the surrounding water before ingestion, allowing appendicularians to feed on bacteria and phytoplankton.

Locomotion: The tail used to generate the feeding current also provides locomotion.

Ecology: Appendicularians are found in coastal and oceanic environments, primarily in near-surface waters. The house used for feeding can quickly become clogged with small particles – for example, the genus *Oikopleura* discards its

Box 8.4 Distinguishing appendicularian families Oikopleuridae from Fritillariidae

Oikopleuridae and Fritillariidae can be readily distinguished based on a suite of characters (Fig. 8.18). Oikopleuridae are small and have a trunk that is oval-shaped, dense and dark. The tail attaches to one side of the trunk (giving a hammer shape) and the tail is relatively thick at the attachment point. By contrast, Fritillariidae are large and have a trunk that is elongated, delicate and relatively transparent. The tail attaches in the middle of the trunk (giving a distinct T-shape) and the tail is relatively thin at the attachment point.

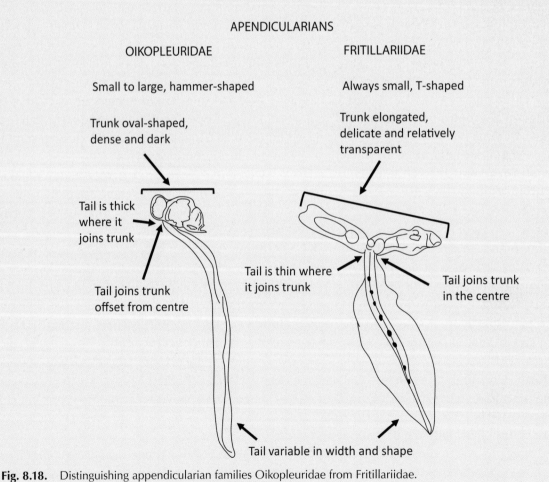

Fig. 8.18.   Distinguishing appendicularian families Oikopleuridae from Fritillariidae.

house and builds a new one every 4 hours. The tail helps to inflate a new house from under its mouth. Discarded houses, together with heavy faecal pellets from appendicularians, are a significant contribution to carbon sinking into deep waters.

Human interactions: Because they eat bacteria and phytoplankton 10 million times smaller than themselves by mass, appendicularians grow faster than any animal on Earth, and because they move slowly, they are important prey for carnivorous zooplankton, and larval and juvenile fish.

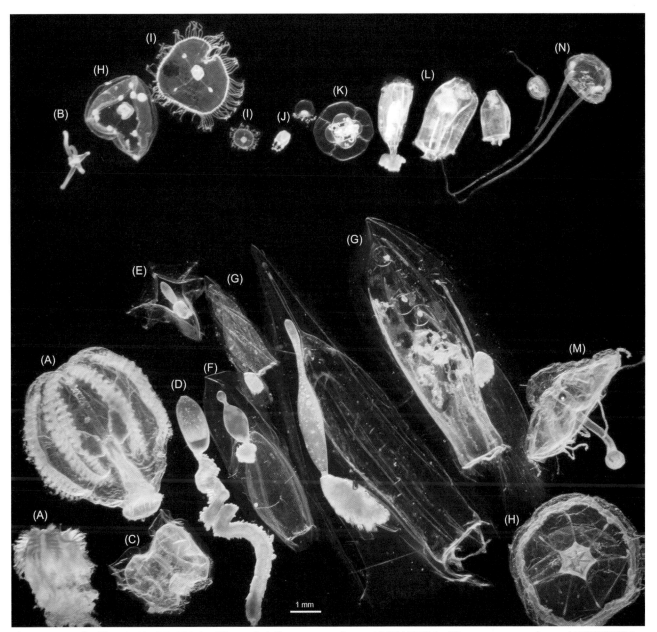

Fig. 8.19.   Jelly-like. CTENOPHORA: **(A)** comb jelly or ctenophore; ANTHOZOA: **(B)** sea anemone larva; HYDROZOA–SIPHONOPHORA: **(C)** physonectid nectophore, **(D)** physonectid pneumatophore, **(E)** calycophoran bract, **(F)** calycophoran eudoxids, **(G)** calycophoran anterior nectophores; HYDROZOA–LEPTOTHECATA: **(H)** medusa of leptothecata, **(I)** *Obelia* sp.; HYDROZOA–ANTHOATHECATA: **(J)** medusa of anthoathecata, **(K)** *Bougainvillia* sp.; HYDROZOA–TRACHYMEDUSAE: **(L)** *Aglaura hemistoma*, **(M)** *Liriope tetraphylla*; HYDROZOA–NARCOMEDUSAE: **(N)** *Solmundella bitentaculata*.

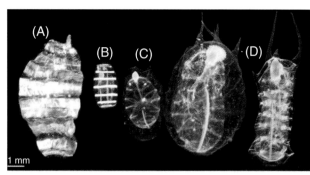

Fig. 8.20. Jelly-like. DOLIOLIDAE: **(A)** doliolid asexual form (Oozooid), **(B)** sexual form (gonozooid); SALPIDAE: **(C)** colonial form, **(D)** solitary forms.

This efficient transfer of energy from lower to higher trophic levels has been termed the 'larvacean shunt'. Appendicularians have been used as a model to understand vertebrate evolution.

8.3.7 Jelly-like

Quick guide
There are four main groups of jelly-like or gelatinous zooplankton, only one of which is a true jellyfish (Figs 8.19, 8.20):

Cnidarians: True jellyfish that have a bell that is round, box-shaped or sometimes triangular, often with tentacles. Some are colonial.

Ctenophores: Usually egg-shaped with eight rows of cilia, usually without tentacles visible. All are solitary.

Salps: A squarish barrel-shape, sometimes with two 'horns' present, and incomplete muscle bands that often fuse. Some are colonial.

Doliolids: Smaller than salps. Rounded barrel-shape with concentric (parallel) muscle bands completely encircling the body like metal bands on a wine barrel. All are solitary.

8.3.7.1 Cnidarians (Phylum Cnidaria)

I remark also in my notes, that having kept a Medusa of the genus Dianaea, till it was dead, the water in which it was placed became luminous.

Charles Darwin, December 1833, made during the voyage of the *Beagle* in Atlantic Ocean

(Darwin 1845) [The genus *Dianaea* is a hydrozoan cnidarian that is bioluminescent and has now been split into several genera]

Identification: Phylum Cnidaria (from Ancient Greek *knídē* meaning 'nettle' referring to the stinging plant) has stinging cells called cnidoblasts: the most complex cells of any animal. Cnidoblasts distinguish cnidarians from all other gelatinous zooplankton, and often from each other. Most jellyfish in plankton samples are hydrozoans (Class Hydrozoa, Fig. 8.19B–N) and are tiny (~1 mm diameter), but some reach tens of metres in length. Hydrozoan cnidarians can be bell-shaped, triangular or squarish, and sometimes have many tentacles. Many cnidarians are difficult to identify in plankton samples because they are juveniles.

Diversity: There are more than 10 000 species of cnidarians in six classes: Hydrozoa (individually small jellyfish, but can form large colonies); Scyphozoa (true jellyfish, with a large medusa stage); Cubozoa (box jellyfish); Anthozoa (almost all are benthic as adults and includes anemones and corals); Staurozoa (stalked, benthic, polyp-like cnidarians); and Myxozoa (very small, obligate parasites). Staurozoa and Myxozoa will not be seen in plankton samples.

Lifespan: The lifespan of cnidarians varies, with larger species likely to survive for longer periods than smaller species. The hydrozoan jellyfish *Turritopsis* is known as the immortal jellyfish because after it reaches sexual maturity as a solitary individual it can revert to a sexually immature colonial stage indefinitely.

Some common species: Most jellyfish you see on the beach or floating at sea are either large scyphozoans (e.g. the blue blubber jellyfish *Catostylus mosaicus* blooms in subtropical Australian bays in huge congregations during warmer months – see Section 3.10.1 on jellyfish blooms) or colonial neustonic (live near the surface) siphonophores (e.g. the Portuguese man o'war or blue bottle *Physalia*, by-the-wind sailor *Velella*, and the blue button *Porpita*). However, most jellyfish that you will capture in a plankton sample are small hydrozoans that are found a bit deeper in the water, particularly siphonophora (Fig. 8.19C–G), leptothecata (Fig. 8.19H,I), anthoathecata (Fig. 8.19J,K), trachymedusae (Fig. 8.19L,M) and narcomedusae (Fig. 8.19N).

Reproduction: Cnidarians have a diverse range of reproductive strategies. There are typically two forms: the sessile asexual hydroid that are permanently attached to a substrate and reproduce asexually by budding off new individuals; and the mobile medusa stage that are free-swimming, bell-shaped or flattened, and reproduce sexually (Fig. 2.5). Some cnidarians have only a hydroid stage in their life cycle, others only a medusa stage, and others both. Meroplanktonic cnidarians thus alternate between the two forms, and holoplanktonic cnidarians do not have a benthic stage. Even though jellyfish can be very large (some colonial siphonophores can reach 50 m in length), they never become part of the nekton because they cannot swim against currents.

Feeding: Most cnidarians obtain the majority of their food from predation. Cnidoblasts shoot nematocysts into prey, which are poisoned and entangled. Some groups rely more on their oral arms to capture prey than on the tentacles on the edge of the bell. Other species rely more on suspended particles for food and these species rely on cilia to produce currents to bring food towards their mouths. Colonial siphonophores rely on specific individuals for feeding, reproduction and locomotion. Some jellyfish have symbiotic relationship with zooxanthellae, particularly medusa from the class Scyphozoa (see Chapter 3).

Locomotion: Scyphozoan jellyfish swim by contracting the umbrella-shaped bell at the top of their body, which forms a vortex behind the animal and propels them forwards. In siphonophore colonies, the nectophores are individuals specialised for swimming.

Ecology: Although cnidarians have historically been viewed as trophic dead ends, over 124 fish species and 34 species of other animals are reported to feed either occasionally or predominantly on jellyfish. Of these, 11 fish species are jellyfish specialists, and some critically endangered animals, such as the leatherback turtle *Dermochelys coriacea*, target jellyfish blooms. Juveniles of some fish species find safe refuge from predation by living on or nearby jellyfish.

Human interactions: Jellyfish outbreaks can have many deleterious consequences for people, including: losses in tourist revenue through beach closures and even the death of bathers (Box 9.1); power outages following the blockage of cooling intakes at coastal power plants; blocking of alluvial sediment suction in diamond mining operations; burst fishing nets and contaminated catches; interference with acoustic fish assessments; killing of farmed fish; reduction in commercial fish abundance through competition and predation; and as probable intermediate vectors of various fish parasites. Human consumption of jellyfish in China dates back to at least 1700 years ago. Jellyfish is usually salted and dried to produce a crunchy texture. In traditional Chinese medicine, there is a long list of purported health benefits from eating jellyfish. Jellyfish are thus the focus of large fisheries, particularly in Asia (Section 3.11). Jellyfish catches are increasing globally and are now estimated to be 1.2 million tonnes in 19 countries, with largest catches in China (Box 3.9). It is not clear, however, whether this increase in catch is because of the development of new fisheries or a sign of environmental degradation. Not only are jellyfish used as food, they are also used in agriculture (livestock feeds, fertilisers), aquaculture (feeds), cosmetics (gelatin), environmental monitoring (pollution detection), fishing (bait), materials science (absorbent polymers, cement additive, nanoparticle filters), and pharmaceuticals (anticoagulants, anti-cancer agents, antimicrobiotics, collagen). The amazing variety of forms of medusae have been popularised by Ernst Haeckel's classic 'Art forms in Nature' in 1904 (https://publicdomainreview.org/collections/ernst-haeckels-jellyfish/).

8.3.7.2 Ctenophores (Phylum Ctenophora)

The structure of the Beroe (a kind of jelly fish) is most extraordinary, with its row of vibratory cilia, and complicated though irregular system of circulation.

Charles Darwin, December 1833, made during the voyage of the *Beagle* in the Atlantic Ocean
(Darwin 1845)

Identification: Ctenophores (from Ancient Greek *kteís* meaning 'comb' and *phóros* meaning 'bearing') are commonly known as comb jellies or sea gooseberries (Fig. 8.19A). Ctenophores are egg-shaped, ranging in size from a pea to an orange. The combs refer to the eight longitudinal rows of cilia. Combs produce a rainbow effect known as iridescence, which is not caused internally by bioluminescence, but by the scattering of light as the cilia move. Ctenophores are distinguished from all other animals by having colloblasts, which are sticky cells that capture prey (they are not stinging cells as in cnidarian jellyfish). However, ctenophores are similar to jellyfish in their structure, as they have only two basic living tissues: an inner layer (endoderm) and an outer layer (ectoderm), separated by a thicker layer of non-living tissue (mesoglea). Sometimes a plankton net may be full of gelatinous slime; this can be the remains of lobate ctenophores.

Diversity: There are only ~150 species of ctenophores and nearly all are holoplanktonic. Although ctenophores have historically been in the same phylum as cnidarians, more recent molecular genetic evidence suggests that ctenophores are more different to cnidarians than are arthropods, echinoderms or molluscs.

Some common species: There are three main groups of ctenophores: the cydippids (e.g. *Pleurobrachia*) that are rounded or egg-shaped; the lobates that have a pair of muscular lobes that project beyond the mouth (e.g. *Mnemiopsis*); and the beroids (e.g. *Beroe*) that have bundles of several thousand large cilia fused into 'teeth' inside their mouth that can be used to 'bite' off pieces of other

ctenophores. Coastal species tend to be robust so they can withstand wave action, whereas oceanic species tend to be more fragile, surviving in a more benign environment. Thus, the most commonly observed species are robust coastal genera – *Pleurobrachia*, *Beroe* and *Mnemiopsis* – and the oceanic species are rarely seen and do not preserve well.

Lifespan: From less than a month to 3 years.

Feeding: Ctenophores are voracious predators, eating over 10 times their bodyweight in zooplankton per day. Ctenophores eat prey ranging from larvae to adults of small crustaceans. The feeding rate seems to be related to tidal turbulence, bringing zooplankton into contact with the sticky cells on their tentacles or lobes. For example, once *Pleurobrachia* senses it has 'fly-papered' a copepod onto a tentacle, it wraps its tentacles around the body, and wipes the copepod across its mouth opening.

Reproduction: Most species are hermaphrodites, with both male and female sex organs. Unusually, some species are capable of reproducing before reaching adulthood. Hermaphroditism, combined with early reproduction, means ctenophore populations can expand rapidly under good food conditions, especially in semi-enclosed bays.

Locomotion: The eight combs of cilia are responsible for locomotion in ctenophores. The cilia beat away from the mouth when feeding, or beat towards the mouth to move away from predators.

Ecology: Ctenophores are solely marine and holoplanktonic. They are major predators of copepods and larval fish. Ctenophores may sometimes bloom in massive numbers and may fill a plankton net making it difficult to retrieve into the boat. Some species are parasitic on salps. Ctenophores have a balance organ known as a statocyst, with a tiny grain (statolith) inside, supported on four bundles of cilia that sense orientation. The statocyst, made from calcium carbonate, could be impacted by ocean acidification.

Human interactions: Ctenophores are often displayed in public aquaria. The ctenophore *Mnemiopsis leidyi* is a devastating alien species. *Mnemiopsis leidyi* was accidentally introduced into the Black Sea and expanded rapidly to nearly 900 000

Box 8.5 Distinguishing salps from doliolids, and sexual from asexual forms

It can be challenging to distinguish salps from doliolids, especially considering that they have both sexual and asexual stages. However, there are three diagnostic features (Fig. 8.21):

Size: Salps are always larger than doliolids.
Muscle bands: The key distinguishing feature is that salps have muscle bands that are incomplete (not continuous) and some touch each other (intersect). By contrast, doliolids have complete muscle bands that remain parallel and never touch.
Shape: Salps have a more irregular shape than doliolids, which are neat rounded barrels.

It is relatively straightforward to distinguish sexual and asexual stages in salps and doliolids. In salps, the sexual stage is colonial (so you often see individual salps together in chains), the wall is thick and gelatinous, and there are no processes (horns). By contrast, the asexual stage of salps is solitary, thin-walled and gelatinous, and some species have thin horns (and none in others). In doliolids, the sexual form is solitary, thin-walled, and there are no processes. Although the asexual form of doliolids is also solitary and thin-walled, it has a diagnostic single appendix that buds off small, genetically identical, individuals.

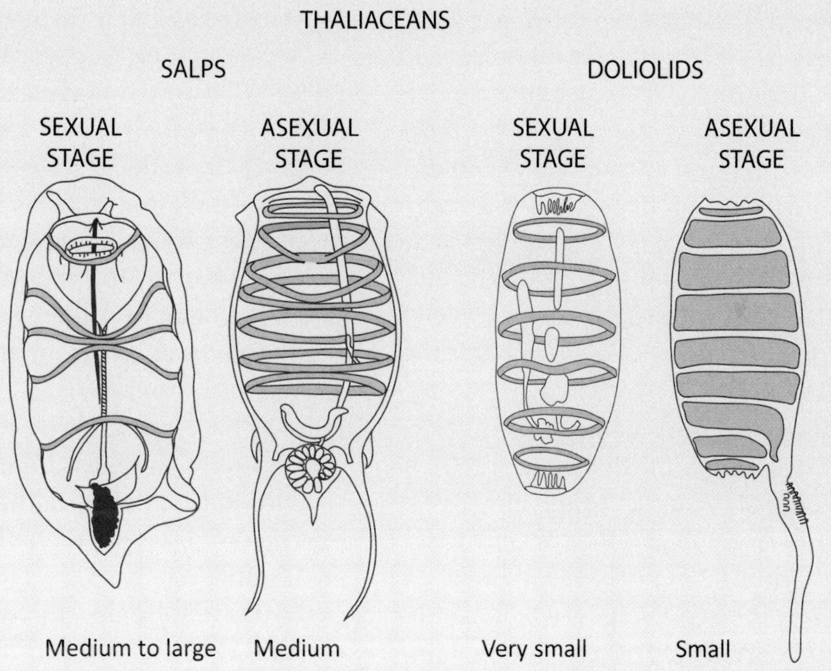

THALIACEANS

	SALPS		DOLIOLIDS	
	SEXUAL STAGE	ASEXUAL STAGE	SEXUAL STAGE	ASEXUAL STAGE
SIZE	Medium to large	Medium	Very small	Small
MUSCLE BANDS	Incomplete, intersect	Incomplete, intersect	Thin, parallel	Thick, parallel
SHAPE	More irregular	More irregular	Barrel	Barrel
FORM	Colonial	Solitary	Solitary	Solitary
WALL	Thick, gelatinous	Thin, gelatinous	Thin	Thin
PROCESSES	No horns	Thin horns in some species	No appendix	Appendix with buds

Fig. 8.21.   Distinguishing thaliacean families Salpidae and Doliolidae, and sexual from asexual forms.

tonnes, presumably because of agricultural runoff that increased plankton prey densities and because of the collapse of the main potential predator, the European anchovy *Engraulis encrasicolus*. The later accidental introduction of the predatory ctenophore *Beroe* helped to mitigate the problem because *Beroe* preyed on *M. leidyi* and reduced its population substantially. *Mnemiopsis leidyi* has now spread along European coastlines in the Baltic and North Seas.

8.3.7.3 Salps (Order Salpida)

Salps (from Ancient Greek *sálpē* meaning 'a species of fish'), together with doliolids, are specialised pelagic tunicates (from Latin *tunicatus* meaning 'to clothe with a tunic'). Tunicates are in the Class Thaliacea, relatives of benthic sea squirts, and indeed ourselves, because larval salps have a notochord (a flexible rod that is the forerunner of the vertebrate backbone) and a rudimentary brain, placing them in the Phylum Chordata (from Latin *chorda* meaning 'chord'). It can be challenging to distinguish salps from doliolids, considering that they both have sexual and asexual stages. Box 8.5 shows the diagnostic distinguishing features between salps and doliolids.

Identification: Salps are small squarish gelatinous barrels, from 0.5 to 190 mm long (Fig. 8.20C,D). They have incomplete muscle bands that often fuse, are generally larger than doliolids, and sometimes have two 'horns' or projections. Salps are found singly or in stringy colonies, which can be several metres in length. They have no limbs, tentacles or eyes.

Diversity: There are 48 known species of salps.

Some common species: At times, salps form massive blooms, including off eastern Australia. One species that is commonly responsible for these blooms is *Thalia democratica*. Other distinctive common species include *Ihlea magalhanica*, *Iasis zonaria* and the huge *Thetys vagina* (up to 190 mm).

Lifespan: A few weeks to possibly over a year.

Feeding: Salps filter feed small particles by pumping water through their inhalant siphon at one end and trapping particles through a continuously renewed mucous net, which is then ingested together with the particles entrapped. Water is then expelled from the exhalant end. Salps normally swim and feed continuously. Salps are remarkable in marine food webs in their ability to feed on particles over three orders of magnitude in size, from bacteria (<1 µm) to nauplii (1 mm), but bacteria and small phytoplankton are their typical prey. Salps have historically been considered 'trophic dead-ends', consuming the same food as copepods, but not transferring this energy up the food chain to economically important fish. However, at least 202 species are now known to feed on salps, including 149 species of fish, as well as other organisms such as corals, ctenophores, molluscs, crustaceans and marine turtles.

Reproduction: Salps have two forms: a solitary (Fig. 8.20D) and an aggregate (colonial) form (Fig. 8.20C). The solitary form reproduces asexually by budding to produce chains of aggregate individuals. These aggregate forms reproduce sexually to produce the solitary forms, which can have 'horns' or projections. Salp populations grow rapidly under favourable food conditions because they produce hundreds of clones per individual through asexual reproduction, they have the fastest growth rates of any multicellular animal, and they have short generation times (2 to 14 days).

Locomotion: Salps move using jet propulsion, but have also adapted for slow continuous swimming.

Ecology: Salps are common in oceans from the equator to the poles, although most species are tropical. They are especially abundant in the Southern Ocean. Because of their rapid production in the asexual stage when food is abundant, salps can form dense swarms, to the near exclusion of nearly all other zooplankton. Swarms are ephemeral, lasting weeks to a few months, and are usually driven by intrusions of cool nutrient-rich water that enhance phytoplankton abundance, commonly in spring. At these times, delicate chains of salps can be seen, composed of two to dozens of individuals (e.g. *Salpa* and *Pegea*). Salps produce large, fast-sinking (up to 2700 m per day), carbon-rich, faecal pellets, contributing disproportionately to carbon flux compared with other zooplankton. Salps thus play a major role in global biogeochemistry, and the episodic swarm events can provide pulses of food that sustain abyssal communities long after the swarm has declined.

Human interactions: Because salp swarms can export large amounts of carbon from the euphotic zone, geoengineering solutions to combat climate change have explored ocean fertilisation to stimulate salp blooms. Salps, as chordates with a primitive concentration of nerves, have been used widely as a model for early vertebrates and the evolution of the central nervous system.

8.3.7.4 Doliolids (Order Doliolida)

Identification: Doliolids (from Latin *dolium* meaning 'large jar') are a rounded barrel shape 0.5–5 mm in size, with eight or nine concentric muscle bands (Fig. 8.20A,B). The muscle bands force water through their body and eject it through the rear siphon in a jet that propels the animal. Like salps, doliolids have no limbs, tentacles or eyes.

Diversity: There are 23 described species of doliolids.

Some common species: Two common species are *Doliolum denticulatum* and *D. nationalis.*

Lifespan: The sexual stage of doliolids lives 10–14 days. With their fast growth rates, pelagic tunicates such as doliolids and salps are adapted to event-scale variation in environmental conditions.

Feeding: Doliolids feed on particles from 2 to 50 μm in size, from bacteria to phytoplankton. Similar to salps, as water passes in an inhalant siphon at one end and out an exhalant siphon at the other, the water passes through a fine mucous filter, which collects particles and the entire filter, with attached particles, is then periodically ingested.

Reproduction: Doliolids have a complex life cycle that includes alternation of sexual and asexual generations. The sexual generation consists of individuals, called gonozooids, with eight muscle bands, each having male or female gonads (Fig. 8.20B). Fertilised eggs produce individuals with nine muscle bands, no gonads, and one large posterior projection that buds off zooids (Fig. 8.20A). These asexual individuals are informally called 'nurses', and each one produces tens of thousands of mature sexual and asexual zooids.

Locomotion: Doliolids propel themselves by contracting the muscular bands around their body, creating a temporary water jet that thrusts them forwards or backwards. They are adapted for single rapid jet pulses that can propel them at velocities equivalent to 50 body lengths per second.

Ecology: Almost all doliolid species are found in the tropics. Doliolids are likely to be a common prey item for many animals, but currently only 10 predators are known. The copepod *Sapphirina nigromaculata* chews through the body of the doliolid *Dolioletta gegenbauri* and then ingests its internal tissues.

Human interactions: The transparency of doliolids means their internal anatomy is easily visible, making them valuable model organisms in developmental biology for studying vertebrate and deuterostome evolution.

8.3.8 Snails (Class Gastropoda)

Snails (from Ancient Greek *gastĕr* meaning 'stomach' and *podós* 'foot') are gastropod molluscs with unsegmented soft-bodies, partially or wholly covered by a mantle, a sheet of tissue exclusive to this phylum (Fig. 8.22). The body is often divided into a head, with eyes or tentacles, a muscular foot used for locomotion that is modified in some species for swimming, and a visceral mass housing the organs. Most have a protective shell, usually external, excreted by the mantle, but in a few species the shell is internal, reduced or absent.

Quick guide

The taxonomy of planktonic molluscs is in a state of flux, making identification problematic. This is exacerbated by many species being gelatinous and delicate, and thus damaged during net sampling. It is possible to identify three major groups relatively easily, and sometimes distinctive taxa within these:

Pteropods: Wing-like foot and shell spirals to the left. Less frequently they have straight shells.

Heteropods: Gelatinous mass with reduced shell or flattened coiled shell with keel (and distinct eyes).

Prosobranchs: Shell spirals to the right.

8.3.8.1 Pteropods

Identification: This was a formal taxonomic entity that was removed, and recently reinstated (Order Pteropoda). The name Pteropods comes from Ancient Greek *pterón* meaning 'winged' and *podós* meaning 'foot' because the foot is divided into two flaps for flying through the water. Pteropods with a spiral shell have the spiral twisting to the left, which means that if you sat the shell on a surface with the whorl point facing up, the shell opening will be on the left. Some pteropod species have a straight shell.

Diversity: There are ~140 holoplanktonic marine gastropods (pteropods and heteropods).

Some common species: Shelled pteropods appear as left-coiled shells (e.g. *Limacina* sp. in Fig.

8.22D) or simple cones (e.g. *Hyalocylis* sp. in Fig. 8.22E and *Creseis* sp. in Fig. 8.22F). Other common pteropod genera are *Clione* and *Cavolinia*.

Lifespan: The pteropod *Limacina* has survived up to 18 months in the laboratory.

Feeding: Pteropods are carnivorous, using their highly developed eyes to locate small crustaceans and gelatinous zooplankton, and using their protrusible radula and hook-like teeth to capture their prey. Pteropods feed at night near the surface and migrate to deeper waters in the day.

Reproduction: Pteropods are protandric hermaphrodites (born male and transform to female at a later stage) and cross fertilisation is the rule, although self fertilisation is believed to occur at times. Entire life cycles are around 1 year, and eggs are gelatinous ribbons or spheres. Veliger larva hatches, metamorphoses, and develops into a juvenile.

Locomotion: Pteropods wave their wing-like foot to move themselves through the water (e.g. *Desmopterus* sp. in Fig. 8.22C, *Hyalocylis* sp. in Fig. 8.22E and *Creseis* sp. in Fig. 8.22F).

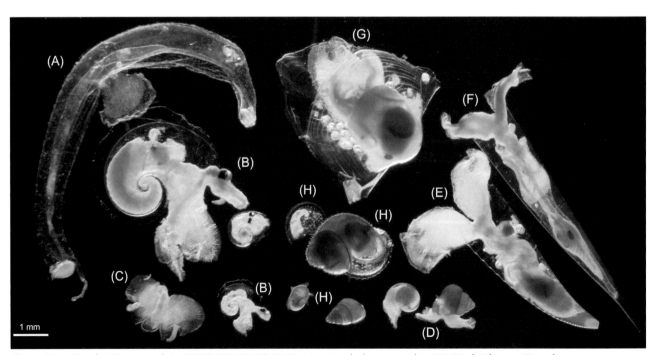

Fig. 8.22.   Snails (Gastropoda). PTEROTRACHEOIDEA (previously heteropods): **(A)** *Firoloida* sp., **(B)** *Atlanta* spp.; PTEROPODA: **(C)** *Desmopterus* sp., **(D)** *Limacina* spp.; **(E)** *Hyalocylis* sp., **(F)** *Creseis* sp., **(G)** *Diacavolinea* sp.; PROSOBRANCHIA: **(H)** species of prosobranch gastropods.

Ecology: Pteropods are common from the poles to the equator, with most being epipelagic, but some are mesopelagic. Although pteropods are occasionally found in neritic waters, there are no truly coastal species. Pteropods are important in ecological and geological research because they are a significant contributor to the oceanic carbon cycle. Pteropod shells are made from aragonite: a crystalline form of calcium carbonate that is particularly susceptible to shell dissolution under low pH conditions. There is some concern about the ability of pteropods to cope with ocean acidification.

Human interactions: Pteropods leave a fossil record important for paleoclimatic, paleoceanographic and paleoecological work.

8.3.8.2 Heteropods (now Pterotracheoidea)

Identification: Heteropods (from Ancient Greek *héteros* meaning 'different' and *podós* meaning 'foot') are commonly called sea elephants because they have a trunk-like proboscis. The term heteropod is not a formal taxonomic entity and has been replaced by the superfamily Pterotracheoidea. They have soft tissue and are laterally compressed, the small transparent shell (<14 mm) is beneath the foot, which is modified into a single ventral fin (i.e. they swim upside down, as do pteropods).

Diversity: There are ~140 holoplanktonic marine gastropods (pteropods and heteropods).

Some common species: A common heteropod is *Firoloida*, where the female possesses a permanent egg filament protruding posteriorly (Fig. 8.22A). Another common heteropod is *Atlanta* (Fig. 8.22B).

Feeding: Heteropods are carnivorous, using their highly developed eyes to locate small crustaceans and gelatinous zooplankton, and using their protrusible radula and hook-like teeth to capture their prey.

Reproduction: Heteropods have separate sexes and show sexual dimorphism. After internal fertilisation, eggs are laid individually or in egg-strings (free floating or attached to the female). Larvae are free floating veligers, with right-coiled shells.

Locomotion: Heteropods use a single ventral fin to propel themselves.

Ecology: Heteropods are holoplanktonic and found primarily in tropical to subtropical open oceans, down to a depth of 250 m.

Human interactions: *Atlanta* has a shell made from aragonite, and there is some concern about the impact of ocean acidification on this taxon.

8.3.8.3 Prosobranchs

Identification: The term prosobranch (from Ancient Greek *pró* 'before' and *bránkhia* meaning 'gills' because the heart is in front of the gills) is not a formal taxonomic entity. Almost all prosobranchs are meroplanktonic, except for the purple snail *Janthina*, a large holoplanktonic snail that builds a raft of bubbles for a float and is part of the neustonic community (animals that live on the sea surface). By contrast with pteropod shells that spiral to the left, prosobranch shells spiral to the right (i.e. looking down on the whorl point the shell opening will be on the right). For example, compare Fig. 8.22H (right-hand specimen, right spiral = prosobranch) with Fig. 8.22D (right-hand specimen, left spiral = pteropod).

Diversity: There are 20 000 species of marine gastropods, and most of these are prosobranchs that are benthic as adults and have planktonic larval stages.

Some common species: Most prosobranchs collected in zooplankton samples are unlikely to be identified beyond prosobranch. *Janthina* may be present as an adult, but are rare in plankton samples, although are sometimes washed up on the beach during summer.

Feeding: Almost all prosobranchs are larvae when they are in the plankton, and feed on small zooplankton.

Reproduction: Prosobranchs typically develop from eggs that are retained in the mantle cavity and then deposited onto a hard substrate or shed into the water column.

Locomotion: Prosobranch larvae use cilia to move, but the holoplanktonic prosobranch *Janthina* uses a raft of bubbles to keep itself afloat.

Ecology: Prosobranch molluscs are found on rocky shores and their larvae are thus only periodically abundant in coastal waters following spawning.

Human interactions: Conchologists collect sea shells, predominantly of prosobranch molluscs. Due to their sensitivity to chemical pollutants, particularly to the anti-biofouling chemical tributyltin (TBT), prosobranchs are used in biomonitoring and bioaccumulation studies.

8.3.9 Spiky

Quick guide

Forams: Most appear as clumps of dark, fused, often spiky balls.

    Radiozoans (radiolarians): Star-shaped, spherical and spiky. Spikes sometimes random.

    Echinoderm larvae: Spiky/thorny, shaped like a capital 'A'. A tissue mass is usually visible in the apex.

    Polychaete trochophore larvae: Young stage of worm with chaeta (bristles).

8.3.9.1 Forams (Phylum Foraminifera)

Protist Protest

Little protists of the sea
How do we treat thee?
As foraminifers, Oh wee beasties of the sea? Or,
shall it be, foraminifera,

for the plural or the singular?
Perhaps we can float the word 'foraminiferan' but
then again,
it still would mean a single cell,
but how in hell
can foraminifera be for one and two,
when so many live in the ocean blue? Please,
please tell me Dr. Foram Man or M'am,
Is it -minifer, -minifera, or -miniferan?
    By Sally Walker in Lipps *et al.* (2011)

Identification: Foraminifera (from Latin *foramen* meaning 'opening' and *fer* meaning 'bearing', because they were originally thought to be gastropods), or forams for short, are single-celled organisms. Planktonic forms have multi-chambered shells (also called tests) that are usually coiled and sometimes spiky (Fig. 8.23B). Some types appear flat and two-dimensional (e.g. *Globorotalia*), whereas others are more three-dimensional and appear as three or four variable-sized spherical chambers fused together (e.g. *Globigerina*). Shells of forams can be made of organic compounds or calcium carbonate. They are usually <1 mm in size.

Diversity: Forams are typically benthic, with ~40 planktonic species.

Some common species: Some well-known planktonic forams are *Globigerina*, *Globigerinoides*, *Neogloboquadrina*, *Orbulina* and *Turborotalia*.

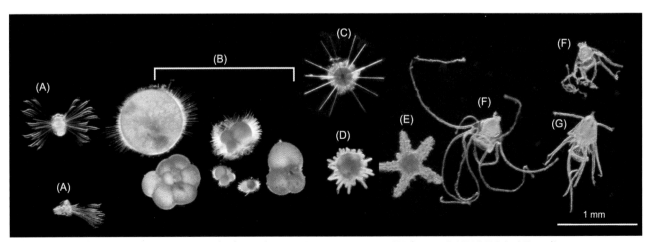

Fig. 8.23. Spiky. POLYCHAETA: **(A)** polychaete larva; FORAMINIFERA: **(B)** foram; RADIOZOA: **(C)** radiozoan (radiolarian); ECHINODERMATA: **(D)** sea urchin larva juvenile, **(E)** brittle star juvenile, **(F)** ophiopluteus larvae, **(G)** echinopluteus larva.

Lifespan: Typical lifespan of forams is 2–4 weeks, but some can live for years.

Feeding: Most forams consume dissolved organic molecules, bacteria, diatoms and other single-celled algae, but they can also ingest stages of copepods that are caught with a network of reticulopodia (thin pseudopodia) that extend from one or more apertures in the shell. Some forams have algal symbionts that provide some nutrition from photosynthesis.

Reproduction: Forams can reproduce either sexually or asexually. There is no alternation of strategies, so that animals produced asexually may again produce offspring asexually.

Locomotion: Forams have limited locomotory abilities and float passively in the water column.

Ecology: The shells of forams are major components of sand in some regions, particularly in the deep ocean. Because the shells of some forams are made from calcium carbonate, there are concerns that foram shells are susceptible to dissolution as oceans acidify.

Human interactions: Because they are abundant, have complex and distinctive test morphology, and continually rain down to the deep ocean where they are buried, pelagic forams are key trace fossils for dating sedimentary rock. Using forams to date sedimentary rock is a critical step in finding oil and gas deposits. Because foram tests are formed from elements found in the ancient seas where they lived, they provide valuable proxies for paleoclimate and paleoceanography. Global temperature and ice volume can be estimated using oxygen isotopes of forams, and the history of the carbon cycle and oceanic productivity by their carbon isotopes. The foram *Neogloboquadrina pachyderma* is an indicator of past ocean temperatures: when the Earth is relatively cool, tests of *N. pachyderma* coil to the left; whereas when the Earth is relatively warm tests of *N. pachyderma* coil to the right. Living foram assemblages are also used as bioindicators in coastal environments such as coral reefs because they provide an integrated measure of water quality.

8.3.9.2 Radiozoans (Phylum Radiozoa)

Identification: Radiozoans were previously known as radiolarians (from Latin *radiolus* meaning 'little sun-beam'). Radiozoans are star-shaped, single-celled organisms that produce intricate silica or strontium sulphate skeletons (Fig. 8.23C). Part of the cell extends out through tiny perforations and along spines for feeding. They can range from 0.02 to 3 mm in diameter, but are usually around 0.1–0.2 mm in size. Radiozoans can be solitary or colonial.

Diversity: There are two major groups of Radiozoa; the Polycystina and the Acantharia. The Polycystina have silica skeletons, whereas the Acantharia have strontium sulphate skeletons.

Some common species: The two most commonly captured radiozoans in plankton samples off the east coast of Australia are *Collozoum inerme* and *Sphaerozoum* sp.

Lifespan: Usually a few weeks to a few months.

Feeding: Radiozoans use pseudopodia to ingest bacteria, algae, protists (especially tintinnids), copepods, appendicularians and other small zooplankton. Some radiozoans have symbiotic microalgae (particularly dinoflagellates) that provide most of their energy; these species live in surface waters.

Reproduction: Asexual reproduction in radiozoans is by budding or fission. When the skeleton divides, each daughter cell produces the missing half. Sexual reproduction is thought to occur but has not been observed.

Locomotion: Radiozoans have limited abilities to move, but are buoyant so can remain in the water column without expending energy.

Ecology: Radiozoans do not have skeletons made from calcium carbonate and thus are not expected to be sensitive to ocean acidification. There is some limited information that copepods and euphausiids can feed on radiozoans. Radiozoans are exclusively marine pelagic organisms, except for one brackish water species. Radiozoans are most common in tropical and subtropical oceans.

Human interactions: Their diversity and long geological history (going back to the Cambrian

Period) make radiozoans important sources of information for evolutionary, stratigraphic and paleoecological research. Radiozoa are particularly important for dating sedimentary rocks that lack calcareous fossils such as forams. The intricate skeletons of radiozoans have an amazing variety of forms, and were popularised by Ernst Haeckel's classic 'Art forms in Nature' in 1904 (https://publicdomainreview.org/collections/ernst-haeckels-radiolaria-1862/).

8.3.9.3 Echinoderm larvae (Phylum Echinodermata)

Identification: Spiky echinoderm (from Ancient Greek *ekhînos* meaning 'hedgehog' and *derma* meaning 'skin') larvae belong to sea urchins (echinoids) and brittle stars (ophiuroids). Echinoderm larvae look like a capital 'A', because many are shaped like a building truss (i.e. triangular with cross-members) (Fig. 8.23F–G). As juveniles, they look more like their adults (Fig. 8.23D–E). Echinoderms are a solely marine group and none are holoplanktonic.

Diversity: There are ~7000 species of echinoderms and most have pelagic larvae.

Some common species: It is not generally possible to identify echinoderm larvae to species, but merely to groups. Characteristic larval stage in the brittle stars (Ophiuroidea, Fig. 8.23F) and sea urchins (Echinoidea, Fig. 8.23G) is the pluteus.

Lifespan: Pluteus larvae that feed remain in the plankton for weeks to several months, whereas non-feeding larvae are in the plankton for a much shorter period.

Feeding: Pluteus larvae have ciliated arms that they use to capture unicellular algae.

Reproduction: Most echinoderms shed their gametes into the water where they are fertilised, but some echinoderms do not have a larval stage and develop directly. The pluteus larvae develops rapidly from the embryo.

Locomotion: Pluteus larvae swim with their arms upwards; a band of cilia over each arm creates a current that propels the larvae.

Ecology: Adults are relatively sedentary on the sea floor (e.g. sea stars, sea urchins, sand dollars and sea cucumbers), but their larvae are often common in the plankton, particularly in shelf and coastal waters. Larvae of deep-water species are unlikely to be caught in plankton tows.

Human interactions: In 2012, a total of 109 222 tonnes of sea cucumbers, sea urchins and other echinoderms were harvested from the ocean. Sea cucumbers are considered a delicacy in many countries in South-East Asia, where it is boiled, then dried, and finally smoked. Gonads of male and female sea urchin are eaten in Japan, Peru, Spain and France. Echinoderm larvae, particularly sea urchins, are commonly used in short-term, sub-lethal toxicity tests. The ability of starfish to regenerate lost body parts is being studied to understand regrowth and cloning processes that may provide insights into the potential for regeneration of human limbs and organs.

8.3.9.4 Polychaete trochophore larvae (Class Polychaeta)

Larvae of polychaetes are common in the plankton, particularly in shallow habitats. They are variable in appearance, but generally have chaeta (spines) and/or cilia (Fig. 8.23A). See Section 8.3.6.1 for information on pelagic polychaete adults.

8.3.10 Fish and squid

Fish larvae have great diversity in morphology, ranging from miniature versions of their adults to larvae that undergo a significant metamorphosis from the larval to the juvenile and adult stages. Nevertheless, they are identifiable as fish. Squid are cephalopods, and we include here the paralarvae of squid, octopus and cuttlefish: all are miniature versions of their adults and are thus unmistakable. Young stages of cephalopods are not termed larvae but rather paralarvae, because they do not go through true metamorphosis (they are miniature adults).

Quick guide

Squid, octopus and cuttlefish paralarvae: Miniature versions of their adults.

Fish larvae: Variable body shape with large eyes, fins, variable numbers of myomeres, pigmentation and head spine patterns.

8.3.10.1 Squid, octopus and cuttlefish paralarvae (Class Cephalopoda)

Identification: Cephalopods (from Ancient Greek *kephalḗ* meaning 'head' and *podós* 'foot') are in the Phylum Mollusca and include squid, octopus and cuttlefish (Fig. 8.24A). The only other animals with similar looking eyeballs to cephalopod paralarvae are fish larvae (Fig. 8.24B,C). Squid paralarvae have two small fins on their body, and in octopus the fins are absent.

Diversity: Cephalopod paralarvae are found in all oceans, most commonly in surface waters but also throughout the mesopelagic to abyssal depths. There are over 800 species currently described. Diversity is greatest closer to the equator.

Some common species: Cephalopod paralarvae are generally not identifiable to species.

Lifespan: Cephalopods grow very fast and reach full size quickly. Many species of squid die after spawning and probably only live for a year or two. Even the largest squid may only live for 5 years, but some species, such as the vampire squid, can live for close to 10 years. Cephalopods remain as paralarvae for weeks to several months.

Feeding: As paralarvae, cephalopods eat zooplankton, particularly crustaceans.

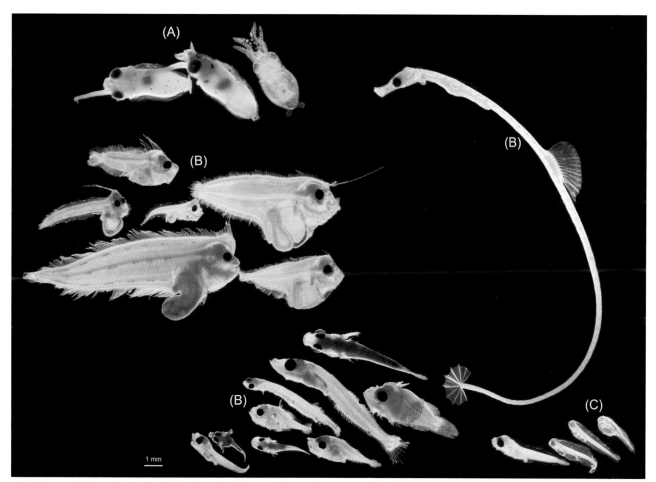

Fig. 8.24.   Fish and squid. CEPHALOPODA: **(A)** squid early stages; PISCES: **(B)** Fish larvae of different fish species, **(C)** early stages of fish development, some with yolk-sac visible.

Reproduction: Cephalopods have separate sexes. After mating, females attach eggs to the substrate. The paralarvae that hatch from the eggs resemble miniature versions of the adults.

Locomotion: Cephalopods swim using a combination of jet propulsion and propulsion developed by undulations of their fins; different species rely more or less on each type of propulsion.

Ecology: Cephalopods grow quickly and so are an important food source for other species, including tuna, myctophids and toothed whales.

Human interactions: Squid, octopus and cuttlefish fisheries have been increasing globally and now total 4.5 million tonnes annually. Shells of nautilus are used in decorations, and the internal shell of a cuttlefish is used as a calcium source for pet birds.

8.3.10.2 Fish larvae (Class Actinopterygii, previously Class Osteichthyes)

Identification: The large distinctive eyes of fish larvae catch your attention (Fig. 8.24B). Newly hatched fish larvae still have their yolk sac attached, which they rely upon as a food reserve before they start feeding (visible as a globule in Fig. 8.24C). A recently hatched larva is ~2–3 mm long and reaches the settlement/transformation stage at ~10–30 mm. During this short larval development period, morphological features change rapidly, making it difficult to construct keys for larval identification. The early life of fish is normally divided into five stages: egg, yolk-sac larva, larva, settlement/transformation and juvenile. Fish larvae have variable body shapes with large eyes, fins, variable numbers of myomeres and pigmentation and head spine patterns. Accurate identification of fish larvae is difficult and is based on a variety of morphological characters: body shape, number and position of melanophores, head spination, relative position of fins, shape and size of fin rays, myomere/vertebra counts and meristic characters.

Diversity: There are more than 32 000 fish species worldwide, with over 4500 species from Australian waters. Most marine fish, whether pelagic or demersal, spawn pelagic eggs and larvae that spend their time in the upper layers of the water column.

Some common species: Fish have specific habitats (e.g. estuarine, coastal, oceanic, reef dwellers), so the common species you will encounter depend upon where you sample. Larval fish usually have different morphology than their pelagic or demersal adults, making them difficult to identify. Recently, several larval fish identification guides have been produced that illustrate the development of larvae from different geographical regions. Because of their importance in commercial and recreational fishing, here we describe in more detail the development of larval fish.

Developmental stages of larval fish: One of the most commonly used terminologies to describe the development of larval fish is based on that used by Ahlstrom and Moser (1980). The larval stage is defined as the development stage between hatching (or birth) and the attainment of full external meristic complements (i.e. the number of fin rays and scales) and loss of specialisation for planktonic life. The larval stage is divided into preflexion, flexion and postflexion stages that are related to the development of the caudal fin and the corresponding upward (dorsal) flexion of the notochord.

The length and stage of development are important features in classification. The gas bladder (Table 8.1), which is present in many larvae is absent in adults (such as gobies). During the day, the gas bladder may be small in larvae, but strongly inflated and conspicuous in larvae caught at night, which can result in larvae of the same species appearing different depending on when they were caught. The inflation of the swim bladder is related to the diel vertical migration that larvae of many species undertake.

The most common method of identification of unknown fish larvae is the series method. This involves identifying the largest available larval or juvenile specimen in the samples, based on adult characteristics such as fin meristics and vertebral number (equivalent to the number of myomeres or muscle blocks – in larvae). The largest specimen is

Table 8.3.   Key identification of features of fish larvae in estuaries and marine coastal waters

Family	Estuary (E) or Marine (M)	Features
Engraulidae (anchovy)	E/M	38–47 myomeres; body very elongate; gut very long; lightly pigmented (Fig. 8.25A)
Clupeidae (herring)	E/M	41–55 myomeres; body very elongate; lightly pigmented; gut very long (Fig. 8.25B)
Mugilidae (mullet)	M	24–25 myomeres; body depth moderate; heavily pigmented; gut long and coiled; small preopercular spines (Fig. 8.25C)
Atherinidae (hardyhead)	E	35–47 myomeres, body very elongate, moderately pigmented, gut coiled and compact (Fig. 8.25D)
Hemiramphidae (garfish)	E	51–57 myomeres; body very elongate; moderately to heavily pigmented; gut very long (Fig. 8.25E)
Syngnathidae (pipefish/seahorse)	E	Elongate body; prominent dermal plates; moderately to heavily pigmented (Fig. 8.25F)
Platycephalidae (flathead)	M	27 myomeres; body depth moderate; moderately pigmented; gut moderate to long and coiled; large and early forming pectoral fins; extensive head spination (Fig. 8.25G)
Scorpaenidae	M	24–28 myomeres; body depth moderate; moderately pigmented; gut moderate to long and coiled; large early forming pectoral fins; extensive head spination (Fig. 8.25H)
Sillaginidae (whiting)	M	32–45 myomeres; body elongate; lightly pigmented; gut moderate to long and coiled; very small preopercular spines (Fig. 8.25I)
Carangidae (scad/trevally)	M	24–25 myomeres; body depth moderate to deep, lightly to heavily pigmented; gut coiled and moderate; preopercular spines and occipital crest (Fig. 8.25J)
Gerreidae	M	24–25 myomeres; body depth moderate; lightly pigmented; gut moderate coiled and compact; prominent ascending premaxillary process; small preopercular spines (Fig. 8.25K)
Sparidae (bream, porgy)	M	24–25 myomeres; body depth moderate; lightly pigmented; gut moderate, coiled and compact; small to large preopercular spines (Fig. 8.25L)
Monodactylidae (moonfish)	M	24 myomeres; body deep and laterally compressed; moderately to heavily pigmented; gut moderate, coiled and compact; large early forming pelvic fins; large preopercular spines (Fig. 8.25M)
Girellidae (blackfish)	M	26–27 myomeres; body depth moderate; lightly to moderately pigmented; gut moderate, coiled and compact; small preopercular spines (Fig. 8.25N)
Ambassidae (glass perchlet)	E/M	24–25 myomeres; body depth moderate; lightly pigmented; gut coiled and compact; conspicuous gas bladder; small preopercular spines (Fig. 8.25O)
Terapontidae (trumpeter)	M	25 myomeres; body elongate; lightly pigmented; gut coiled and moderate; small preopercular spines (Fig. 8.25P)
Labridae (wrasse)	M	23–28 myomeres; body elongate to moderate and laterally compressed; lightly pigmented; gut initially straight and later coils (Fig. 8.25Q)
Gobiidae (goby)	E/M	24–34 myomeres; body elongate to moderate; lightly to heavily pigmented; gut moderate and slightly coiled; conspicuous gas bladder (Fig. 8.25R)
Blenniidae (blenny)	E	Typically, 30–40 myomeres; body elongate; lightly to moderately pigmented; gut short and coiled; moderate to large teeth; none to large preopercular spines (Fig. 8.25S)
Callionymidae (dragonets)	M	20–22 myomeres; body robust and moderately deep; heavily pigmented; gut coiled and moderate to long; one large preopercular spine (Fig. 8.25T)
Monacanthidae (leatherjacket)	M	19–20 myomeres; body deep and laterally compressed; moderately to heavily pigmented; gut moderate, coiled and compact; prominent dorsal and pelvic spine with barbs (Fig. 8.25U)
Bothidae (flounder)	M	33–61 myomeres; body moderately deep and laterally compressed; lightly pigmented; gut coiled and moderate to long (Fig. 8.25V)
Paralichthyidae (flounder)	M	33–39 myomeres; body moderately deep and laterally compressed; moderately pigmented; gut coiled and moderate to long; small preopercular spines (Fig. 8.25W)
Cynoglossidae (tongue sole)	M	42–66 myomeres; body moderate to elongate and laterally compressed; lightly pigmented; gut coiled and pendulous (Fig. 8.25X)

Based on information from Leis and Carson Ewart (2001) and Neira *et al.* (1998).

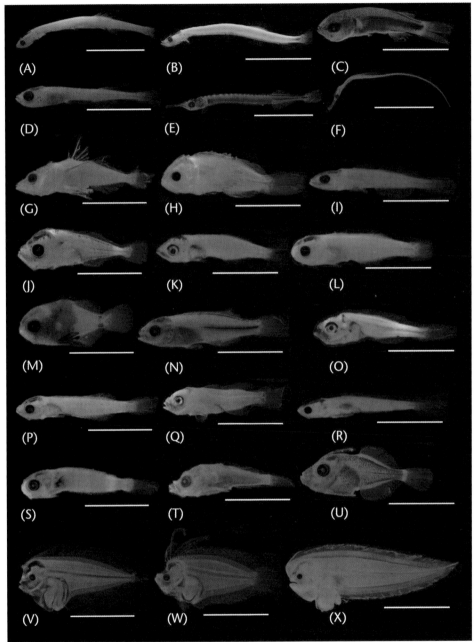

Fig. 8.25. The postflexion stages of some dominant families of fish in estuarine and coastal marine plankton samples. **(A)** ENGRAULIDAE: *Engraulis australis* (anchovy); **(B)** CLUPEIDAE: *Sardinops sagax* (pilchard); **(C)** MUGILIDAE: *Liza argentea* (mullet); **(D)** ATHERINIDAE: *Atherinostoma microstoma* (hardyhead); **(E)** HEMIRAMPHIDAE: *Hemiramphus* sp. (garfish): **(F)** SYNGNATHIDAE: *Urocampus carinirostris* (pipefish): **(G)** PLATYCEPHALIDAE: *Platycephalus fuscus* (flathead): **(H)** SCORPAENIDAE: *Centropogon australis* (fortesque); **(I)** SILLAGINIDAE: *Sillago ciliata* (whiting): **(J)** CARANGIDAE: *Trachurus novozelandiae* (scad): **(K)** GERREIDAE: *Gerres subfaciatus* (silver biddy), **(L)** SPARIDAE: *Acanthopagrus australis* (bream); **(M)** MONODACTYLIDAE: *Monodactylus argenteus* (moonfish): **(N)** GIRELLIDAE: *Girella tricuspidata* (blackfish); **(O)** CHANDIDAE (AMBASSIDAE): *Ambassis jacksoniensis* (perchlet); **(P)** TERAPONTIDAE: *Pelates sexlineatus* (trumpeter); **(Q)** LABRIDAE: *Acheorodus viridis* (wrasse); **(R)** GOBIIDAE: *Pseudogobius* sp. (goby); **(S)** BLENNIIDAE: *Omobranchus anolius* (blenny); **(T)** CALLIONYMIDAE: *Repomucenus calcaratus* (dragonet); **(U)** MONACANTHIDAE: (leatherjacket); **(V)** BOTHIDAE: *Lophonectes gallus* (flounder): **(W)** PARALICHTHYIDAE: *Pseudorhombus* sp. (flounder): **(X)** CYNOGLOSSIDAE: *Cynoglossus* sp. (tongue sole). Note the 5 mm scale bar beside each image.

then linked to smaller specimens in the series by using morphological and pigment characteristics. A variety of characters can be used to identify fish larvae including general morphology, such as body shape, gut length and degree of coiling, number of myomeres, pigmentation patterns (melanophores, Table 8.1), the sequence of development of fins and the pattern of head spination (Table 8.3).

Despite these characters, identification of fish larvae to species is difficult, even for experienced workers, because species descriptions are based on adult characteristics, which are not evident in early developmental stages. In addition, the eggs and larval development of many species are still undescribed. A technique used for identification of adult fish and now used for identification of fish larvae is DNA barcoding using the cytochrome c oxidase gene subunit I (COI; Lewis *et al.* 2016; Ardura *et al.* 2016).

Lifespan: The larval stage usually lasts 3–4 weeks. For adult fish, there is considerable variation in their life span, with some such as orange roughly living up to 149 years and reaching 75 cm in length while mahi mahi live only 2 years but reach 1.8 m long. Despite this variation, ultimate adult size is generally indicative of longevity.

Feeding: Once fish larvae have consumed the yolk from their yolk sac, they eat small plankton. Fish larvae generally start eating phytoplankton and then move onto zooplankton when they are larger. As the jaw length and mouth gape increase with growth, this allows larvae to ingest larger and more diverse prey items.

Reproduction: Most fish species have separate sexes but some are sequential hermaphrodites (first one sex and later in life the other), and a few are simultaneous hermaphrodites. Many marine

Box 8.6 Fish larvae in coastal waters and estuaries

Larval fish communities in coastal waters vary considerably in time and space. The variability in community structure is a result of interactions among adult fish spawning patterns, larval behaviour, and oceanic processes such as dominant currents that transport larvae. Many larval fish species vary in abundance seasonally, and other species occur all year round.

Based on their life history, the majority of fish larvae caught in estuaries can be categorised as estuarine or marine opportunists, with low numbers of freshwater or marine straggler species. Estuarine species, which are usually small as adults, spawn and spend their whole life cycle in the estuary. By contrast, estuarine opportunist species spawn in coastal waters, with their larvae entering estuaries where the juveniles settle into nursery habitats such as seagrass beds and mangroves. Adults of these species remain in estuaries, leaving the estuary to spawn or permanently migrate out of the estuary to coastal reefs. There may also be small numbers of larvae of freshwater and marine straggler species in estuaries.

Estuarine-spawning species have a short life cycle of only 1–2 years and have several reproductive strategies to reduce the mortality of eggs and larvae. These strategies include being live bearers with juveniles hatching at an advanced stage of development, benthic eggs that are attached to seagrasses or other hard substrates, or spawning pelagic eggs in the upper reaches of the estuary to minimise the chance of the eggs being washed out to sea. Such strategies mean that larvae hatch at an advance stage of development, which allows them to be retained within estuaries.

The abundance of larvae of estuarine species usually shows a seasonal pattern, with highest abundances in summer and the lowest in winter. The seasonal variation in abundance closely follows the cycle of water temperatures. The abundance of larvae of marine opportunist species entering estuaries also shows a similar, but less-marked, seasonal variation.

Species diversity of fish larvae generally decreases with increasing distance upstream, away from the mouth of the estuary. Although samples from estuarine stations usually have a much lower diversity compared with marine stations, abundances of larvae of estuarine species are usually higher than for larvae of marine opportunist species.

species produce hundreds of thousands of eggs ~1 mm in diameter (see Section 8.3.11.4).

Locomotion: Early larvae can only swim poorly and drift with currents, but their swimming ability increases as fin elements develop. Once fish larvae have developed scales and complete fin complements, they are referred to as juveniles.

Ecology: Nearly all fish have a pelagic larval stage and thus are an interesting component of zooplankton samples. Most fish larvae hatch from pelagic or demersal eggs (i.e. attached to hard substrates or seaweed), but there are a few live-bearing species. Fish diversity is highest in tropical waters and decreases towards the poles. Most species of fish larvae are found in estuarine and coastal marine waters and these larval fish communities vary considerably in time and space and have adaptations to maintain their coastal existence (see Box 8.6).

Human interactions: Fish and the fisheries they support have huge economic, social, cultural and health benefits for people. About 70 million tonnes of fish are harvested from the oceans each year, and 90% of all fisheries landings is derived from the continental shelf. Virtually all commercial fish species have larvae that rely on eating phytoplankton and small zooplankton for survival during early life. The magnitude of fish recruitment each year is considered to be primarily regulated by differential survival of fish larvae (Box 8.7), but is difficult to predict, making fisheries management more challenging. Enhancing the spawning of resident fish is one reason for the establishment of marine parks.

Recent genetic research shows that the amount of spill over of eggs and larvae from reserves into neighbouring areas can supplement local fisheries, although the extent is of some debate. The presence of fish larvae can indicate important spawning areas, and therefore larval diversity may have greater relevance than mere presence of a transient adult.

8.3.11 Eggs

Marine species that do not have direct development have an egg stage. Most eggs in plankton samples are small (~0.1 mm) because they are the eggs of crustaceans such as copepods. Eggs are spherical and often reddish, brownish or clear, and can be distinguished from similar-sized phytoplankton, which is normally greenish and flat (Fig.

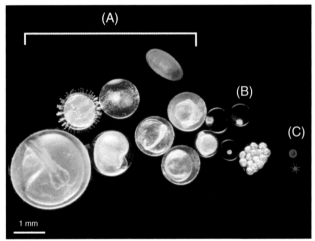

Fig. 8.26. Eggs. PISCES: **(A)** fish eggs with embryos inside; **(B)** recently spawned eggs without embryos; COPEPODA: **(C)** copepod eggs.

Box 8.7 Larval fish condition, fish recruitment and water quality

The larval stage of fish is considered a bottleneck for fisheries, through starvation of the larvae (insufficient food), predation (from jellyfish, ctenophores, krill or fish) and dispersal due to unfavourable currents. More than 99% of all fish eggs and larvae do not survive, owing to predation from krill, jellyfish or other fish, and therefore rapid growth may enhance survival by reducing the duration of the vulnerable larval stages. Consequently, fisheries biologists estimate age and larval growth from the daily growth increments of the ear-stone or otolith (analogous to tree rings). Body width or weight are also useful indicators of larval condition (fatter fish are healthier), in response to environmental conditions (food, water temperature and pollution). Fish larvae are very delicate – without scales – so they are susceptible to poor water quality from urban run-off, acidic water or sewage effluent.

8.26). The finer the mesh net used, the more eggs will be seen.

8.3.11.1 Copepod eggs

The most common eggs seen in plankton samples are typically ~0.1 mm in diameter and are from copepods (Fig. 8.26C). The majority are smooth spheres, although a few such as the eggs of the common calanoid copepod *Centropages* are spiky. These spikes may help keep the eggs in suspension or may reduce predation. In unfavourable conditions, some species of copepods produce thick-shelled resting or dormant eggs. These eggs are less likely to be seen in plankton samples because they slowly sink to the substrate where they remain until environmental conditions are more favourable, which may be months.

8.3.11.2 Krill eggs

A little larger than copepod eggs at 0.2–0.5 mm, with a large perivitelline space. They are often seen in plankton samples. Krill eggs are negatively buoyant, so they slowly sink to the sea floor. In some species, eggs can last up to 10 days until the nauplius hatches.

8.3.11.3 Decapod eggs

A little larger than copepod eggs at 0.2–0.5 mm, decapod eggs are rarely seen. The only decapods to spawn eggs into the water column are sergestoids and some penaeid prawns. Most penaeids are probably on the bottom when they spawn, so their eggs never enter the water column. Penaeid eggs are also short-lived (<48 hours for some species) and negatively buoyant, so sink. The penaeid genus *Trachypenaeus* probably has positively buoyant eggs because they have a very large perivitelline space.

8.3.11.4 Fish eggs

Fish eggs are typically between 0.5 mm and 1.5 mm in diameter, and are larger but rarer than crustacean eggs (0.1 mm). Fish eggs are translucent, whereas invertebrate eggs are often dark and <0.5 mm diameter. Almost all fish eggs are perfectly spherical, making the elliptical anchovy egg distinctive when seen in zooplankton samples. Inside a young fish egg is a ball of yolk and maybe oil globule(s) (Fig. 8.26B). Older eggs contain an embryo delicately suspended inside (Fig. 8.26A). Eggs hatch into larvae with large and distinctive pigmented eyes and only fin folds (Fig. 8.24, Section 8.3.10.2). Compared with fish larvae, there has been little work undertaken on the identification of fish eggs. Characters that can be used to identify eggs are: their size and shape; the number, position and pigmentation of oil globules; the degree of yolk segmentation; the presence and size of the perivitelline space; and embryonic characteristics. Fish eggs are a major food source in the ocean for even the largest predators. For example, whale sharks in Qatari waters of the Arabian Gulf feed on eggs released by a large spawning aggregation of mackerel tuna.

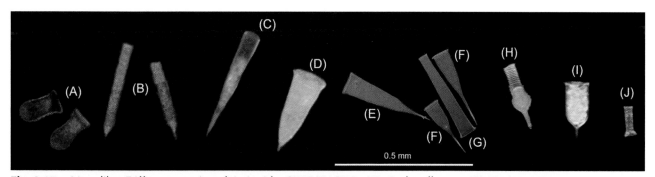

Fig. 8.27.   Vase-like. Different species of tintinnids. CILIOPHORA: **(A)** *Codonella* spp., **(B)** *Tintinnopsis* spp., **(C)** *Xystonellopsis heros*, **(D)** *Cyttarocylis cassis*, **(E)** *Xystonella treforti*, **(F)** *Rhabdonella* spp., **(G)** *Eutintinnus* sp., **(H)** *Codonellopsis orthoceras*, **(I)** *Favella* sp., **(J)** *Tintinnopsis nordqvisti*.

8.3.12 Vase-like

8.3.12.1 Tintinnids (Phylum Ciliophora)

Identification: The term tintinnid is from the Latin *tintinnabulum* meaning 'a small tinkling bell': a reference to the shape of many species. Other species can be shaped like miniature vases, cones or pipes (Fig. 8.27). Tintinnids are ciliates (previously placed in the protozoa) that, when undisturbed, extend a crown of cilia around the open end to capture small cells.

Diversity: Tintinnids are complex single-celled eukaryotes (having a cell nucleus with a membrane). Characteristics of their lorica, or shells, are classically used to distinguish the ~1000 described species.

Some common species: Tintinnids have distinctive shapes: some have rounded ends (*Codonella* spp. Fig. 8.27A); others are thin with a pointy end and parallel sides (*Tintinnopsis* spp. Fig. 8.27B) or have tapering sides (*Xystonellopsis heros* Fig. 8.27C, *Cyttarocylis cassis* Fig. 8.27D and *Xystonella treforti* Fig. 8.27E); while others have tapering sides and a long point (*Rhabdonella* spp. Fig. 8.27F) or have a bulb before the point (*Codonellopsis orthoceras* Fig. 8.27H); some are fat and have parallel sides with a point (*Favella* sp. Fig. 8.27I); and some are like pipes (*Eutintinnus* sp. Fig. 8.27G and *Tintinnopsis nordqvisti* Fig. 8.27J).

Lifespan: There is nothing published about the longevity of tintinnid species under natural conditions or about the viability of tintinnid cysts. Their lifespan is probably short because their doubling time is only a few days.

Feeding: The cilia of the cell project out of the top of the shell and generate a water flow across the mouth of the cell, bringing bacteria and phytoplankton into the mouth, and also propelling the tintinnid through the water. Tintinnids such as *Favella* feed on other tintinnids.

Reproduction: Tintinnids commonly reproduce through binary division, but sexual reproduction has been described for some species, with the exact mechanisms of gamete production and fusion being variable. Some tintinnid species can produce two generations per day when conditions are favourable, and some species produce cysts when conditions are unfavourable.

Locomotion: The cilia that are used for feeding also provide locomotion for tintinnids. The action of the cilia causes tintinnids to have a jerky swimming pattern.

Ecology: Tintinnids are most abundant in marine waters, but also found in freshwaters. Tintinnids are the top predators in microbial food webs and are thus a vital link from bacteria and nanophytoplankton to copepods and fish larvae.

Human interactions: Tintinnid loricas – the outer coverings made from protein and often reinforced with sand grains – have an amazing variety of forms that have been popularised by Ernst Haeckel's classic 'Art forms in Nature' in 1904 (https://en.wikipedia.org/wiki/Kunstformen_der_Natur#/media/File:Haeckel_Ciliata.jpg).

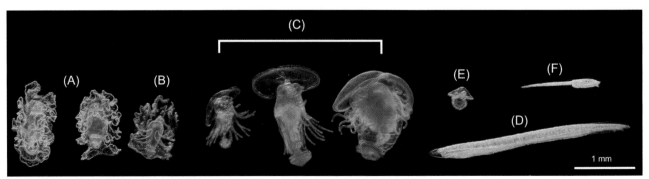

Fig. 8.28.   Oddities. ECHINODERMATA: **(A)** brachiolaria larva; HEMICHORDATA: **(B)** tornaria larva; PHORONIDA: **(C)** actinotroch larva; CEPHALOCHORDATA: **(D)** *Branchiostoma* (juvenile); ANNELIDA/MOLLUSCA: **(E)** trochophore larva; CHORDATA: **(F)** ascidian larva.

8.3.13 Oddities

After identifying the common and larger specimens in a plankton sample, you are left with an odd assortment of rare groups, covering a range of sizes (Fig. 8.28).

Brachiolaria larvae: These are second stage of development in many sea stars (Asteroidea) that look amorphous and jelly-like (Fig. 8.28A). The brachiolaria stage follows the bipinnaria stage, which is short-lived and has ciliated folds. Bipinnaria would only be seen in shallow, coastal plankton samples. Older larvae of sea stars (Asteroidea) resemble adults.

Hemichordate tornaria larvae: The tornaria larva of the peanut worms (Sipunculida) larvae occasionally appear in the plankton (Fig. 8.28B).

Phoronid actinotroch larvae: Phoronids are unusual, almost hydroid-like annelids (Fig. 8.28C). The adults are horseshoe worms. Phoronids are only rarely encountered.

Cephalochordates: *Branchiostoma* has very small eyes and no visible fin rays (Fig. 8.28D).

Trochophore larvae: Not only polychaetes produce trochophore larvae (see Fig. 8.23), but so do bivalves and snails, although these are less spiky and more ciliated (Fig. 8.28E). Trochophore larvae are tiny (0.2 mm), free-swimming and top-shaped, with a mouth opening just below an equatorial ring of cilia and with an apical tuft of cilia. Beating of the cilia spins them and propels them through the water. These are difficult to identify and usually only appear for a short time before further development. As they develop, trochophore larvae tend to become more like adults: for example, annelid larvae become more elongate and segmented with age. Trochophore larvae use their cilia bands for locomotion. The trochophore stage is followed in the gastropods and bivalves by a veliger larva, with a foot and shell. The veliger then settles to the sea floor as a young adult.

Depending where you are sampling, you might also see freshwater rotifers, particularly in inshore waters following floods. You might also see insects. Most are blow-ins, but the remarkable *Halobates*, known as sea skaters or water striders, can be found on the surface of tropical and subtropical oceans, far out to sea. Insects may dominate the land, but their arthropod cousins the crustaceans dominate the sea. Another animal that you see on the surface of the water is the blue sea slug (nudibranch) *Glaucus*, which uses surface tension to float upside down. *Glaucus* eats other surface-dwelling organisms such as the siphonophore Portuguese man o'war *Physalia physalis*. Amazingly, the undischarged stinging nematocysts of the Portuguese man o'war can be stored in the tissue of *Glaucus* and be used for its own defence!

8.3.14 Phytoplankton

Although zooplankton are collected with a net, there can be many phytoplankton species present in a sample, especially when a finer mesh (<0.1 mm) is used. The larger phytoplankton species and those that form chains are the ones captured in nets. Phytoplankton can be distinguished from smaller zooplankton by their greenish tinge from their photosynthetic pigments and their lack of appendages (although they can have projections, spikes and flagella, but these are never jointed) (Fig. 8.29). Some groups such as dinoflagellates are mixotrophic – they are both autotrophic (photosynthesise) and heterotrophic (ingest particles) – and particles of food can sometimes be seen inside them.

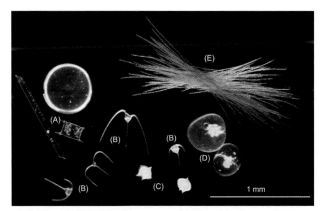

Fig. 8.29. Phytoplankton. **(A)** BACILLARIOPHYCEAE: Diatoms including *Coscinodiscus* (top right), *Proboscia* (left) and *Odontella* (bottom right); **(B)** DINOFLAGELLATA: *Tripos* spp.; **(C)** *Protoperidinium* spp.; **(D)** *Noctiluca scintillans*; **(E)** CYANOBACTERIA: *Trichodesmium* spp.

Coscinodiscus (**diatom**): You might see flat greenish discs that look like hockey pucks 0.1–0.2 mm in size: these are *Coscinodiscus* (Fig. 8.29A), but be careful to confirm these are flat rather than spherical copepod eggs.

Proboscia (**diatom**): Proboscia (formerly *Rhizosolenia*) have long straight strands with a thin protuberance ('nose') at the end (Fig. 8.29A). This genus can form large rafts.

Odontella (**diatom**): You might notice thin green-tinged strips – these are diatoms and if you look closely they are chains. Chains formed by squarish cells with thin straight protuberances are often *Odontella* (Fig. 8.29A).

Chaetoceros (**diatom**): Diatom chains with long lateral spines are usually *Chaetoceros*; the spines help with defence from grazers and reduce sinking.

Tripos (**dinoflagellate**): You might also see small anchors: these are the dinoflagellate *Tripos*, formerly called *Ceratium* (Fig. 8.29B). Dinoflagellates are protists having two flagella, so they can swim. Some dinoflagellate species are solely photosynthetic (like plants), some only eat small particles (like animals) and some are mixotrophs (both).

Protoperidinium (**dinoflagellate**): *Protoperidinium* spp. are ~0.1 mm in size and have a distinctive polygonal shape (Fig. 8.29C). *Protoperidinium* is a large genus of ubiquitous marine heterotrophic dinoflagellates. This genus typically follows diatom blooms and are generally found in coastal regions. Blooms often have pinkish to yellowish hues. *Protoperidinium* is an armoured cell and feeds on diatoms via extracellular digestion. *Protoperidinium* can make a neurotoxin called azaspiracid.

Noctiluca scintillans (dinoflagellate):

The sea was luminous in specks & in the wake of the vessel, of a uniform, slightly milky colour. – When the water was put into a bottle, it gave out sparks for some minutes after having been drawn up. – When examined both at night and next morning, it was found full of numerous small (but many bits visible to the naked eye) irregular pieces of (a gelatinous?) matter.

Charles Darwin, 1831–1836, made during the voyage of the Beagle (Darwin 1845)

This account from Charles Darwin was most likely of dinoflagellate bioluminescence, probably *Noctiluca*. You can see the large (0.5 mm diameter), translucent, reddish, apple-shaped cells of *Noctiluca* with a stereo microscope (Fig. 8.29D), but you might not see its striated flagella. This unarmoured (naked) dinoflagellate feeds on diatoms and other plankton. *Noctiluca* forms red tides and are entirely carnivorous, having no photosynthetic pigments, and thus rightly is covered in a zooplankton chapter! They can bloom in the estuary or coastal ocean in response to diatom blooms, which in turn have bloomed in response to upwelling of nutrients or sewerage. *Noctiluca* tend to bloom within a critical temperature range around 20°C and can numerically dominate the zooplankton in some inshore areas.

Trichodesmium:

The Sea in many places is here cover'd with a kind of a brown scum, such as Sailors generally call spawn; upon our first seeing it, it alarm'd us, thinking we were among Shoals, but we found the same depth of Water where it was as in other places.

Captain James Cook, 28 August 1770, made during the voyage of the Endeavour (Cook 1770)

These were observations of the floating cyanobacterium *Trichodesmium* outside the Great Barrier Reef off the east coast of Australia. *Trichodesmium*, also called sea sawdust, is found in subtropical and tropical areas (Fig. 8.29E). Individual cells are strung together in filaments, and then filaments are often clumped together in colonies, and colonies can form large mats, which are easily visible to people. *Trichodesmium* is a cyanobacterium, and fixes atmospheric nitrogen dissolved in sea water, and therefore is a vital primary producer in warm oligotrophic waters.

8.4 Top tips for identifying zooplankton

The approach adopted in this chapter for the identification of zooplankton first involves choosing which body type a specimen belongs to (see Section 8.2, Fig. 8.5), and then proceeding to the relevant section on the particular zooplankton body types (Section 8.3) where there is more information to distinguish particular taxonomic groups and sometimes species. An alternative approach is to use the approximate taxonomic tree (Fig. 8.2) to go more directly to a particular taxonomic group, bypassing the zooplankton body type. This might be more useful for researchers who have some experience in zooplankton identification.

We all find zooplankton taxonomy challenging to start with, and perseverance is the key to success. For your own sanity, it is important to remember that you will not be able to identify everything in your Petri dish, especially to species. Some specimens are damaged, especially those belonging to delicate groups such as jellyfish or molluscs, so diagnostic features might be absent. Many specimens are likely to be juveniles, and these often lack the diagnostic characters of adults for species-level identification. Molecular genetics offers some hope in this regard – offering the ability to identify damaged specimens and juveniles – something that is usually impossible with microscopy. Molecular genetics is revolutionising the identification of zooplankton, but there remain challenges associated with species coverage, validity of voucher specimens in GenBank, the ability to distinguish sexes, the ability to distinguish stages and, perhaps most importantly, obtaining abundance estimates. Nevertheless, molecular genetics is increasingly being used for zooplankton identification and this tool perfectly complements traditional taxonomy based on morphology. It is not routinely used for long-term time series monitoring, but is likely to be in the near future.

The level of identification that you require in your study depends on the question being asked. For example, if you are investigating the food environment of planktivorous fish (e.g. sardine or manta rays), then maybe you do not need to identify all zooplankton to species. Putting them into general categories such as copepods, decapod larvae might be sufficient, especially combined with identifying the one or two very abundant species. However, if you are interested in ecological indicators of water masses or climate change, identifying specimens to species is much more important because each species has its own environmental niche. Your question will thus determine the taxonomic resolution you require.

There is a plethora of resources out there to help you identify zooplankton; the references at the end of this chapter lists many of them, but we wanted to mention a few of the most useful ones here. The classic beginner's guide on plankton is by Newell and Newell (1977). For copepods, the 'Banyuls' website (www.copepodes.obs-banyuls.fr/en/) provides a comprehensive resource summarising the literature on the diversity and geographic distribution of all marine planktonic copepods. For non-copepod zooplankton, Boltovskoy (1999) is the most comprehensive. A perpetual challenge is to keep up with the changing taxonomic status and relationships among marine organisms. The World Register of Marine Species (WoRMS, www.marinespecies.org) is the definitive resource for up-to-date taxonomy.

Keep in mind that local identification guides can be better for your waters than global ones, primarily because a regional guide will have a much smaller number of species than a global guide. This is akin to identifying birds in different regions – you would always use a local field guide (maybe with hundreds of species) rather than a global one with all 10 000 species. In the Australian region, the *Australian Marine Zooplankton: Taxonomic Guide and Atlas* (www.imas.utas.edu.au/zooplankton) is a free online guide for zooplankton that is useful for Australian and Southern Hemisphere waters. This guide has taxonomic sheets for a couple of hundred copepod species and other common and distinctive zooplankton, with colour images highlighting diagnostic features of both males and females.

One approach that makes identification easier is the use of a Lucid key (http://www.lucidcentral.

com). A Lucid key uses diagnostic features of specimens to narrow down possible groups and species, rather than a traditional taxonomic key that uses dichotomous decisions. When a user says that a specimen has a specific character, all the taxa that have that character are retained, and all the taxa that do not are discarded, so that a limited subset of possible species is presented at each step. This process is repeated with multiple characters, until the user is left with a small number of (or one) species. This makes the identification process much easier. Few Lucid keys are available currently for zooplankton, but there are likely to be more in the future.

Finally, try not to work in isolation – seek out friendly experts or taxonomists. The internet makes this easier, so you can reach out across the ocean. We have found it very helpful to have a mentor: someone nearby or on the internet who can provide guidance. As your taxonomic confidence and experience grows, you will be able to identify more of what you see, and you might need multiple mentors for particular taxonomic groups. We have found that our mentors have often become our collaborators. If you are interested in keeping up with some of our taxonomic and ecological work on zooplankton, our Facebook page (www.facebook.com/imosaustralianplanktonsurvey) has regular discussions about interesting plankton we see and features awesome colour images.

8.5 References

When writing this chapter, the authors referred to the references listed here, but to improve readability most are not specifically cited in the text.

8.5.1 General references

Animal Diversity Web (2018) *Online Encyclopedia.* Website. University of Michigan, Michigan IL, USA, <https://animaldiversity.org>.

Barnes RD (2006) *Invertebrate Zoology: A Functional Evolutionary Approach.* Thomson Press, Calcutta, India.

Boltovskoy D (Ed.) (1999) *South Atlantic Zooplankton, Volumes 1 and 2.* Leiden, Backhuys, Netherlands.

Conway DVP (2012) *Marine Zooplankton of Southern Britain – Part 1: Radiolaria, Heliozoa, Foraminifera, Ciliophora, Cnidaria, Ctenophora, Platyhelminthes, Nemertea, Rotifera and Mollusca.* Occasional Publication of the Marine Biological Association 25. Marine Biological Association of the United Kingdom, Plymouth, UK.

Conway DVP (2012) *Marine Zooplankton of Southern Britain – Part 2: Arachnida, Pycnogonida, Cladocera, Facetotecta, Cirripedia and Copepoda.* Occasional Publication of the Marine Biological Association 26. Marine Biological Association of the United Kingdom, Plymouth, UK.

Conway DVP (2015) *Marine Zooplankton of Southern Britain – Part 3: Ostracoda, Stomatopoda, Nebaliacea, Mysida, Amphipoda, Isopoda, Cumacea, Euphausiacea, Decapoda, Annelida, Tardigrada, Nematoda, Phoronida, Bryozoa, Entoprocta, Brachiopoda, Echinodermata, Chaetognatha, Hemichordata and Chordata.* Marine Biological Association of the United Kingdom, Plymouth, UK.

Darwin C (1845) *Journal of researches into the natural history and geology of the countries visited during the voyage of the H.M.S.* Beagle *round the world, under the command of Capt. Fitz Roy, R.N.* 2nd edn. John Murray, London, UK. [See John van Wyhe (Editor), 2002, The Complete Work of Charles Darwin Online, <http://darwin-online.org.uk>]

Horton T, Kroh A, Ahyong S, Bailly N, Boury-Esnault N, Brandão SN, *et al.* (2018) World Register of Marine Species (WoRMS). Online database, <http://www.marinespecies.org>.

Huys R, Boxshall GA (1991) *Copepod Evolution. The Ray Society Volume 159.* Gresham Press, Old Woking, Surrey, UK.

Martin JW, Olesen J, Høeg JT (Eds) (2014) *Atlas of Crustacean Larvae.* Johns Hopkins University Press, Baltimore MD, USA.

Mauchline J (1998) The biology of calanoid copepods. *Advances in Marine Biology* **33**, 1–710.

Newell GE, Newell RC (1977) *Marine Plankton – A Practical Guide.* 5th edn. Hutchins, London, UK.

Razouls C, de Bovée F, Kouwenberg J, Desreumaux N (2018) *Diversity and Geographic Distribution of Marine Planktonic Copepods.* Sorbonne Université, CNRS, Paris, France, <http://copepodes.obs-banyuls.fr/en>.

van Couwelaar M (2003) *Zooplankton and Micronekton of the North Sea* (CD-ROM). Expert Centre for Taxonomic Identification, Amsterdam, Netherlands.

Van Dam RA, Harford AJ, Houston MA, Hogan AC, Negri AP (2008) Tropical marine toxicity testing in Australia: a review and recommendations. *Australasian Journal of Ecotoxicology* **14**, 55–88.

Young C (2006) *Atlas of Marine Invertebrate Larvae.* Elsevier, Amsterdam, The Netherlands.

Wiktionary (2018) *The Free Dictionary* [Greek and Latin names]. Website, <https://en.wiktionary.org>.

8.5.2 References on zooplankton sampling

Harris RP, Wiebe PH, Lenz J, Skjoldal HR, Huntley M (2000) *ICES Zooplankton Methodology Manual.* Academic Press, San Diego CA, USA.

Steedman HF (Ed.) (1976) *Zooplankton Fixation and Preservation.* UNESCO Press, Paris, France.

Wiebe PH, Benfield MC (2003) From the Hensen net toward four-dimensional biological oceanography. *Progress in Oceanography* **56**, 7–136. doi:10.1016/S0079-6611(02)00140-4

8.5.3 Specific references for taxonomic groups

Anderson DT (1998) *The Cnidaria and Ctenophora. Invertebrate Zoology.* Oxford University Press, Oxford, UK. [Jellyfish]

Armstrong AO, Armstrong AJ, Jaine FRA, Couturier LIE, Fiora K, Uribe-Palomina J, *et al.* (2016) Prey density threshold and tidal influence on reef manta ray foraging at an aggregation site on the Great Barrier Reef. *PLoS One* **11**, e0153393. doi:10.1371/journal.pone.0153393 [Calanoid copepods]

Bondad-Reantaso MG, Subasinghe RP, Josupeit H, Cai J, Zhou X (2012) The role of crustacean fisheries and aquaculture in global food security: past, present and future. *Journal of Invertebrate Pathology* **110**, 158–165. doi:10.1016/j.jip.2012.03.010 [Decapod larvae]

Boyko CB, Bruce NL, Hadfield KA, Merrin KL, Ota Y, Poore GCB, *et al.* (Eds) (2018). *World Marine, Freshwater and Terrestrial Isopod Crustaceans Database.* Website. Flanders Marine Institute, Ostend, Belgium, <www.marinespecies.org/isopoda>. [Isopods]

Bouchet P, Rocroi JP, Hausdorf B, Kaim A, Kano Y, Nützel A, *et al.* (2017) Revised classification, nomenclator and typification of gastropod and monoplacophoran families. *Malacologia* **61**, 1–526. doi:10.4002/040.061.0201 [Snails]

Brotz L, Pauly D (2017) Studying jellyfish fisheries: toward accurate national catch reports and appropriate methods for stock assessments. In *Jellyfish: Ecology, Distribution Patterns and Human Interactions.* (Ed. GL Mariottini) pp. 313–329. Nova Publishers, New York, USA. [Jellyfish]

Cook J (1770) *From Torres Strait to Batavia. August 1770.* Chapter 9. Captain Cook's Journal during his first voyage round the world made in H.M. Bark "Endeavour" 1768–71. [Trichodesmium]

Dahms H-U, Qian P-Y (2004) Life histories of the Harpacticoida (Copepoda, Crustacea): a comparison with meiofauna and macrofauna. *Journal of Natural History* **38**, 1725–1734. doi:10.1080/0022293031000156321 [Harpacticoids]

Dakin WJ, Colefax AN (1940) The plankton of the Australian coastal waters off New South Wales. Part I. With special reference to the seasonal distribution, the phytoplankton, and the planktonic Crustacea, and in particular, the Copepoda and crustacean larvae, together with an account of the more frequent members of the groups Mysidacea, Euphausiacea, Amphipoda, Mollusca, Tunicata, Chaetognatha, and some references to the fish eggs and fish larvae. *Publications of the University of Sydney. Department of Zoology Monographs.* **1**, 1–209. [Mysids, Radiolarians]

Dall W, Hill BJ, Rothlisberg PC, Staples DJ (1990) *Biology of Penaeidae.* Advances in Marine Biology 27. Academic Press. London, UK. [Decapod larvae]

Deibel D, Lowen B (2012) A review of the life cycles and life-history adaptations of pelagic tunicates to environmental conditions. *ICES Journal of Marine Science* **69**, 358–369. doi:10.1093/icesjms/fsr159 [Doliolids]

Eberl R, Cohen S, Cipriano F, Carpenter EJ (2007) Genetic diversity of the pelagic harpacticoid copepod *Macrosetella gracilis* on colonies of the cyanobacterium *Trichodesmium* spp. *Aquatic Botany* **1**, 33–43. doi:10.3354/ab00002 [Harpacticoids]

Gallienne CP, Robins DB (2001) Is *Oithona* the most important copepod in the world's ocean? *Journal of Plankton Research* **23**, 1421–1432. doi:10.1093/plankt/23.12.1421 [Cyclopoids]

Gold K (1971) Growth characteristics of the mass-reared tintinnid *Tintinnopsis beroidea*. *Marine Biology* **8**, 105–108. doi:10.1007/BF00350925 [Tintinnids]

Greenwood JG, Greenwood J, Skilleter GA (2002) Comparison of demersal zooplankton in regions with differing extractive-dredging history, in the subtropical Brisbane River estuary. *Plankton Biology and Ecology* **49**, 17–26. [Mysids]

Hamano T, Morrissy NM, Matsuura S (1987) Ecological information on *Oratosquilla oratoria* (Stomatopoda, Crustacea) with an attempt to estimate the annual settlement date from growth parameters. *Journal of Shiminoseki University of Fisheries* **36**, 9–27. [Stomatopods]

Henschke N, Everett JD, Richardson AJ, Suthers IM (2016) Rethinking the role of salps in the ocean. *Trends in Ecology & Evolution* **31**, 720–733. [Salps]

Holland LZ (2016) Tunicates. *Current Biology* **26**, R146–R152. doi:10.1016/j.cub.2015.12.024 [Appendicularians]

Huber M (2010) *Compendium of Bivalves. A Full-color Guide to 3,300 of the World's Marine Bivalves. A Status on Bivalvia after 250 Years of Research.* ConchBooks, Hackenheim, Germany. [Bivalves]

Jackson CJ, Rothlisberg PC, Pendrey RC, Beamish MT (1989) A key to the genera of Indo-Pacific penaeid larvae and early postlarvae and descriptions of *Atypopenaeus formosus* Dall and *Metapenaeopsis palmensis* Haswell (Decapoda: Penaeoidea) reared in the laboratory. *Fishery Bulletin* **87**, 703–733. [Decapod larvae]

Jamieson AJ, Malkocs T, Piertney SB, Fujii T, Zhang Z (2017) Bioaccumulation of persistent organic pollutants in the deepest ocean fauna. *Nature Ecology and Evolution* **1**(3), 51. [Amphipods]

Jennings RM, Bucklin A, Ossenbrügger H, Hopcroft RR (2010) Species diversity of planktonic gastropods (Pteropoda and Heteropoda) from six ocean regions based on DNA barcode analysis. *Deep-sea Research. Part II, Topical Studies in Oceanography* **57**, 2199–2210. doi:10.1016/j.dsr2.2010.09.022 [Gastropods]

Kiorboe T (2011) What makes pelagic copepods so successful? *Journal of Plankton Research* **33**, 677–685. doi:10.1093/plankt/fbq159 [Copepods]

Koski M, Kiørboe T, Takahashi K (2005) Benthic life in the pelagic: Aggregate encounter and degradation rates by pelagic harpacticoid copepods. *Limnology and Oceanography* **50**, 1254–1263. doi:10.4319/lo.2005.50.4.1254 [Harpacticoids]

Lee CS, O'Bryen P, Marcus NH (2005) *Copepods in Aquaculture.* Blackwell Publishing, Ames IA, USA. [Copepods]

Lipps JH, Finger KL, Walker SE (2011) What should we call the foraminifera? *Journal of Foraminiferal Research* **41**, 309–313. doi:10.2113/gsjfr.41.4.309 [Forams]

Lisenkova AA, Grigorenko AP, Tyazhelova TV, Andreeva TV, Gusev FE, Manakhov AD, *et al.* (2017) Complete mitochondrial genome and evolutionary analysis of *Turritopsis dohrnii*, the "immortal" jellyfish with a reversible life-cycle. *Molecular Phylogenetics and Evolution* **107**, 232–238. doi:10.1016/j.ympev.2016.11.007 [Jellyfish]

Murphy EJ (2001) *Krill*. In *Encyclopedia of Ocean Sciences*. 2nd edn. (Ed. JH Steele) pp. 349–357. Academic Press, Oxford, UK. [Euphausiids]

Nimmo DR, Hamaker TL (1982) Mysids in toxicity testing – a review. *Hydrobiologia* **93**, 171–178. doi:10.1007/BF00008110 [Mysids]

Nishibe Y (2005) The biology of oncaeid copepods (Poecilostomatoida) in the Oyashio region, western subarctic Pacific: its community structure, vertical distribution, life cycle and metabolism. PhD dissertation, Hokkaido University, Sapporo, Japan. [Poecilostomatoids]

Poore GCB, Bruce NL (2012) Global diversity of marine isopods (except Asellota and Crustacean symbionts). *PLoS One* **7**(8), e43529. doi:10.1371/journal.pone.0043529 [Isopods]

Richardson AJ, Bakun A, Hays GC, Gibbons MJ (2009) The jellyfish joyride: causes, consequences and management actions. *Trends in Ecology & Evolution* **24**, 312–322. doi:10.1016/j.tree.2009.01.010 [Jellyfish]

Schminke HK (2007) Entomology for the copepodologist. *Journal of Plankton Research* **29**(Suppl. 1), i149–i162. doi:10.1093/plankt/fbl073 [Copepods]

Soo P, Todd PA (2014) The behaviour of giant clams (Bivalvia: Cardiidae: Tridacninae). *Marine Biology* **161**, 2699–2717. doi:10.1007/s00227-014-2545-0 [Bivalve larvae]

Suzuki N, Not F (2015) Biology and Ecology of Radiolaria. In *Marine Protists*. (Eds S Ohtsuka, T Suzaki, T Horiguchi, N Suzuki and F Not) pp. 179–222. Springer, Tokyo, Japan. [Radiolarians]

Tafe DJ, Greenwood JG (1997) The Bodotriidae (Crustacea: Cumacea) of Moreton Bay, Queensland. *Oceanographic Literature Review* **3**, 235–236. [Cumaceans]

Tait RD, Maxon CL, Parr TD, Newton FC, III (2016) Benthos response following petroleum exploration in the southern Caspian Sea: Relating effects of nonaqueous drilling fluid, water depth, and dissolved oxygen. *Marine Pollution Bulletin* **110**, 520–527. doi:10.1016/j.marpolbul.2016.02.079 [Cumaceans]

Uchima M, Hirano R (1988) Swimming behavior of the marine copepod *Oithona davisae*: internal control and search for environment. *Marine Biology* **99**, 47–56. doi:10.1007/BF00644976 [Cyclopoids]

Vezzulli L, Grande C, Reid PC, Hélaouët P, Edwards M, Höfle MG, *et al.* (2016) Climate influence on *Vibrio* and associated human diseases during the past half-century in the coastal North Atlantic. *Proceedings of the National Academy of Sciences of the United States of America*

113, E5062–E5071. doi:10.1073/pnas.1609157113 [Cyclopoids]

Wang K (2014) The life cycle of the pteropod *Limacina helicina* in Rivers Inlet (British Columbia, Canada). PhD dissertation, University of British Columbia, Vancouver, Canada. [Pteropods]

Watling L, Gerkin S (2018) *Cumacea World Database*. Website, <http://www.marinespecies.org/cumacea>. [Cumaceans]

Wirtz KW (2012) Who is eating whom? Morphology and feeding type determine the size relation between planktonic predators and their prey. *Marine Ecology Progress Series* **445**, 1–12. doi:10.3354/meps09502 [Larvaceans]

Yancey P (2014) *Exploring the Mariana Trench. The World's Deepest Living Animals*. Schmidt Ocean Institute, Palo Alto CA, USA <https://schmidtocean.org/cruise-log-post/the-deepest-living-animals/>. [Amphipods]

York T, Powell SB, Gao S, Kahan L, Charanya T, Saha D, *et al.* (2014) Bioinspired polarization imaging sensors: from circuits and optics to signal processing algorithms and biomedical applications. *Proceedings of the IEEE* **102**, 1450–1469. doi:10.1109/JPROC.2014.2342537 [Stomatopods]

8.5.4 References for fish eggs and larvae

Asch RG (2015) Climate change and decadal shifts in the phenology of larval fishes in the California Current ecosystem. *Proceedings of the National Academy of Sciences of the United States of America* **112**, E4065–E4074. doi:10.1073/pnas.1421946112

Ahlstrom EH, Moser HG (1980) Characters useful in identification of pelagic marine fish eggs. *California Cooperative Oceanic Fisheries Investigations Reports* **21**, 121–131.

Ardura A, Morote E, Kochzius M, Garcia-Vazquez E (2016) Diversity of planktonic fish larvae along a latitudinal gradient in the Eastern Atlantic Ocean estimated through DNA barcodes. *PeerJ* **4**, e2438. doi:10.7717/peerj.2438

Baldwin CC, Brito BJ, Smith DG, Weigt LA, Escobar-Briones E (2011) Identification of early life-history stages of Caribbean Apogon

(Perciformes: Apogonidae) through DNA barcoding. *Zootaxa* **3133**, 1–36.

Fahay MP (2007) *Early Stages of Fishes in the Western North Atlantic Ocean (Davis Strait, Southern Greenland and Flemish Cap to Cape Hatteras). Volumes 1 and 2.* Northwest Atlantic Fisheries Organisation, Dartmouth, Nova Scotia, Canada.

Gray CA, Miskiewicz AG (2000) Larval fish assemblages in south-east Australian coastal waters: seasonal and spatial structure. *Estuarine, Coastal and Shelf Science* **50**, 549–570. doi:10.1006/ecss.1999.0595

Harada AE, Lindgren EA, Hermsmeier MC, Rogowski A, Terrill E, Burton RS (2015) Monitoring spawning activity in a Southern California marine protected area using molecular identification of fish eggs. *PLoS One* **10**(8), e0134647. doi:10.1371/journal.pone.0134647

Hubert N, Espiau B, Meyer C, Planes S (2015) Identifying the ichthyoplankton of a coral reef using DNA barcodes. *Molecular Ecology Resources* **15**, 57–67. doi:10.1111/1755-0998.12293

Kendall AW, Jr (Ed.) (2011*) Identification of Eggs and Larvae of Marine Fishes.* Tokai University Press, Tokyo, Japan.

Kimerling N, Zuqert O, Amitai G, Gurevich T, Armoza-Zvuloni R, Kolesnikov I *et al.* (2017) Quantitative species-level ecology of reef fish larvae via metabarcoding. *Nature Ecology and Evolution* **2**, 306–316.

Ko H, Wang Y, Chui T, Lee M, Leu M, Chang K, *et al.* (2013) Evaluating the accuracy of morphological identification of larval fishes by applying DNA barcoding. *PLoS One* **8**, e53451. doi:10.1371/journal.pone.0053451

Leis JM, Carson Ewart BM (2000) *The Larvae of Indo-Pacific Coastal Fishes: An Identification Guide to Marine Fish Larvae.* Fauna Malesiana Handbooks, Brill, Leiden, Netherlands.

Lewis LA, Richardson DE, Zakharov EV, Hanner R (2016) Integrating DNA barcoding of fish eggs into ichthyoplankton monitoring programs. *Fishery Bulletin* **114**, 153–165. doi:10.7755/FB.114.2.3

Moser HG, Richards WJ, Cohen DM, Fahay MP, Kendall AW, Jr, Richardson SL (Eds) (1984) *Ontogeny and Systematics of Fishes.* Special Publication 1. American Society of Ichthyologists and Herpetologists, Lawrence KS, USA.

Moser HG (1996) *The Early Stages of Fishes in the California Current Region.* CalCOFI Atlas no. 33. California Cooperative Oceanic Fisheries Investigations, La Jolla, CA, USA.

Neira FJ, Miskiewicz AG, Trnski T (Eds) (1998) *Larvae of Temperate Australian Fishes. Laboratory Guide for Larval Fish Identification.* University of Western Australia Press, Perth, Australia.

Olivar MP, Fortuño JM (1991) Guide to ichthyoplankton of the southeast Atlantic (Benguela Current region). *Scientia Marina* **55**, 1–383.

Okiyama M (Ed.) (2014) *An Atlas of the Early Stages Fishes in Japan.* 2nd edn. Tokai University Press, Tokyo, Japan.

Ozawa T (Ed.) (1986) *Studies on the Oceanic Ichthyoplankton in the Western North Pacific.* Kyushu University Press, Fukuoka, Japan.

Potter IC, Beckley LE, Whitfield AK, Lenanton RCJ (1990) The roles played by estuaries in the life cycles of fishes in temperate Western Australia and southern Africa. *Environmental Biology of Fishes* **28**, 143–178. doi:10.1007/BF00751033

Richards WJ (Ed.) (2006) *Early Stages of Atlantic Fishes: An Identification Guide for the Western Central North Atlantic. Volumes I and II.* CRC Press, Boca Raton FL, USA.

9

Educating with plankton

Timothy Roe, Anthony J. Richardson and Iain M. Suthers

Plankton studies are a superb vehicle for education generally, and scientific education in particular. Plankton are accessible for students and show interesting changes in species composition in time and across space in a range of freshwater and marine environments, providing students with a great range of taxonomic groups. The sheer variety and strangeness of this largely hidden world provides a great engagement tool for exciting students to become learners. To engage students, they need to see the connection between their theme of study and their lives. With regard to plankton, there are connections students understand but probably are unaware of, such as: sustaining fish and fishing; jellyfish stings at the beach; young lobster, crabs and prawns; the food and the young of oysters and barnacles on the rocky shore; and brine shrimp in aquaria. Because plankton can also be used to infer ecosystem health – through phytoplankton blooms, harmful algal blooms, jellyfish blooms, the impacts of pollution – studying them is a way to connect students to their local place.

Studying plankton gives students and their communities an opportunity to do interesting citizen science in their local area. A survey of 122 citizen science project leaders (Pecl *et al.* 2015) found that educational outputs were rated as almost as important as research outputs. If many schools take up the challenge of plankton studies and collaborate to standardise their methods, the resultant dataset created over time will also be of great interest to the scientific community and aquatic managers.

This chapter is about bringing plankton into the classroom, and bringing the class to the water's edge. A good place to start is with some fun facts (Section 1.2) about how much we rely on plankton, choosing the facts that are age-appropriate. Not many of us know that plankton fuel our cars, inspire movie aliens, produce the oxygen in every second breath we take, can be 40 m long, are used to help architects design buildings, and even change our weather and climate. To open their eyes to the often hidden world of plankton, you might like to show students the images of marine zooplankton – there are dozens of striking images that highlight the exquisite variety of forms. If you want to show students that many of the common animals that we know and eat as adults – such as fish, crabs, prawns, lobsters, mussels, oysters, squid – start their life as plankton, then Fig. 8.3 will be valuable. And if you want a quick guide to all the marine groups you are likely to see, then Fig. 8.2 could be helpful.

Students in Years 1–3 may find microscopes difficult to use, so try displaying a digital camera image of live *Artemia* (brine shrimp, sea monkeys) on a screen. Older students will be enthralled looking at a plankton sample with a properly set up (Fig. 4.8, Box 8.1) binocular dissecting microscope. They can compare their knowledge of insect life cycles with those of plankton (Chapter 2). More senior students in environmental science can consider the practical uses of plankton (Chapter 3). There are many useful plankton teaching resources online for all ages: for example, search for the 'great plankton race' or 'plankton planet' or 'blue planet'.

We provide some practical approaches to engaging students in aquatic science through project-based learning (e.g. how to build a plankton net) and collaborative learning (e.g. what bug is that?). Although we make specific references to the Australian curriculum (ACARA 2014), these are quite broad and generally applicable to curricula in other countries.

9.1 Plankton: for education, science inquiry and learning

In Australia, the *Melbourne Declaration* (MCEETYA 2008) is the document outlining the outcomes desired for education for the decade and what is required to achieve them; other countries will have their own declaration for learning outcomes, but it will undoubtedly be similar. It includes the following educational goals for students to 'become successful learners, confident and creative individuals, and active and informed citizens' [p. 8]. To create successful learners, it states that successful learners will:

- 'play an active role in their own learning'
- be 'creative and productive users of technology'
- 'think deeply and logically, and obtain and evaluate evidence in a disciplined way'
- are 'creative, innovative and resourceful, and are able to solve problems in ways that draw upon a range of learning areas and disciplines'
- will 'plan activities independently, collaborate, work in teams and communicate ideas'
- 'be able to make sense of their world and think about how things have become the way they are'.

These goals are easily supported by the use of plankton and science inquiry throughout a student's learning (Fig. 9.1), but they should be supported by appropriate pedagogy to enhance student learning. The opportunities for learning with plankton include covering the key ideas of science, and the fifth pedagogy that was developed by Ballantyne and Packer (2009) as an extension to the productive pedagogies (Lingard *et al.* 2001), which detailed the particular pedagogies and efficacy of learning outside the classroom. In Australia, this is supplemented by science inquiry skills from the Australian Curriculum Science (ACARA 2014).

ACARA (2014) states 'that students participating in science should 'experience the joy of scientific discovery and nurture their natural curiosity about the world around them'. Rantala and Määttä (2012) also value joy as an important emotion to encourage in educational settings. They put forward 10 theses that supported successful student learning. These included success, play, freedom to explore, and being an expert, which link well to science inquiry skills and which can be easily achieved by students during the study of plankton.

Studying plankton is a useful tool for environmental monitoring as changes in plankton populations indicate changes in water quality. Therefore local plankton studies fit well into the citizen science frame with the ensuing benefits of student engagement, rich discussion, student driven questions and critical thinking (Cigliano *et al.* 2015).

When students observe plankton for the first time, there is a transformation of a student's understanding of the world that is exciting and leads to them wanting to know more. Plankton studies are complex, with plankton populations in a particular area affected by: tidal, diurnal, lunar and seasonal cycles; predator–prey relationships; proximity to land masses, distance from the equator; turbidity; flood events; and nutrient enrichment including human impacts. This gives great scope for students to start by simply identifying plankton and their body parts, through identification and classification of specimens and on to ecological studies identifying and assessing the significance of several variables. The goal is that students will ultimately conduct their own studies to practice five key science inquiry skills (ACARA 2014):

Questioning and predicting: Identifying and constructing questions, proposing hypotheses and suggesting possible outcomes.

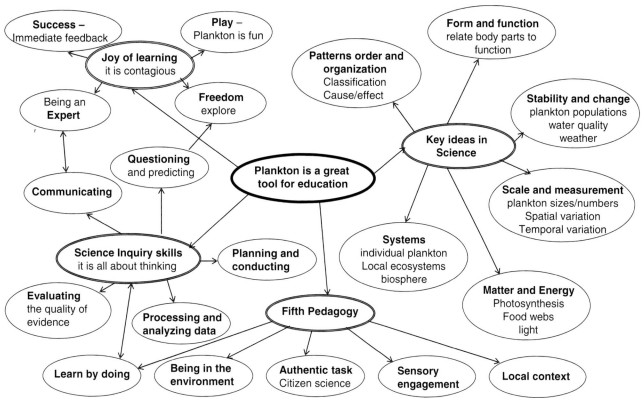

Fig. 9.1. The utility of plankton for education generally and scientific education in particular. The interrelationships between four pedagogical approaches are shown: the 'key ideas in science' and the 'science inquiry skills' from the Australian Curriculum Science (ACARA 2014), the 'fifth pedagogy' of Ballantyne and Packer (2009) and the 'joy of learning' from Rantala and Määttä (2012). The bold circles highlight these four domains that each have several characteristics linked to them. The number of interconnections demonstrate how plankton can be used to integrate these pedagogical approaches to achieve significant outcomes for students.

Planning and conducting: Making decisions about how to investigate or solve a problem and carrying out an investigation, including the collection of data.

Processing and analysing data and information: Representing data in meaningful and useful ways; identifying trends, patterns and relationships in data, and using this evidence to justify conclusions.

Evaluating: Considering the quality of available evidence and the merit or significance of a claim, proposition or conclusion with reference to that evidence.

Communicating: Conveying information or ideas to others through appropriate representations, text types and modes.

9.2 Studying plankton: where and how

Often, we think about plankton studies based around marine habitats (see Chapters 6 and 8) but lakes, ponds and rivers will also hold interesting populations of algae, rotifers, cladocerans and copepods that can be discovered by students (see Chapter 7). There may be some in your household water tank – which reinforces that a healthy water supply is not sterile and still contains plankton.

To study plankton, a few basic items are needed to allow you to collect and observe plankton: a net, a microscope, identification resources and other resources such as pipettes and Petri dishes. These suggestions may be modified to match your needs and experience (see Box 8.1).

A plankton net is basically a fine-mesh bag that can be pulled through the water and has a bottle at the bottom end to collect the critters. Nets typically have the following components: towing ring and bridle, tapered mesh sides and a removable capture jar. A weight may be required to sample deeper water, or a float to sample surface water (Fig. 9.2).

Nets can also be made by students out of a stocking, some wire such as a wire coat-hanger, tape and string (Fig. 9.3A). Bend the wire coat-hanger into a circle and flatten the hook onto the circle as well, covering up any sharp areas with tape. To make it stronger, twist the circle into a fig-ure-of-eight and then fold into a double hoop. Then stretch the leg opening of the stocking over the wire circle (Fig. 9.3B), and make several wraps with the stocking and secure with duct tape (Fig. 9.3C). Cut off the toe of the stocking, place the small

Fig. 9.2. A plankton net being rinsed down (and see Fig. 4.5D for another method). Insert shows the net ready for retrieval. Note the Secchi disc dangling from the block in top right (see Fig. 4.2F).

opening over a plastic jam collecting jar (~200 mL) and firmly tie on with several wraps of string and possibly tape if necessary (Fig. 9.3D).

To make the tow bridles, tie three lengths of cord 60 cm each, at equal intervals around the cir-cumference of the net opening (Fig. 9.3E). You will need to puncture through the mesh and tape (with a pencil or scissors) to tie around the wire, and tie securely (this could be a good time to teach stu-dents about types of knots). Draw the three ends together to form a towing loop, where you may attach the tow rope. Put on a lifejacket and tow the net from the jetty or the dinghy, at about walking pace. Ensure that one of the students records the duration of the tow and water speed – even if they do not intend to quantify the number of plankton per litre, it is an important practice to know that it can be quantified and how (see later in this chapter and Box 4.9).

When designing a plankton net to estimate numbers, the mesh size, front opening area and length need to be considered. A useful starting point would be a net with a 400 mm diameter opening, 200 µm (0.2 mm) mesh and a length of around 1.8 m. If the students are more interested in phytoplankton, then consider a smaller mesh around 100 µm (0.1 mm) to capture more phytoplankton.

It is ideal to have the students involved in the capturing of your plankton sample to further their understanding of the process (note that if you are working outside normal classroom settings, we have provided some suggestions in Box 9.2). You can challenge the students to explain how the net concentrates plankton into a small volume of the capture jar. They could achieve this by examining just a few millilitres of unfiltered sea water and comparing with to the sample in the capture jar. The life of a copepod can be a lonely one!

The best method to collect a plankton sample is to slowly tow it behind a boat, or tow it by walking along a jetty or pontoon. If a member of your school community is a regular fisher, they may tow a net for 5 minutes as they finish for the day and place the sample in the fridge for the students to look at the next day. If sampling in faster flowing water such as a river or tidal estuary, you can tie the net to a structure and allow the moving water to carry plankton into your net.

Depending on your set up, a weight may be a useful addition to keep your net working efficiently and allow you to sample different depths (especially between day and night – see Section 9.7).

Fig. 9.3. Construction of a simple plankton net from a wire coat hanger and half a nylon stocking: **(A)** the coat hanger is stretched into a circle ~30 cm diameter; **(B)** the circle is twisted into a figure-of eight and folded over, **(C)** to make a double wire ring; **(D)** the top of the pantihose is stretched over the wire circle and taped in place around the leading edge; **(E)** the three bridle strings are attached by puncturing through the tape with a pencil and tying the string around the wire circle at three equally spaced points on the circumference of the net frame; **(F)** the pantihose toes are the cut and a jar is then firmly tied on the end, ensuring that it can be undone (or access the sample by pushing the jar up through the net to the mouth); **(G)** the completed net; **(H, I)** gently tow the net from a small boat in a circle, or from a jetty or pontoon.

Ensure that your net doesn't bump along the bottom and capture sand and mud as well as plankton. Having your weight hanging from a short strop below your net can make this easier to achieve, because the weight can skim the sediment and the disturbance does not enter the net. You will need to set the duration of your tow dependent upon how much plankton is in the water and how much you want to capture. If there is a lot of plankton, shorter times are recommended. In most estuaries, a tow of 5–8 minutes at 1 to 2 knots (0.5–1 m/s) is sufficient to gather a rich plankton sample.

More advanced students could consider how to quantify the abundance of plankton, in terms of a concentration (number per m^3). This requires estimating the amount of water that the net has filtered. Flow meters provide an estimate of how much water has flowed through your net, but can be expensive (Fig. 4.6A). Flow meters are commonly used in university-level plankton practicals, but are not generally needed for younger students.

The cheapest and easiest method to estimate the volume of water filtered by a net in cubic metres (m^3) is to multiply the mouth area of the net (m^2) by the duration of the tow (seconds) and by the tow speed (m/s) of the net (Box 4.9). The easiest way to estimate boat speed is by counting the velocity of flotsam passing from the bow to stern. Thus, the filtered volume of a net of ~15 cm mouth diameter has a mouth area of ~0.0175 m^2 (Fig. 9.4), such that when towed at walking speed of 0.5 m/s for ~2 minutes, you filter ~1000 L (or 1 m^3).

An important part of the scientific process is to ensure a systematic process of recording information about your samples. Because there are many variables that may affect plankton populations, we recommend that students complete a research data card that includes location, water temperature, salinity, time of day, tide height and Secchi depth (Fig. 9.4). These data are needed to enable students to analyse their samples with respect to the environment. For example, one plankton tow could be made on the flood tide with oceanic water, while later another tow could be made on the ebb tide

with estuarine water; the difference in zooplankton could be bewildering without the environmental data. The data card could be modified to suit your needs or the equipment you have available, but a minimum of date and location should be recorded.

Once your net is recovered, you need to wash down the outside of the net to wash the plankton into the cod-end. This is achieved using a deck wash hose or dipping the net and collection jar repeatedly into the water and raising it, and/or splashing water up onto the end of the net with your hand (Figs 9.2, 4.5D). The cod-end (capture jar) can be unscrewed and the plankton poured into a clear container for initial observations, or the capture jar of your hand-made stocking net can be pushed up from the bottom and through the interior of the net up to the mouth, to pour off the contents.

MICRO – Plankton trawl - Research Card

Date		Site - Waypoint	
Start time		Finish time	
Secchi Depth	cm	Water depth	m
Water Temp	°C	Tide Height at Start (m)	m
Ebb or Flood		Salinity	
Wind Speed (knots)		Wind direction	
Start Position	Lat.		
	Long.		
Finish Position	Lat.		
	Long.		

Net mouth size	..........cm	Net mesh size	.... µm .... mm
Notes:	Tow speed?..........Tow duration?........
	....
School: Group etc	

Fig. 9.4.   Research data card for each tow.

For qualitative studies, it is recommended that students use pipettes to suck up plankton from the sample and place small drops either on a glass slide or in the middle of a Petri dish for observation with a microscope. Plastic pipettes can be cut off shorter to make a larger opening for larger organisms. After 5–10 minutes, many of the crustaceans will swim to the corners and be easier to capture. For quantitative studies, the sample will have to be sub-sampled volumetrically (Section 4.8.2).

9.3 How to observe plankton

Binocular dissecting microscopes are the simplest for students to use and a magnification of 20–40 times is suitable. Modern microscopes with a rechargeable LED light source greatly increase the flexibility of your plankton observations. One microscope for every two students is great if you can. This allows students to collaborate and provide feedback on their discoveries to each other (Rantala and Määttä 2012). Having one larger trinocular dissecting microscope with a camera (i.e. the microscope has a camera tube) able to record images and providing a live feed to a display is a useful teaching tool allowing the whole class to see the specimen under discussion. The camera can also be used to capture images of the plankton that can be used for creating a local identification resource (Fig. 9.5).

Students enjoy looking at live samples with a properly set up microscope (much more than the live digital image). If the live organisms are too fast, they can be slowed down by cooling in a fridge or cooler, or adding some soda water (the CO_2 slows them down). Identification is an important part of plankton studies, but, particularly when you are just beginning, it can be somewhat daunting. This gives another opportunity for students to enjoy learning when they take on the role of expert with their peers. If possible, encourage discussion about differing identifications to foster the ability of students to use evidence to back up their suggested identifications.

For identifying zooplankton, take a look at Chapter 8 for zooplankton, particularly: Fig. 8.5 on typical zooplankton body types; Fig. 8.2 on a simple taxonomic tree of zooplankton; and Section 8.3.14 on some common phytoplankton seen in plankton nets. Occasionally non-planktonic animals (mosquito larvae, insects) will be seen (Fig. 7.1). Other identification resources can be accessed from the internet, but often the best guides are ones produced locally using photographs from students checked by an expert from a museum or research institution.

In the next sections, we provide a description of plankton practicals that teachers can do with students, with reference to the Australian Curriculum (ACSSU).

9.4 Opportunities for primary students

9.4.1 Year 1 to 3 Learning Intent

- Living things have a variety of external features (ACSSU017)
- Living things live in different places where their needs are met (ACSSU211)

Tasks: The teacher and/or students make some plankton jigsaw pieces out of scissors and paper, such as a prawn or a brine shrimp. A whole prawn is a useful model of a krill, to explain zooplankton to students. If it is cooked then it is easier to handle, and a useful demonstration of form and function for Years 4 to 6 (ensure there are no allergies to seafood in your group). Then identify the head, antennae and eyes, abdomen, tail-fan, walking limbs and swimming limbs.

Or students draw and identify the external features of a plankton observed with a microscope (or from the digital camera and screen), such as body, head, legs, tail, eyes, antennae and tentacles.

Thinking activity: Students match their identified body part to a function (e.g. eyes versus antennae; legs used for walking versus swimming; and the tail fan for a prawn or krill for rapid escape from predators).

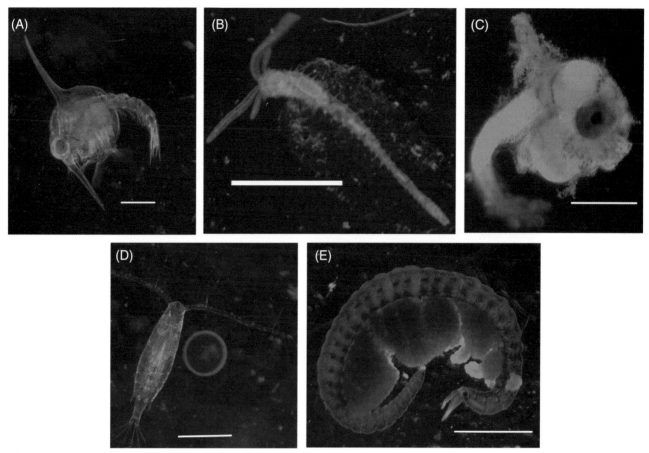

Fig. 9.5.   Pictures taken by primary school students with a microscope and digital camera: **(A)** crab zoea (scale bar = 1 mm); **(B)** terebellid polychaete worm *Lanice* (scale bar = 1 mm), which lives in that transparent mucus cylinder and it is common in coastal shallow bays of south-east Queensland; **(C)** a larval leatherjacket (Monacanthidae with a little dorsal barb buried in the amorphous blob (scale bar = 1 mm); **(D)** calanoid copepod and a centric diatom (scale bar = 0.5 mm); **(E)** the posterior, reproductive part (stolon) of a polychaete worm (Syllidae, Autolytinae) (scale bar = 0.5 mm) breaks free from the anterior part, to make a hazardous journey into the plankton, carrying the eggs in a ventral sac (see Section 8.3.6.1).

9.4.2 Year 4 to 6 Learning Intent

- Living things have life cycles (ACSSU072)
- Living things depend on each other and the environment to survive (ACSSU073)
- Living things have structural features and adaptations that help them to survive in their environment (ACSSU043)
- The growth and survival of living things are affected by physical conditions of their environment (ACSSU094)

Tasks: Identify the stages in the life cycle of crabs within a plankton sample e.g. zoea, megalopa, adult (Fig. 2.5). Make your own plankton net (Fig. 9.3). Distinguish between phytoplankton and zooplankton (Chapter 2). Introduce concepts of plankton buoyancy, density, shape and surface area.

Thinking activity: Analyse and evaluate the advantages and disadvantages of this life cycle that produces hundreds of thousands of young to produce just a few individuals in the next generation. A comparison of relative abundance of young and old crab larvae (zoea and megalopa) is a useful tool to start the students thinking about predation and ability to colonise favourable habitats. Contrast this life cycle with a more familiar pet or farm animal – such as a chicken.

9.5 Opportunities for lower secondary study

9.5.1 Year 7 Learning Intent

- Classification helps organise the diverse group of organisms (ACSSU111)

Task: Students identify plankton using keys or other methods to discriminate various major shapes and taxa (e.g. Figure 8.5, 8.7 of typical zooplankton body shapes). They then use identifiable features to group plankton together to create a dichotomous key. Students describe and evaluate the benefits of using a hierarchical classification system.

Thinking activity: Students evaluate the usefulness of using observable physical features to create a key. Students analyse the difficulties of using a single key for all organisms and create a solution.

Notes: Students can save digital photos of plankton identified to make a reference collection (Fig. 9.5). Start students with just four zooplankton that have relatively obvious differences. Then ask them to expand their key to cope with any new plankton identified.

Have extension students classify the four major groups of copepods (Fig. 8.10).

- Interactions between organisms, including the effects of human activities, can be represented by food chains and food webs (ACSSU112).

Task: Students create a food web including the plankton they have identified and extend it from phytoplankton to humans.

Thinking activity: Students identify and rank the benefits and detriments provided to humans by plankton (consider for example oxygen production; larval fish and fisheries; jellyfish stings; harmful algal blooms; environmental indicators of ocean acidification and climate change, Chapter 3).

Notes: To ensure capture of phytoplankton, a net with a fine mesh of around 100 µm may be useful. Among the many small zooplankton, you will likely find large centric diatoms, and armoured dinoflagellates (e.g. *Tripos*, Fig. 8.29).

9.5.2 Year 9 Learning Intent

- Ecosystems consist of communities of interdependent organisms and abiotic components of the environment; matter and energy flow through these systems (ACSSU176)

Task: Students compare plankton samples between ebb tide and flood tide, or before and after rainfall, or taken at different times of day or year to evaluate variability within populations (Chapter 4, and Figs 2.4, 2.7).

Thinking activity: Students will identify variables such as sunlight availability, rainfall, water temperature, time of day and food availability that may affect plankton populations then evaluate the significance of these to predict what plankton are expected at different times of the year.

Notes: Students must evaluate which variables can be controlled and ensure those that cannot be controlled are recorded at the time of sampling. This will require multiple sampling throughout the year and standardisation of collection methods. A multi-year dataset would be useful for students to analyse. An online 'Plankton Club' that could share data with other schools around the country would be a great goal. An example of a research data card is given in Fig. 9.4. Have a look in your country whether there is online plankton information used for ecosystem monitoring programs available that the students could use.

9.5.3 Year 10 (or Year 11 or 12) Learning Intent

- Global systems, including the carbon cycle, rely on interactions involving the biosphere, lithosphere, hydrosphere and atmosphere (ACSSU189)

Task: Students quantify phytoplankton from jar collections, or from chlorophyll extractions to highlight the importance of phytoplankton for the carbon cycle, the biosphere and atmosphere.

Thinking activity: Students identify phytoplankton as the pastures of the ocean (Chapter 1, fun facts); 50% of our oxygen is from plankton; all seafood relies on phytoplankton to a greater or lesser extent. Consider the huge atmospheric change that planktonic photosynthesis brought to our atmosphere, over 2 billion years ago.

Notes: Collect 1 or 2 L jar of water and let phytoplankton settle in a fridge; after >3 hours (or overnight) then use a pipette to suck up the bottom sediment to observe on a slide (this step could be quantified, to get counts per litre).

Make a wet mount of the sediment in the jar (Fig. 4.7B) and view under a compound microscope at 40× (100× is more difficult with a wet mount, but rewarding). If an inverted microscope is available, this process is avoided by looking up through the bottom of the container.

9.6 Opportunities for upper high school study

9.6.1 Year 11 and 12 Learning Intent

- Biological classification is hierarchical and based on different levels of similarity of physical features, methods of reproduction and molecular sequences (ACSBL016)
- Biological classification systems reflect evolutionary relatedness between groups of organisms (ACSBL017)

Task: Students identify plankton using keys or other methods. They then use identifiable features to group plankton together into the hierarchical classification system. This is a great way to quickly become familiar with the major phyla and their characteristics.

Thinking activity: Students debate the level of evolutionary relatedness of the organisms they find. For example, the various crustaceans (arthropods), molluscs and annelids, relative to sea urchin larvae or fish larvae is a good start.

Notes: Evolutionary relatedness can be checked using Fig. 8.2.

9.6.2 Year 11 and 12 Learning Intent

- Biodiversity includes the diversity of species and ecosystems; measures of biodiversity rely on classification and are used to make

comparisons across spatial and temporal scales (ACSBL015)
- Conduct investigations, including using ecosystem surveying techniques, safely, competently and methodically for the collection of valid and reliable data (ACSBL003)

Task: Design and conduct an investigation to show variations in plankton across spatial and temporal scales.

Thinking activity: Students evaluate the impact of different variables on the plankton population and use evidence collected to justify their conclusions.

The students will need to: represent data in meaningful and useful ways; organise and analyse data to identify trends, patterns and relationships; qualitatively describe sources of measurement error, and uncertainty and limitations in data; and select, synthesise and use evidence to make and justify conclusions (ACSBL004). This sort of plankton study can fulfil the requirements for an extended experimental investigation (EEI).

Students could form an online 'Plankton Club' to compare similarly collected data, arranged in a similar format from different areas.

9.7 Opportunities for undergraduate study

9.7.1 University Undergraduates Learning Intent

- Marine ecology and ecosystem understanding.

Task: Classify plankton by trophic level, or functional group such as prey or predator, herbivore or carnivore. Estimate a trophic pyramid of biomass for terrestrial systems (plants and trees; kangaroos; dingoes and eagles). Then compare it with the trophic pyramid of biomass of marine ecosystems where there are more trophic levels (biomass of phytoplankton; zooplankton, herring, tuna, shark, killer whale).

Thinking activity: Students identify and evaluate the food security implications of the differing trophic pyramids on land and sea.

Notes: There are many more trophic levels in the sea than on land. The marine pyramid is balanced on a tiny column of phytoplankton biomass because of rapid turnover of cells (production). Challenge the students how they can determine biomass (mg/m^3) versus production ($mg/m^3/year$) and the units of measurement. Diel vertical migration (DVM, diel refers to 24 hour cycles, compared with diurnal or nocturnal over ~12 hours).

- Plankton ecology.

Task: Tow the plankton net during the day and night; students could collect samples every 2–4 h through the day and night on a jetty; and collect corresponding benthos (sand or mud) with a benthic grab or with a jar, in day and night (an hour after dusk is sufficiently dark).

Thinking activity: Students create an hypothesis pertaining to the cause of DVM. For example, the low abundance in the day may be due to net avoidance whereas at night zooplankton cannot sense the net as well. Alternatively, the high abundance at night is due to plankton at the surface feeding on phytoplankton because visual predators such as fish cannot see them (but jellyfish and chaetognaths can sense them!). Or perhaps in the day zooplankton remain at depth or live on seagrass, or in the sediment. Students should design an experiment that could be used to test or disprove their hypothesis.

Notice that the night-time zooplankton in estuaries has many benthic species in it. Consider the massive vertical migrations by zooplankton in the open ocean: every night from 500 m deep to the surface. The benthic material is best sorted by vigorously shaking inside mesh bags (similar mesh to plankton net), and if possible stain the invertebrates with rose bengal.

- Jellyfish biology.

Task: Quantify abundance between morning and afternoon. Survey surface jellyfish abundance with a speedboat (e.g. 2 m either side of boat) and GPS using measured transects with respect to wind.

Deploy jars along shoreline in sunlight, containing a small *Catostylus* (or *Aurelia*) and a small *Phyllorhiza* (or chunks of the bell), and control jars. *Phyllorhiza* has zooxanthellae and will produce oxygen. Measure oxygen with DO probe after 1–2 hours.

Stinging cells of *Catostylus* – gently scrape a small amount of tissue from tentacles onto a glass slide, stain with methyl blue, apply a coverslip (Fig. 4.7B) and examine under high power reveal stinging cells (cnidocytes);

Thinking activity: What is the life cycle of jellyfish? How does this lifecycle compare with a ctenophore, and with a salp? Why is *Phyllorhiza* rather like a carnivorous plant? (hint: consider the need for nitrogen rather than fixed carbon). Why do jellyfish sometimes bloom? Consider the effect of wind on distribution of jellyfish. And what do we know about box jellyfish and Irukandji and climate change?

Trophic ecology. Carnivores and/or and symbiosis with zooxanthellae. Discuss the importance of the control jar result; and the activity of the jellyfish.

Sometimes small fish hide among the tentacles of *Catostylus* – how is this possible? (think of an anemone fish).

What are the predators of jellyfish? (look these up the internet).

Notes: Students should be careful with jellyfish stings but *Catostylus*, *Aurelia* or *Phyllorhiza* are not dangerous to most people (Box 9.1).

9.8 Plankton studies and the student

If there's ever a moment of realisation with students and plankton (that 'light-bulb' moment), it's usually when they have a properly set up dissecting microscope with eye-pieces the correct distance apart (Fig. 4.8B). Without squinting, they should have a focused binocular view of live zooplankton. However, getting to this point may take time, and patience could be in short supply. They will likely be sharing a microscope, and there may be 10 or 20 other students in the room. Suddenly the dirt under a fingernail becomes as interesting as the plankton! Therefore an old-fashioned hand-lens or a digital camera and TV display is a great start, and drawing the animal makes them observe the morphological features. Or they can connect with crayon various larval and adult animals on pre-prepared printed sheets. Having a squeezable soda-pop bottle to make a Cartesian diver (an eye-dropper) go up and down is another way to communicate density and the planktonic life.

For older students, plankton studies enable the important introduction to technology – from satellite images of ocean colour, to video, optical and acoustic counting methods. The projects outlined in this chapter give them ownership of the question and the answers. Hopefully the group dynamic among the students will enable them to share their new expertise of recognising a larval fish or crab zoea, but there's always the reality of some students not engaging (they may just look at their phones). These students could look up the many plankton identification tools available on the web, or search for the influence of plankton on the modern world (Chapter 1). It is also possible to use the camera on a smart phone to take a decent picture down the microscope's ocular (eye-piece), but this takes patience and eye-hand coordination.

Plankton studies enable students to cover many aspects of the curriculum, including biology, ecology, chemistry, mathematics and physics, but the main benefit is that by studying plankton students take on the role of a scientist. The students may

Box 9.1 Handling jellyfish: a note on safety

Most jellyfish stings are not lethal, but a few are. Many more cause rashes, swelling and other symptoms such as nausea, sweating, muscle and joint pain and difficulty breathing. Wear rubber gloves when handling jellies and avoid water into which stinging cells might have been released. Wear a rash-vest (with gloves, booties and hood if possible) if swimming with them. Be especially careful to protect the eyes, nose and mouth. If you need to capture jellyfish, avoid taking them out of water, because they are fragile. Instead, capture and move them in bags and buckets.

First aid is regionally and species specific. The recommended procedure is DRABCD (danger; response; send for help; airway clear; breathing; CPR; defibrillator).

In cooler, temperate waters an effective and practical treatment for pain from bluebottle (*Physalia*) stings and from jimbles (*Carybdea*) is to carefully **remove any remaining tentacles** with tweezers or your fingertips and thoroughly **rinse** the sting area with **sea water**. Then treat with immersion in warm to **hot water** (45°C for 20 minutes) or with **ice** or ice packs. Calm the patient (distract them from the pain), and do not rub the area with sand, or other materials. Treatment is similar for the brown blubber *Catostylus*, although it is usually not as aggravating.

If stung in eye or mouth, then seek medical advice (e.g. lifeguards).

In summary: remove tentacles; rinse with sea water; treat with hot water or ice; keep calm.

Both cold or heat treatments only alleviate the pain, which should disappear in 1–2 hours. Keep the patient calm and provide distraction from the pain. The venom can be denatured at 60°C, which is too hot for most people. Take care to thoroughly rinse the area with sea water to remove any undischarged cells, because the hot (fresh) water could discharge more cells, and the hot water may dilate surface capillaries to spread the toxin – especially relevant for treating box jellyfish stings.

In warmer tropical waters (and these days in subtropical waters), the recommended treatment for very painful box jellyfish stings is DRABCD; and douse the area with vinegar. If none available then **rinse with plenty of sea water (not fresh)**, then treat with ice (for pain).

Useful links

http://lifesaving.com.au/wp-content/uploads/Marine-Stinger-Fact-Sheets-all-2018.pdf

http://www.ambulance.nsw.gov.au/Media/docs/090730bluebottle-eee3bc83-ce7c-4281-a095-b427eb01e6d0-0.pdf realise this as they use or make the nets, collect samples using standardised procedures, record the data and discuss environmental variables. For teachers preparing to go into the field there is some preparation (Box 9.2), but it is worth the effort. Older students find it empowering to realise they are researching a specific question in their lake or estuary that no-one else has done before and that their persistence should reveal an answer.

It is wonderful to share with students when they see plankton for the first time, or find their first megalopa if they are exploring crab life cycles. Or they find a tiny squid and can see the chromatophores pulsing and understand how these animals change their appearance so quickly.

9.9 References

When writing this chapter, the authors referred extensively to the references listed here, but to improve readability most are not specifically cited in the text.

ACARA (2014) *Foundation to Year 10 Curriculum: Science (ACELA 1428)*. Australian Curriculum,

---

Box 9.2 Suggestions for teachers in risk management of day-trips

Most schools and community groups will need to prepare some paperwork before any field trip, especially for those involving students and boats. Not only does it display the necessary duty of care, the forms facilitate planning and serve as a checklist. **The main hazards are sunburn, heat exposure, bites and stings, falls and tripping hazards, allergens, and water hazards.** See Chapter 4 for other aspects of safety and permits such as if working in a marine park. Consider the following points and enter as appropriate into your risk assessment or activity plan.

- Provide pre-visit information to the students, outlining risk management, safety considerations and school requirements. Schools should ensure that their students arrive at the program with adequate sun protection, closed walking shoes, food and water. The 'pre-excursion school visit information' details the required drinking water, clothing, shoes and sun safety protection required by the students and teachers.
- Include details of the study site and/or details of the vessel (e.g. if drinking water and a toilet is available).
- Ensure the all staff understand the overall management plan and have the supervisors' and demonstrators' phone numbers, who can communicate back to the school and parents if necessary.
- Ensure the class list is returned before the program date. This form should outline any physical or special needs of the students. The form should also include relevant medical information pertaining to students, any non-swimmers, as well as a copy of any Anaphylaxis Action Plans.
- Provide safety briefings to students at appropriate times, to alert students to any hazards, and to outline appropriate student behaviour.
- Ensure the staff are trained in safe boarding procedures and emergency drills (and provide vessel's safety documents to teachers). Teachers or group leaders should have current First Aid and CPR certificates.

On the day:

- Before departing, supervisors should be informed of their role, potential hazards and precautions to be taken. A minimum of two adults should be present in all conditions. In calm conditions, an adult present should have the ability to recover a student from water, perform first aid and cardio-pulmonary resuscitation, and use signalling devices needed in a distress situation (including marine radio and flares).
- Prepare to modify the program to use more protected sites in the event of wind or waves.
- Plan for the students to have access to toilets during the day, and hand-washing facilities before eating. Teachers leading activities should inspect public space areas and infrastructure immediately before the students are given access.
- Students should handle only creatures considered safe by teachers, and taught how to correctly and ethically handle them. Control the group numbers for the various displays and collections.

Assessment and Reporting Authority (ACARA), Sydney, Australia, <https://www.australian-curriculum.edu.au/f-10-curriculum/science/>.

Ballantyne R, Packer J (2009) Introducing a fifth pedagogy: experience-based strategies for facilitating learning in natural environments. *Environmental Education Research* **15**, 243–262. doi:10.1080/13504620802711282

Brynjolfsson E, A McAfee (2016) *The Second Machine Age: Work, Progress, and Prosperity in a Time of Brilliant Technologies*. WW Norton & Co., New York, USA.

Cigliano JA, Meyerb R, Ballard HL, Freitag A, Phillips TB, Wasserf A (2015) Making marine and coastal citizen science matter. *Ocean and Coastal Management* **115**, 77–87. doi:10.1016/j.ocecoaman.2015.06.012

Lingard RL, Ladwig JG, Mills M, Bahr MP, Chant DC, Warry M (2001) *The Queensland School Reform Longitudinal Study*. Education Queensland.

MCEETYA (2008) *Melbourne Declaration*. Ministerial Council on Education, Employment, Training and Youth Affairs (MCEETYA), Melbourne, Australia.

Pecl G, Gillies C, Sbrocchi C, Roetman P (2015) Building Australia through Citizen Science. Office of the Chief Scientist – Occasional Paper Series, Issue 11 July 2015.

Rantala T, Määttä K (2012) Ten theses of the joy of learning at primary schools. *Early Child Development and Care* **182**, 87–105. doi:10.1080/03004430.2010.545124

Renshaw P, Tooth R (2018) *Diverse Pedagogies of Place*. Routledge, London, UK.

Epilogue

Most people have little experience or understanding of plankton, and this has led to the perception of them not being relevant. In this book we have endeavoured to show that plankton have many practical applications in management (see Table 1.1). Plankton are essential indicators in water quality assessments of the health of our waterways, such as report cards. Active management is needed to minimise the risks of harmful algal blooms, and the impacts of their associated toxins on shellfish and people. Following the decay of harmful algal blooms, deoxygenated waters can cause fish kills. Plankton are so sensitive to water quality they are used in toxicity testing to detect acute and sublethal impacts of pollutants.

A common question we get asked – from students and experienced researchers – is 'when can a computer identify all the plankton for us?' In some ways, there is exasperation to that question because it misses the point of studying plankton in the first place. Although we believe that the inherent beauty and charm of plankton under the microscope is reason alone to study it, the truth is that microscopy is time consuming, labour intensive and highly specialised. However, recent technological advances have provided a suite of modern approaches that not only deliver data rapidly but provide novel insights into the world of plankton. These modern approaches can be categorised into three main groups: imaging technology (including size), molecular approaches, and identification based on phytoplankton pigments.

Although imaging technology has been used for decades – in the laboratory based on bottle or net samples or towed at sea – the complexity of shapes and orientations overwhelmed traditional image analysis of plankton. Consequently, many of the first technologies did not image plankton at all but measured the size of particles, such as the Coulter Counter for phytoplankton and forerunners of the Laser Optical Plankton Counter for zooplankton. Such size-based technologies led to the development of the theory underlying size-structured ecosystems in the 1970s and 1980s (Herman and Harvey 2006), and the modern size spectrum models of today (Everett *et al.* 2017).

The simplest true imaging systems are used in the laboratory and are either a camera or a high-resolution flatbed scanner (e.g. Zooscan, Section 4.9), where the preserved sample is poured into a tray for imaging. The great advantage of imaging systems is that they can automatically identify plankton using machine-learning techniques, without time-consuming microscopy. The key is to have a good library of classified images for the machine-learning algorithm to recognise to a basic level of taxonomy, and also provide the area of the particle. This is very useful because the area can be used to estimate biovolume, which can indicate the physiology, longevity and consumption of an individual particle.

More recent imaging technology can take stunning images of plankton *in situ* while being towed behind a ship, and use machine-learning algorithms to automatically identify them. One such system is ISIIS (*In Situ* Ichthyoplankton Imaging System, McClatchie *et al.* 2012; Cowen *et al.* 2013). This instrument is towed behind a ship, profiling from the surface to, say, 100 m depth, recording images of plankton ~1 mm in size and larger. The on-board computer retains those images that are in focus and sends them back up the optic fibre to the ship for preliminary analysis. Back on land, a machine-learning algorithm uses a library of known images to classify plankton into groups – even to genus level in some cases. The ability to photograph zooplankton *in situ*, including delicate jellyfish and ichthyoplankton that can be damaged with conventional net sampling, is a great innovation. Weaknesses of all imaging technologies include that the instrument is relatively expensive and fragile, so must be deployed at slow speeds aboard research vessels, and that machine-learning algorithms are good for separating easily distinguishable, morphologically different, functional groups but not for species. Phytoplankton with their simple geometric shapes are thus easier to identify to species with machine-learning techniques than are zooplankton such as copepods, where species-level identification is often based on the morphology of the fifth leg.

The second major group of approaches for plankton identification is based on the molecular revolution. DNA metabarcoding and environmental DNA (eDNA) use certain genetic sequences (e.g. mitochondrial DNA) to genetically identify or fingerprint certain taxa (i.e. order, family, genus or species). Benefits of molecular approaches include that: you can identify specimens to species, even those that are damaged or juveniles that are unable to be identified with microscopy or imaging systems; they are fast so you can potentially process many samples quickly; and they are continually getting cheaper. Metabarcoding uses samples collected by bottles or nets, whereas eDNA uses a sample of water, and requires a large library of genetic sequences from bacteria to whales. This large library is now a reality due to the Census of Marine Life over the past decade and massive databases such as GenBank. However, estimates based on molecular approaches are generally limited to presence or absence and not abundance (as microscopy or imaging systems can) and there is extensive work needed to develop comprehensive sequence libraries (some genetic sequence entries in GenBank are incorrect). Molecular approaches do not identify the sex or age of individuals (but this is standard using microscopy).

The last class of identification methods is based on phytoplankton pigments. Ocean colour satellites can estimate chlorophyll-*a* (an index of phytoplankton biomass) over vast expanses of oceans and inland water bodies. Algorithms can also be used to identify the median size of the phytoplankton community and even some common functional groups based on their different absorption signatures. Pigment analyses using HPLC identify different phytoplankton functional groups based on their accessory pigments (those in addition to chlorophyll-*a*). Fluorometers can be towed *in situ* and typically deliver a pulse of blue light to estimate chlorophyll fluorescence and thus phytoplankton biomass. Multi-spectral fluorometers produce light at multiple wavelengths and can estimate the different phytoplankton functional groups present, based on their different excitation signatures. These approaches based on phytoplankton pigments are excellent at estimating phytoplankton biomass and functional groups, particularly over large time and space scales, but provide little to no information on the species present.

Although these innovative approaches provide new insights into plankton dynamics, no single sampling and identification approach is a panacea. Commonly, aquatic managers and researchers will use multiple approaches to take advantage of the benefits of each. For example, water quality programs commonly use bottle samples to collect samples for chlorophyll-*a* at specific points in time and space, and partner these with estimates of chlorophyll-*a* from fluorometers towed horizontal and vertically. Researchers studying zooplankton dynamics commonly use a Laser Optical Plankton Counter to describe the zooplankton size spectrum, but also collect samples using nets to identify the zooplankton present using microscopy. And researchers pioneering molecular approaches will typically ground-truth their presence–absence data with quantitative abundance data on species from microscopic identification from net-collected samples.

We began this book with 'plankton are effectively our aquatic 'canaries-in-a-coalmine', accumulating the effects of hourly changes in water quality over days, weeks and years'. The future for students of plankton ecology seems bright as the growing human population turns to lakes, rivers, estuaries and the ocean for food and development. The important message is to remember that plankton samples can reveal far more to an observant scientist than electronic, genetic or imagery signals.

References

Cowen RK, Greer A, Guigand C, Hare JA, Richardson DE, Walsh HJ (2013) Evaluation of the In Situ Ichthyoplankton Imaging System (ISIIS): comparison with the traditional (bongo net) sampler. *U.S. Fishery Bulletin* **111**, 1–12. doi:10.7755/FB.111.1.1

Everett JD, Baird ME, Buchanan P, Bulman C, Davies C, Downie R, *et al.* (2017) Modelling what we sample and sampling what we model: challenges for zooplankton model assessment. *Frontiers in Marine Science* **4**, 77. doi:10.3389/fmars.2017.00077

Herman AW, Harvey M (2006) Application of normalized biomass size spectra to laser optical plankton counter net intercomparisons of zooplankton distributions. *Journal of Geophysical Research: Oceans* **111**, C05S05. doi:10.1029/2005JC002948

McClatchie S, Cowen RK, Nieto KM, Greer A, Luo JY, Guigand C, *et al.* (2012) Resolution of fine biological structure including small narcomedusae across a front in the Southern California Bight. *Journal of Geophysical Research: Oceans* **117**, C04020. doi:10.1029/2011JC007565

Glossary of general terms

(**Note:** glossaries of specialised taxonomic terms appear in Table 7.2 and Table 8.1)

alga (plural **algae**):   Chlorophyll-containing plants such as phytoplankton or seaweed that lack roots, stems and leaves.

anthropogenic: A process caused by humans, typically referring to pollution.

aphotic zone: The depth beneath the euphotic zone in which respiration exceeds photosynthesis.

aquatic invertebrates: A general term for animals without backbones (i.e. not in the Phylum Chordata), living in water, including zooplankton.

benthos/benthic: A community of plants and animals living in (or on) the bottom.

billabong (or **oxbow lake**):   A type of wetland found on river floodplains, formed when river meanders are cut off from the main river channel.

bioaccumulation: The gradual concentration of pollutants as they move through the food chain from one trophic level to the next.

bioindicator: An organism characteristic or abundance that changes proportionally to environmental measurements, such as salinity, nutrients or heavy metals.

bioluminescence: The production of light (often greenish) by plankton and bacteria and other marine life.

biomanipulation: Artificially altering or enhancing biological and ecological processes, by adding or removing plants or animals to meet particular management needs.

biomass: The mass of organisms, often expressed as wet weight, dry weight or carbon weight.

biota: The plants and animals of an environment.

bloom: A sudden growth of plankton resulting in a distinctive biomass (such as salps); a bloom of phytoplankton may result in a red tide or a HAB.

blue-green algae: *See* cyanobacteria.

bollard: A sturdy loop or post on the boat to attach an anchor line or tow row.

boundary currents: Large ocean currents adjacent to continental shelves, driven by massive ocean basin gyres (e.g. western boundary currents are poleward).

chlorophyll: The main photosynthetic pigment of plants and often used to quantify phytoplankton abundance (noting that chlorophyll is <5% the biomass of a phytoplankton cell). Chlorophyll-*a* (Chl-*a*) is the main type of chlorophyll.

chitin: A structural chemical compound of the exoskeleton of arthropods such as copepods.

ciliates: A group of protozoans having cilia (small hair-like structures) in rows.

colonial: Organisms of the same species that live closely with each other and interact (*see also* solitary).

community: Groups of plants and/or animals sharing the environment; assemblage (*see also* population).

cosmopolitan: Worldwide.

counting chamber (or **counting cell**):   A small recessed chamber used to contain sample of water for microscopic viewing and counting of zooplankton or phytoplankton.

cyanobacteria: A group of photosynthetic bacteria whose cells lack nuclei, also called blue-green algae (examples are *Microcystis* and *Dolichospermum* (*Anabaena*) and *Trichodesmium*).

cyst: A capsule-like covering that encloses a small organism when in a dormant state.

depressor: A heavy, streamlined weight suspended under a towed plankton net to hold it deep in the water.

detritus: Dead organic matter derived from plants and animals.

diapause: A paused or arrested stage of development (*see also* resting eggs).

diel: A behaviour or phenomenon occurring over a full 24-hour period (as distinct from being simply nocturnal or diurnal).

DNA barcoding (or metabarcoding):   A method of species identification based on nucleotide sequences of a mitochondrial gene (cytochrome oxidase I).

dorsal: On the back of an animal, opposite to the ventral side where the limbs and mouth/anus typically occur.

eddy (plural eddies):   A large gently rotating body of water in the open ocean; either cyclonic or anticyclonic.

e-DNA: Traces of environmental DNA that exist in water for a few hours after shed by an organism, indicating species diversity (presence or absence).

epibenthic: The habitat on the benthic surface, or just above it.

epipelagic: The habitat in the upper well-lit part of the lake or ocean.

euphotic zone: The depth zone from the surface down to where light levels are ~1% of that at the surface and photosynthesis exceeds respiration; the aphotic zone is deeper.

eurytherms: Plankton that can grow and reproduce well over a wide temperature range (*see also* stenotherms).

eutrophic: Water bodies rich in nutrients and productive by plants due to runoff from agriculture or sewage (*see also* oligotrophic, mesotrophic).

exuviae (singular exuvium):   The outer shell or skin of crustaceans (such as copepods or barnacles) discarded during growth by moulting.

flagella (singular flagellum):   Fine, long whip-like appendages that are used for movement.

formaldehyde: A solution of formaldehyde gas used in diluted quantities for preserving biological samples (see Box 4.8).

fresh water: Water with less than 5% sea water or less than 3 g/L of dissolved salt.

fusiform: Spindle-shaped; broad at the middle and narrowing towards the ends.

gelatinous: Jelly-like, such as salps, ctenophores and jellyfish, and some phytoplankton.

genus (plural genera):   A major subdivision of a Family of organisms, comprising one or more species.

HAB: Harmful algal bloom.

holoplankton: Organisms that spend the entire life as plankton (such as copepods or salps); in contrast to meroplankton.

invasive species: A species introduced via human activity such as ballast water or climate change, which increases and dominates the ecosystem.

larva (plural larvae):   The young of invertebrates or larval fish, which are usually different in form to adult.

limnetic zone: *See* pelagic zone.

limnology: The study of inland waters and their ecology.

littoral zone: The shore of river or lake inundated for some or all of the time; the shallow water region extending from the shore to a depth where light is sufficient for rooted aquatic plants to growth.

macrophyte: A large aquatic plant such as a reed or seagrass that grow in or around water.

meroplankton: Organisms that spend only part of their life as plankton (usually as larvae, such as fish or crabs).

metazoan: A multicellular animal.

mesotrophic: Water bodies with a middle level of nutrients and moderately productive (*see also* eutrophic, oligotrophic).

micron: One micrometre (0.000001 m); a thousandth of a millimetre. Also shown as μm.

mixed layer: The warmer, upper layer of the lake or ocean above the thermocline, where most planktonic activity takes place (*see also* stratification).

nauplius: The earliest stage of crustacean larva (such as a copepod) after hatching from an egg.

nekton: Small animals with a good swimming ability, such as krill or jellyfish.

neritic: Coastal; in relatively shallow water over the continental shelf.

neuston: Plankton that occur at or just underneath the surface.

oligotrophic: Water bodies that are poor in nutrients and least productive; many undisturbed highland lakes are oligotrophic (*see also* eutrophic, mesotrophic).

omnivorous: Eating both plants and animals.

otolith(s):   Small calcified bone(s) in the inner ear of fish (and other vertebrates) for balance; in larval fish these three tiny pairs of bones grow by daily increments (analogous to tree rings).

paedamorphosis: The retention of larval or juvenile morphologies in the mature adult (such as in larvaceans and even humans).

parthenogenesis: A mode of reproduction in which eggs do not require fertilisation by male sperm.

pedagogy: The practice of teaching framed and informed by a shared and structured body of knowledge.

pelagic zone: The upper waters of the ocean or open-water region of lakes; also termed limnetic zone in lakes.

pH: The concentration of hydrogen ions in water; the higher the pH value, the fewer hydrogen ions; water is called acidic if the pH values are 1–7, neutral (pH value near 7) or alkaline if the pH values are 7–14.

phylum (plural **phyla):**   A major subdivision of a Kingdom (there are approximately 34 phyla of animals on the planet today, including our own phylum – Chordata). A phylum is made up of one or more classes of animals, which in turn are made up of orders, families, genera and species.

planktivorous: Plankton-eating.

population: A group of individuals of the same species that may interbreed (*see also* community).

primary production: The rate of organic production by photosynthesis.

pseudopodia (singular **pseudopodium):**   The temporary foot-like protrusions of a protozoan.

red (or **brown) tide:**   A bloom of phytoplankton with reddish (or brown) pigments.

rostrum: The pointed part of the carapace, extending between the eyes.

sessile: Attached to the surface of an object (e.g. rock or boat).

setae: Small bristles or spines (also chaetae).

solitary: Organisms that remain as individuals (*see also* colonial).

species: Basic classification of a group of organisms capable of interbreeding (sometimes abbreviated as sp. or spp. when only the genus is known).

stalk: A stem-like structure connecting the body of a protozoan to other animals or substrates.

statocyst: A balance organ to sense gravity (equivalent to an otolith in fish).

stenotherms: Organisms that tolerate only a narrow temperature range; can be warm stenotherms (requiring warm water) or cold stenotherms (requiring cold water) (*see also* eurytherms).

stratification: The layering of density gradients in lakes or oceans caused by solar warming or by rainfall, and broken down by mixing due to wind, tides or thermohaline circulation.

suspended solids: The very small particles of inorganic and organic material in a water body.

symbiotic: Living together in more or less close association or even union.

taxonomy: The science of classifying organisms.

test: A rigid shell covering the body of a protozoan or invertebrate.

thermocline: A narrow depth zone where the gradient in temperature steeply changes with depth.

thermohaline circulation: Density-driven flows caused by temperature and salinity.

thoracic: The middle segment(s) of invertebrate body to which limbs are attached.

thwart: The strong structural beam of the boat, for seats or bollards.

transom: The stern of the boat where the outboard engine or bollards are often attached.

tychoplankton: 'Accidental' plankton such as benthic invertebrates, or mosquito wrigglers (i.e. non-planktonic organisms that appear as plankton due to habitat disturbance).

unicellular: Single-celled (as opposed to multicellular: many cells).

upwelling: An oceanographic process that brings deeper water to the surface (**uplifting** refers to the same process, but not reaching the surface).

vector: An organism that transmits germs or other agents of disease.

ventral: On the abdominal (front) side.

zooplanktivorous: Zooplankton-eating.

INDEX